METEOROLOGICAL MEASUREMENT SYSTEMS

METEOROLOGICAL MEASUREMENT SYSTEMS

Fred V. Brock
Professor Emeritus of Meteorology
University of Oklahoma

Scott J. Richardson
Cooperative Institute for
Mesoscale Meteorological Studies
and
Oklahoma Climatological Survey
University of Oklahoma

2001

OXFORD
UNIVERSITY PRESS

Oxford   New York
Athens   Auckland   Bangkok   Bogotá   Buenos Aires   Calcutta
Cape Town   Chennai   Dar es Salaam   Delhi   Florence   Hong Kong   Istanbul
Karachi   Kuala Lumpur   Madrid   Melbourne   Mexico City   Mumbai
Nairobi   Paris   São Paulo   Shanghai   Singapore   Taipei   Tokyo   Toronto   Warsaw

and associated companies in
Berlin   Ibadan

Copyright © 2001 by Oxford University Press, Inc.

Published by Oxford University Press, Inc.
198 Madison Avenue, New York, New York, New York 10016

Oxford is a registered trademark of Oxford University Press.

All rights reserved. No part of this publication may be reproduced,
stored in a retrieval system, or transmitted, in any form or by any means,
electronic, mechanical, photocopying, recording, or otherwise,
without the prior permission of Oxford University Press.

Library of Congress Cataloging-in-Publication Data
Brock, Fred V.
   Meteorological measurement systems / Fred V. Brock and Scott J. Richardson.
      p. cm.
   Includes bibliographical references and index.
   ISBN 0-19-513451-6
      1. Meteorological instruments. 2. Meteorology. I. Richardson, Scott J. II. Title.
QC876.B76 2000
551.5′028′7–dc21   00-028551

9 8 7 6 5 4

Printed in the United States of America
on acid-free paper

Preface

This material has been used in connection with courses in the School of Meteorology at the University of Oklahoma. The first course in instrumentation is a junior level course with prerequisites of mathematics through calculus and ordinary differential equations and one year of calculus-based physics. The second course is first-year graduate level.

The objective of the courses is to examine the physical principles of meteorological sensors, to develop static and dynamic performance concepts, and to explore the concepts of meteorological measurement systems.

SI units are used throughout this text along with commonly accepted units such as °C for temperature and mb for pressure.

The first twelve chapters are presented in the order that F.V.B. has found useful in presenting lecture material. After discussing barometry (chap. 2), it is somewhat easier to present the material on static characteristics (chap. 3) as the students have the pressure sensor examples to consider. Dynamic performance characteristics are presented as needed. Chapter 8, Dynamic Performance Characteristics, Part 2, may be skipped in an undergraduate course. Chapter 13, Sampling and Analog-to-Digital Conversion, may be inserted almost anywhere in the course.

Some fundamental concepts, such as static and dynamic performance, are presented after discussion of sensors, so that students will be able to immediately apply these concepts to sensors they have discussed and extend their knowledge of these sensors.

Throughout, in situ or immersion sensors are discussed with only a few exceptions. There is a brief mention of radar in connection with rain gauges because radar–rain gauge comparisons are done so often. It is useful to include some treatment of radar so the student can see what the problem is. Visibility and cloud height sensors are included because they are part of some surface observing systems.

Contents

1 Overview 1
   1.1 Instrument Design and Selection 1
      1.1.1 Performance Characteristics 2
      1.1.2 Functional Model 2
      1.1.3 Sources of Error 4
   1.2 Standards 6
      1.2.1 Calibration 6
      1.2.2 Performance 6
      1.2.3 Exposure 6
      1.2.4 Procedural 7
   1.3 System Integration 8
      1.3.1 Instrument Platforms 8
      1.3.2 Communication Systems 8
      1.3.3 Power Source 10
   1.4 Human Aspects of Measurement 11
      1.4.1 Human Perception versus Sensor Measurements 11
      1.4.2 Reasons for Automation 11
      1.4.3 Design, Implementation, and Maintenance of Measurement Systems 12
      1.4.4 Interpretation of Sensor Specifications 12
      1.4.5 Interpretation of Results 13
      1.4.6 Human Judgment 14
   1.5 Quality Assurance 15
      1.5.1 Laboratory Calibrations 16
      1.5.2 Field Intercomparisons 17
      1.5.3 Data Monitoring 17

viii   Contents

   1.5.4 Documentation 18
   1.5.5 Independent Review 18
   1.5.6 Publication of Data Quality Assessment 19
  1.6 Scope of this Text 19
  Questions 19
  Bibliography 20
  General Instrumentation References 21

2 Barometry 22
  2.1 Atmospheric Pressure 22
  2.2 Direct Measurement of Pressure 23
   2.2.1 Mercury Barometers 24
   2.2.2 Aneroid Barometers 29
  2.3 Indirect Measurement of Pressure 33
   2.3.1 Boiling Point of a Liquid 34
  2.4 Comparison of Barometer Types 37
   2.4.1 Mercury Barometers 38
   2.4.2 Aneroid Barometers 38
   2.4.3 Hypsometer 38
  2.5 Exposure Error 39
  2.6 Laboratory Experiment 40
  2.7 Calibration of Barometers 42
  Questions 43
  Bibliography 45

3 Static Performance Characteristics 47
  3.1 Some Definitions 47
  3.2 Static Calibration 49
   3.2.1 Definition of Terms Related to the Transfer Plot 50
   3.2.2 Calibration Procedure 52
  3.3 Example of a Static Calibration 56
  3.4 Multiple Sources of Error 57
  3.5 Significant Figures 58
  Questions and Problems 60
  Bibliography 61

4 Thermometry 62
  4.1 Thermal Expansion 63
   4.1.1 Bimetallic Strip 63
   4.1.2 Liquid-in-Glass Thermometer 65
  4.2 Thermoelectric Sensors 67
  4.3 Electrical Resistance Sensors 70
   4.3.1 Resistance Temperature Detectors 70
   4.3.2 Thermistors 74
  4.4 Comparison of Temperature Sensors 76
  4.5 Exposure of Temperature Sensors 77
  Questions 82
  Bibliography 84
  Notes 85

5 Hygrometry 86
  5.1 Water Vapor Pressure 86
  5.2 Definitions 89
  5.3 Methods for Measuring Humidity 93

      5.3.1 Removal of Water Vapor from Moist Air 93
      5.3.2 Addition of Water Vapor to Air 93
      5.3.3 Equilibrium Sorption of Water Vapor 96
      5.3.4 Measurement of Physical Properties of Moist Air 100
      5.3.5 Attainment of Vapor–Liquid or Vapor–Solid Equilibrium 104
      5.3.6 Chemical Reactions 107
  5.4 Choice of Humidity Sensor 107
  5.5 Calibration of Humidity Sensors 108
  5.6 Exposure of Humidity Sensors 110
  Questions 111
  Laboratory Exercises 114
  Bibliography 114
  Notes 116

6 Dynamic Performance Characteristics, Part 1 117
  6.1 First-Order Systems 118
      6.1.1 Step-Function Input 119
      6.1.2 Ramp Input 122
      6.1.3 Sinusoidal Input 123
  6.2 Experimental Determination of Dynamic Performance Parameters 125
  6.3 Application to Temperature Sensors 126
  Questions 127
  Bibliography 128

7 Anemometry 129
  7.1 Methods of Measurement 129
      7.1.1 Wind Force 130
      7.1.2 Heat Dissipation 140
      7.1.3 Speed of Sound 141
  7.2 Calibration 144
  7.3 Exposure 144
  7.4 Wind Data Processing 144
  Questions 145
  Bibliography 147

8 Dynamic Performance Characteristics, Part 2 151
  8.1 Generalized Dynamic Performance Models 151
  8.2 Energy Storage Reservoirs 152
  8.3 Second-Order Systems 154
      8.3.1 Step Function Input 154
      8.3.2 Ramp Input 157
      8.3.3 Sinusoidal Input 158
  8.4 Application to Sensors 159
  8.5 Experimental Determination of Dynamic Performance Parameters 162
  Questions 164
  Bibliography 165

9 Precipitation Rate 166
  9.1 Definitions 166
  9.2 Methods of Measurement 167
      9.2.1 Point Precipitation Measurement 167
      9.2.2 Radar Rain Measurement 182
  Questions 185
  Bibliography 187

10 Solar and Earth Radiation 189
    10.1   Definitions   189
    10.2   Methods of Measurement   192
        10.2.1   Pyrheliometers   193
        10.2.2   Pyranometers   194
        10.2.3   Pyrgeometers   197
        10.2.4   Pyrradiometers   197
    10.3   Measurement Errors   198
    10.4   Exposure   199
    Questions   200
    Bibliography   201

11 Visibility and Cloud Height 202
    11.1   Definitions   202
    11.2   Measurement of Visibility   204
        11.2.1   Transmissometer   205
        11.2.2   Forward Scatter Meters   206
    11.3   Measurement of Cloud Height   207
        11.3.1   Rotating Beam Ceilometer   207
        11.3.2   Laser Ceilometer   208
    Questions   210
    Bibliography   211

12 Upper Air Measurements 213
    12.1   Methods for Making Upper Air Measurements   213
        12.1.1   Remote Sensing   214
        12.1.2   In-Situ Platforms   214
    12.2   Balloons   215
    12.3   Wind Measurement   219
        12.3.1   Theodolites   220
        12.3.2   Radar   223
        12.3.3   Navigation Aids   223
    12.4   Radiosondes   225
    12.5   Exposure Error   228
    Questions   228
    Bibliography   229

13 Sampling and Analog-to-Digital Conversion 231
    13.1   Signal Path   232
    13.2   Drift   233
    13.3   Sampling   233
    13.4   Analog-to-Digital Conversion   235
    13.5   Information Content of a Signal   239
    Questions   241
    Bibliography   243

A Units and Constants 245
    International System of Units (SI)   245
    Numerical Values   246
    Bibliography   247

B Thermistor Circuit Analysis 248
    B.1   A Thermistor   248
    B.2   A Circuit   250
    B.3   An Alternative Calibration Equation   252

C A Data Logger 253
    C.1 The Data Logger 253
    C.2 Application in a Measurement System 255

D Circuits 258
    D.1 Fundamentals 258
    D.2 Simple Circuits 261
    Questions 263

E Geophysical Coordinate System 266
    E.1 Geophysical versus Mathematical Coordinate System 266
    E.2 Mathematical Coordinates 266
    E.3 Geophysical Coordinates 267

F Instrumentation Glossary 269

Index 285

METEOROLOGICAL MEASUREMENT SYSTEMS

1
Overview

Measurements are required to obtain quantitative information about the atmosphere. Elements of a good measurement system, one that produces high-quality information, are briefly described in the following sections. All of these items are, or should be, of concern to everyone who uses data. None may be safely delegated, in their entirety, to those who have little or no interest in the ultimate use of the data.

1.1 Instrument Design and Selection

An instrument is a device containing at least a sensor, a signal conditioning device, and a data display. In addition, the instrument may contain an analog-to-digital converter, data transmission and data storage devices, a microprocessor, and a data display. The sensor is one of the essential elements because it interacts with the variable to be measured (the measurand), and generates an output signal proportional to that variable. At the other end of this chain, a data display is also essential, for the instrument must deliver data to the user.

To understand a sensor, one must explore the physics of the sensor and of sensor interaction with the measurand. There is a wide variety of sensors available for measuring pressure, temperature, humidity, and so on, and this text discusses each individually. Therefore, each chapter must deal with many different physical principles.

1.1.1 Performance Characteristics

Sensor performance can be described by reference to a standardized set of performance definitions. These characteristics are used by manufacturers to describe instruments and as purchase specifications by buyers.

1.1.1.1 Static

Static characteristics (chap. 3) are those obtained when the sensor input and output are static (i.e., not changing in time). Static sensitivity is an example of a static characteristic and is particularly useful in sensor analysis. When raw sensor output is plotted as a function of the input, the slope of this curve is called the static sensitivity. Relating static sensitivity to fundamental physical parameters is a systematic way of revealing sensor physics and leads to an understanding of the sensor and of how to improve the design.

1.1.1.2 Dynamic

Dynamic characteristics are a way of defining a sensor response to a changing input. The most widely known dynamic performance parameter is the time constant, discussed in chap. 6. This parameter is appropriate to systems whose dynamic performance can be modeled with a simple first-order, ordinary differential equation. More complex dynamical systems are described in chap. 8. Again, the goal is to relate these dynamic characteristics to physical parameters.

1.1.2 Functional Model

A measurement system interacts with the atmosphere and delivers data (information about the desired atmospheric variables) to the users. Common features of a measurement system are shown in fig. 1-1 in functional form. A measurement system may comprise some or all of these blocks, plus many more in a complex system. The blocks essential to any measurement system are 1, the sensor, 2, analog signal conditioning (ASC), and 7, the display. ADC is analog-to-digital converter, and DSC is digital signal conditioning. Raw input is X_i, the measurand (for example, air temperature), and final output is Y_7 (for example, temperature in degrees Celsius).

Block 1 is the sensor with input X_i, called the measurand, and raw output Y_1. A sensor is a transducer, i.e., a device that converts energy from one form to another. An instrument may contain several transducers to convert the energy from the measurand through several steps to a useful form such as electrical voltage. Here the sensor is the primary transducer, the one that interacts with the atmosphere, and block 1 contains only this sensor. Another transducer, if used, will be modeled in the next block. The output of block 1 is the primary transducer raw output in units appropriate to that device.

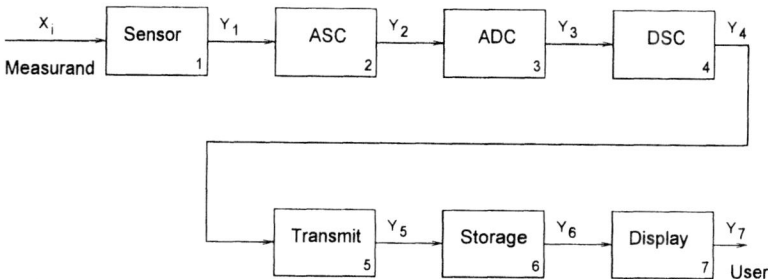

Fig. 1-1 Functional model of a simple measurement system.

EXAMPLE

A mercury-in-glass thermometer may be modeled with blocks 1, 2, and 7. In the sensor, block 1, heat energy is converted into a change in volume of the mercury in the bulb and thus into the height of the mercury column relative to some arbitrary index. The diameter of the column relative to the volume of the bulb is a form of signal conditioning, block 2, which sets the amplification of the thermometer. A smaller column diameter forces the mercury to rise further for a given temperature increase. The input into block 1 (X_i) is air temperature in Kelvin or degrees Celsius while the raw output, Y_1, is the volume of the mercury. After amplification in block 2, the raw output becomes Y_2, the height of the mercury column. The scale etched into the glass, the display (block 7), provides the calibration information that allows the user to translate the raw height, Y_2, into temperature, Y_7.

We may deduce, through calibration, a polynomial to relate the raw sensor output, Y_1, to an estimate of the measurand; thus $X_1 = c_0 + c_1 Y_1$ (a first-order polynomial is used in this example but any order polynomial can be used). The quantity X_1 is an estimate of X_i, based on knowledge of the signal Y_1. If an instrument has seven functional blocks, as in fig. 1-1, we must have a calibration for each block in order to estimate X_7 from Y_7 or, alternatively, we could have obtained the calibration for the complete system in one step using $X_7 = c_0 + c_1 Y_7$. The measurand is truly unknowable because all instruments extract some energy from the measurand and add some noise to the output signal. Therefore, X_i can only be estimated and never known exactly.

X_i and Y_n are signals, that is, information-bearing quantities such as temperature, wind speed, shaft rotation rate, voltage, current, resistance, frequency, and so on, X_i, Y_1, and Y_2 are always analog signals, that is, signals whose information content is continuously proportional to the measurand. Block 2 contains analog signal conditioning (ASC) which may include secondary transducers, an amplifier to provide gain and offset, and filters to reduce high-frequency noise.

EXAMPLE

A cup anemometer is a sensor that converts horizontal wind speed to angular rotation rate of a shaft that is connected to the cup wheel. The sensor input is wind speed in m s-1 and the output is shaft rotation rate in radians s^{-1}. There is usually a secondary transducer, such as a dc generator, connected to the shaft to convert rotation rate to a voltage. This voltage is continuously proportional to the shaft rotation rate which in turn is proportional to the wind speed. Instead of a dc generator, magnets can be used to produce an ac pulse each time the shaft rotates (several pulses can be generated for each shaft rotation, using multiple magnets). In this case, the raw sensor output is shaft rotation rate in Hz.

Block 3 is an analog-to-digital converter (ADC) and is present in most modern measurement systems to convert analog signals to discrete values, i.e., digital. The output signal of an ADC is a stream of numbers representing the value of the input signal. Conversions are usually done at discrete time intervals; thus the output stream is discrete in both value and time.

EXAMPLE

A voltage proportional to wind speed is the input to an ADC. It is set to sample this voltage at 3-second intervals and convert the voltage value of each sample to a binary number that can be read by a microprocessor. These binary numbers can be thought of as integers (as in the programming languages Basic, FORTRAN, or C). Then the ADC output stream is an integer every 3 seconds. There is some scaling applied, so that an integer value of 0 might represent 0 m s^{-1} whereas 500 stands for 25 m s^{-1} and 1000 stands for 50 m s^{-1}.

After the signal is in digital form, it may be manipulated by digital processing elements (DSC = digital signal conditioning), most commonly a microprocessor, represented in block 4. It is convenient to apply the calibration equation here, to correct for nonlinearities, compensate for secondary inputs, format the output, and, in some cases, to drive the output display.

Blocks 5 and 6 represent the operations of data transmission and storage that may not be present in simpler systems but are common in larger measurement systems. Data may be transmitted via a hard-wired connection, telephone lines, a direct radio link, or satellite relay. Data storage could involve anything from holding temporary data to final archiving.

The data display, block 7, is required in even the simplest system as it is the mechanism for user access to the data. It may be a simple analog meter indicator (temperature scale on a mercury-in-glass thermometer) or a complex CRT graphical presentation.

In fig. 1-1, the system input is shown as only a single input, the measurand. In reality, most sensors have some sensitivity to other, unwanted, signals referred to as secondary inputs. In a well-designed instrument, secondary inputs are minimized or controlled but they can seldom be removed completely or ignored. As discussed below, secondary inputs are sources of error in a measurement system.

1.1.3 Sources of Error

There are four basic categories of errors (observed minus actual) in a meteorological measurement system: static, dynamic, drift, and exposure.

1.1.3.1 Static

Static errors are measured when the input is held steady and the output becomes essentially constant. These are the errors remaining after applying a calibration equation. They may be deterministic (e.g., hysteresis, residual nonlinearities, and sensitivity to unwanted inputs such as temperature) or random (noise). See chap. 3 for additional information on static performance characteristics and static errors.

1.1.3.2 Dynamic

Dynamic errors, defined in chaps. 6 and 8, are those due to changing inputs. By definition, dynamic errors disappear when the input is held constant long enough for the output to become constant. Thus dynamic effects are not present during static testing. Every sensor exhibits some time lag and may also produce more complex error modes. See chaps. 6 and 8 for a complete treatment of dynamic errors.

1.1.3.3 Drift

Drift is due to physical changes that occur in a sensor over time. This is a special category of error because these errors are not truly static, nor are they considered to be dynamic, because they are independent of the rate of change of the input. Drift errors are difficult to account for in most measurement systems; the most direct way to compensate for them is frequent calibration. Sensors that drift linearly with time can in principle be corrected, but this can lead to additional uncertainty in the final measurement. And there are many cases where drift does not change linearly with time; sometimes it changes abruptly.

1.1.3.4 Exposure

This is a very special category of errors. They are due to imperfect coupling between the sensor and the measurand. For example, consider the case of using a thermometer to measure air temperature. The sensor will never be at exactly the same temperature as the air because of dynamic errors. Steps can be taken to minimize the differences between the air temperature and the temperature of the sensor, for example, by blowing air on the sensor and shielding it from radiation and conduction sources. However, a temperature sensor will respond to radiative energy exchanges with the sun or other objects and to conductive heat transfer through mechanical supports as well as to the desired convective heat transfer to or from air moving over the sensor. The magnitude of these errors will be a function of global solar radiation, shielding, and the efficiency of convective heat transfer with the air that is strongly dependent upon the rate of air flow over the sensor. These sources of error are not present in the calibration laboratory and are not included in sensor specifications. Therefore, statements about instrument errors assume no exposure error. In a well designed, properly calibrated, and properly maintained measurement system, exposure error can easily exceed all other error sources.

In general, instruments report their own state, which is not necessarily the state of the atmosphere unless great care is taken to provide good exposure. A cup anemometer actually reports the rotation rate of its cup wheel, not the wind speed. If the cup wheel bearings are in good condition and the wind speed is steady, then there is a known relationship between the rate of rotation and the wind speed, determined by the calibration.

Static and dynamic errors are measured during laboratory testing and can be well documented, certainly better than drift or exposure error.

1.2 Standards

There are several kinds of standards that are relevant to meteorological measurement systems: calibration, performance specification, exposure, and procedural. All must be considered in system design and evaluation.

1.2.1 Calibration

Calibration standards are maintained by standards laboratories in each country, such as by the National Institute of Standards and Technology (NIST) in the United States. Standards for temperature, humidity, pressure, wind speed, and for many other variables are maintained. The accuracy of these standards is more than sufficient for meteorological purposes. Every organization attempting to maintain one or more measurement stations must have some facilities for laboratory calibration, including transfer standards. These are standards used for local calibrations that can be sent to a standards laboratory for comparison with the primary standards. This is what is meant by traceability of sensor calibration to NIST standards. Ideally, the calibration of all sensors can be traced back to such a standards laboratory.

1.2.2 Performance

Performance specification standards refer to the terminology, definitions of terms, and the method of testing static and dynamic sensor performance. The American Society of Testing and Materials (ASTM) has been active in establishing these standards. We must agree on use of terms such as time constant, response time, sensor lag, and so on, and the definition of the chosen terms. It is essential that there be a standard method of testing sensors to determine their performance characteristics. Without these standards, vendor performance specifications would be difficult to interpret.

1.2.3 Exposure

Exposure standards are necessary to define what is meant by adequate exposure for certain classes of applications. For example, what is meant by surface wind speed and direction on the synoptic scale? Is it acceptable to mount the anemometer beside a building? How about on the roof of a building? For synoptic observations, we want the measurement to be representative of a large area. An anemometer mounted beside a building or on its roof provides measurements strongly influenced by the building and therefore are not representative of a large area. At what height above ground should measurements be made? The mean wind speed approaches zero near the ground, so an anemometer should be mounted at a standard height above ground. Norment (1992) and Oost (1991) have shown that the shape of the sensor itself, and that of the supporting structure, can also perturb wind flow. To make measurements comparable, there should be, at least, a standard mounting height and some standards about the proximity of obstructions.

The WMO (World Meteorological Organization) specifies a standard mounting height for wind instruments of 10 m above level, open terrain. The distance between the anemometer and an obstruction (buildings, trees, etc.) must be at least ten times the height of the obstruction. This precludes mounting an anemometer on the roof of a building.

Temperature sensors, according to WMO recommended practice, should be exposed in a radiation screen, with or without forced ventilation, at a height of 1.25 m to 2.00 m above a level ground surface. The screen must not be shielded by or close to trees, buildings or other obstructions. A measurement site must not be on a steep slope or in a depression where thermal conditions might not be representative of the larger scale. Exposure on top of buildings is not recommended because of the vertical temperature structure in the atmosphere and the perturbation caused by buildings. Where snow is persistent, it is acceptable to maintain the sensor at a constant height above the snow surface.

Precipitation measurements are best made in clearings surrounded by brush and trees to reduce the wind effect. There is frequently a requirement to locate the rain gauge, with other sensors, close to the data logger. One way to resolve this conflict is to equip each rain gauge with a wind screen. This is a screen designed to minimize the effect of wind on the gauge catch. See chap. 9 for additional details.

In the real world, some of these requirements are mutually exclusive and so many sites fail to meet all of the exposure specifications. Therefore, it is necessary to document carefully and completely sites with photographs to show the terrain, especially the wind fetch.

1.2.4 Procedural

Procedural standards refer to selection of data sampling and averaging periods and to simple algorithms for commonly computed quantities. These standards have been evolving slowly without much compliance so far. When data are used only within one network and for narrowly defined goals, these standards are not so important. However, when data from several networks are combined or data are used in more diverse applications, adherence to procedural standards becomes significant. Implementation of these standards requires local processing capability, usually found in the data logger, the data collection platform, or in another local system element.

1.3 System Integration

Design of a measurement system is powerfully affected by considerations other than the choice of sensor and data logger. Selection of the measurement platform, data communication system, and type of power have a profound effect on overall system design. Communication system limitations may dictate the location of remote sites, forcing compromises in site location. Power limitations may prohibit the use of certain types of sensors.

Sensors are typically mounted on a stationary platform (a simple mast or tall tower) or on a moving platform (balloons, planes, ships, etc.). Ideally, data are communicated in real time from the measurement site or platform to a central archiving facility. In some cases, real-time communication is not possible but, instead, data are manually collected at periodic intervals, usually in some electronic form. Availability of electrical power, or the lack of it, may seriously affect the system design.

1.3.1 Instrument Platforms

It is not surprising that virtually every type of instrument platform is used in meteorology because the atmosphere is so extensive and because most of it is quite inaccessible. These platforms include masts, instrument shelters, tall towers, balloons, kites, cars, ships, buoys, airplanes, rockets, and satellites. Synoptic data platforms include balloons and satellites supplemented by buoys and ships over the ocean. In addition, aircraft are used for hurricane observation and some data are collected from commercial flights to fill in gaps in the observation networks. Aircraft are extensively used for research investigations around thunderstorms or wherever high-density upper-air data are needed.

When selecting a platform, consideration should be given to where the measurement is to be made, whether the platform can be permanently fixed or is moving, cost, and exposure. To some extent, any platform, even a simple tower for surface measurements, interacts with the atmosphere and affects instrument exposure. A simple 10 m tower, shown in fig. 1-2, has a wind sensor at 10 m and temperature and a relative humidity (T&RH) sensor at 1.5 m in addition to a radio antenna for data transmission, a solar panel and battery for power, a barometer, and a data logger. These sensors must be mounted with due consideration for exposure to prevailing winds, to minimize tower effects.

1.3.2 Communication Systems

A communications network is a vital part of almost every meteorological measurement system at all scales. Historically, meteorological communications have relied primarily on land-line and radio links. More recently, polar orbiting and geostationary satellites are used for data communications in macroscale or synoptic measurement systems and even in many mesoscale systems. Commercial satellites are used to broadcast data from central points, with sophisticated uplinks, to users equipped with fairly simple antennas and receivers (inexpensive downlinks).

The ideal communications system would reliably transmit data from the remote instrument platform to a central facility and in the reverse direction with little or no

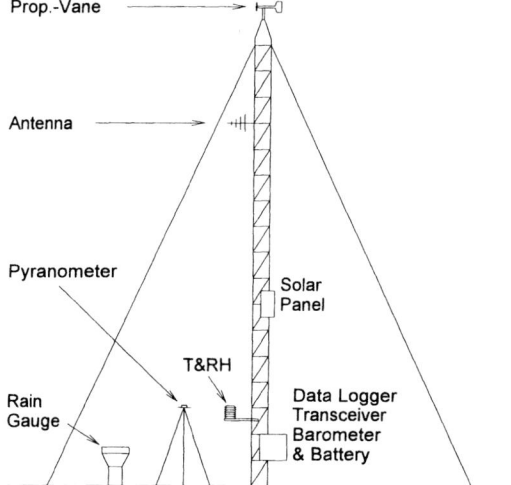

Fig. 1-2 A 10-meter tower for surface measurements.

time delay and without limiting the volume of data to be transmitted. Communication both to and from the remote site is required to synchronize local clocks in the data loggers, to load operating programs into the data loggers, and to make special data requests, to name a few. Two-way communications are not essential but highly desirable.

1.3.2.1 Telephone

Commercial telephone systems provide adequate signal bandwidth, are generally reliable, and cover most land areas. The cost is prohibitive if one must pay for running lines to each station, especially for a short-term project. Even for long-term projects like the Oklahoma Mesonet (Brock et al., 1993) and the ARM Program (Stokes and Schwartz, 1994), phone lines were either avoided entirely or used very sparingly because of the expense involved.

Recent advances in cellular telephone technology coupled with decreasing airtime charges mean that cellular data communications have become a viable alternative to traditional phone lines.

1.3.2.2 Direct Radio

Direct radio links from the remote stations to a central base station are desirable because they offer flexibility, but earth curvature limits line-of-sight links. Figure 1-3 shows the maximum line-of-sight distance between two stations if the remote station antenna is at a height of 10 m. For a base station or repeater antenna height of 200 m, the line-of-sight link is only a little more than 60 km. Direct radio, even

10   Meteorological Measurement Systems

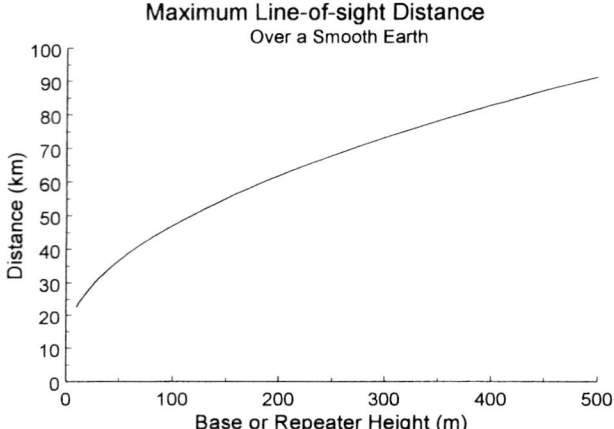

Fig. 1-3 Line-of-sight distance over a smooth earth as a function of height of one end of the link when the other end is fixed at 10 m.

when augmented with repeaters, severely limits the size of a network and causes immense difficulties in complex terrain. For example, if the path of the signal from a remote station to the repeater or to the base station is too close to the ground, the signal could be trapped in an inversion layer and ducted away from the intended destination.

1.3.2.3 Satellite

The first communications satellite that permitted an inexpensive uplink (low power transmitter and simple antenna) was the Geostationary Operational Environmental Satellite (GOES). An inexpensive uplink is essential when a large number of remote stations is involved. In addition, stations may be powered by batteries and solar panels, thereby requiring low-power radio transmitters. As satellite communications technology evolves, communication restraints will be eased, which will lead to vastly improved meteorological networks, especially for the mesoscale.

1.3.3 Power Source

Electrical power consumption of a measurement system is often a vital consideration; the primary concern is cost. Where commercial power is available, cost is not usually a problem. However, many systems are required to be portable or to operate in locations where commercial power is not available. In these cases, the power source is usually batteries, perhaps supplemented with solar panels. Such systems must operate on a severely limited power budget and that constraint affects the selection of components and the overall system design. Battery-powered systems are constrained to select sensors with low power consumption and/or to switch the sensors on only as needed, to conserve power. Heaters generally cannot be used and local computational capability may be severely limited. Therefore, all components must be rated for operation over the expected temperature range.

1.4 Human Aspects of Measurement

Automated instrumentation is used to make measurements that humans cannot, for reasons of safety, cost, or performance (speed or accuracy for example), and the instrumentation must deliver data to the user. Therefore, all measurement systems must be viewed as extensions of the human who wishes to use the data.

1.4.1 Human Perception versus Sensor Measurements

Humans can directly sense a number of variables such as temperature, wind speed and direction, and solar radiation but we have to be calibrated against some standard. We perceive some variables, such as pressure, poorly if at all. In addition, human perception of temperature is affected by wind speed and solar radiation and by humidity in hot weather. Despite all this, the human observer is generally superior to instruments in that we can detect subtle influences and perceive patterns in the weather that would be difficult to discern from the worldview filtered through a measurement system. Humans are superior to instruments for some variables such as precipitation type and reign supreme for visibility that, by definition, is the distance that humans can see objects.

1.4.2 Reasons for Automation

As noted above, the measurement process can be automated with the use of instruments, data loggers, and so on, to perform tasks that humans cannot or will not do.

1.4.2.1 Cost

Very often, a measurement task can be implemented at far lower cost by eliminating people and replacing them with sensors, data loggers, and data communication equipment. Consider a mesoscale network that requires measurement of 10 variables at five-minute intervals 24 hours a day, day after day. Each station would require about five (allowing for shift work, vacations, and sick leave) very dedicated people. Their salary alone would be far more than the cost of the usual equipment.

1.4.2.2 Performance

Instruments can be located in environments hostile to humans and perform tasks difficult or impossible for humans such as recording a number of variables at 0.1

second intervals. In addition, the accuracy of human observations is generally less than that of a well-calibrated and suitably exposed electronic sensor.

1.4.2.3 Eliminating Human Error

We expect instruments to make repetitive measurements with only occasional error. In the same task, a human would be soon bored and, consequently, error prone.

If well done, automation of measurements can reduce costs, improve sensor performance, and reduce human error, but there is a trade-off: the loss of involvement of a human observer. The user of the data may not know what type of sensor was used, how representative its location, magnitude or likelihood of environmentally induced errors, or the response of a sensor to extreme conditions. Automated networks isolate the end user from the sensors and from the people who understand the measurement system.

Automation typically causes a vast increase in the volume of data obtained. Measurements are made more frequently from more sensors in more places. Even with the aid of computers and sophisticated graphics packages it is extremely difficult for a user to examine all of this data and therefore even gross errors may go undetected. Unless automatic data monitoring procedures are implemented for quality control, automation can easily cause a decrease in data quality.

1.4.3 Design, Implementation, and Maintenance of Measurement Systems

While they are important topics, it is beyond the scope of this text to discuss the elements of ergonomic design, or design for easy use and maintenance. The overall measurement system should be designed to facilitate data quality assurance. This could range from local computation and reporting of the standard deviation to installation of redundant sensors.

1.4.4 Interpretation of Sensor Specifications

Reputable vendors offer sensors with a complete set of specifications which can be used to compare sensors from various vendors. What do these specifications tell us about values reported in field use?

EXAMPLE
If the vendor specifies a temperature sensor inaccuracy of $0.2°C$, what can one infer about the reported datum of $30°C$?

(a) that there is complete certainty that the actual air temperature is $30°C \pm 0.2°C$?

(b) that there is a 95% probability that the actual air temperature is $30°C \pm 0.2°C$? (If the errors were randomly distributed in a Gaussian distribution with a standard deviation of $0.1°C$)

(c) the probability statement in (b) is OK provided the user can offer reasonable assurances about drift, dynamic error, and exposure error at the time and place of measurement?

Answer (a) cannot be true; it is impossible to put absolute limits on error as that would require testing of all possible input values throughout the sensor range, an infinite task. In practice, if the sensor range is, say, from $-30°C$ to $50°C$, the sensor will be tested in a calibration laboratory at a finite number of temperature steps. It might be tested at 17 different temperatures in $5°$ steps. Setting absolute bounds also requires testing of all sensors. This can be done, but usually at extra cost. Further, one component of error has a random source that manifests itself as irreproducibility: the property that a sensor, when exposed repeatedly to identical conditions, will not produce identical outputs.

Answer (b) could be correct only in the context of a laboratory calibration where conditions are carefully controlled and there is no possibility of drift, dynamic error, or exposure error.

In field conditions, answer (c) is correct. Given a high-quality sensor that has been properly calibrated and a well-designed data acquisition system where the complete system is properly maintained, the largest source of error will be exposure error. This is not included in the vendor specifications because it is a function of many variables that are not under the vendor's control. Exposure error for a temperature sensor is a function of sensor design, radiation shield design, climatology, solar radiation, wind speed and direction, surface reflectivity, and so on. See sect. 3.2 for additional information.

Interpretation of measurements requires involvement of the user, who must judge whether the instrument was properly exposed and whether unusual conditions exist. A measurement that is unusual or unexpected might result from instrument failure or from unusual atmospheric conditions. Sometimes the cause is obvious. If the reported temperature is $-273°C$, one can reasonably conclude that there has been some kind of instrument failure. The obvious cases should be detected by a data quality assurance system. There will always be some odd data, outliers, that a quality assurance system will pass. The user is the ultimate judge.

1.4.5 Interpretation of Results

Do sensors measure what we want them to measure? A measurement system reports air temperature. Does that mean that the sensor exposure is representative of a large area? That it implies exposure at the WMO standard height somewhere between 1.25 m and 2.0 m? If the data come from a general-purpose mesoscale or synoptic scale measurement system, the presumption is that WMO guidelines have been followed and that the temperature sensor was not installed near a building, a tree, or other obstruction. However, even if WMO guidelines have been followed, either aspirated or unaspirated temperature shields could have been used (see chap. 4). Under certain conditions, shields can induce large errors.

First came the observer who was, at once, the sensor, data acquisition system, and data user. One can sense directly, for example, wind speed and direction and be aware of wind obstacles or shelters that may alter one's perception of the phenomena.

Then came the observer with a hand-held sensor, better equipped to make quantitative measurements if proper operation and calibration of the sensor could be verified. With simple instruments, the observer could readily verify operation and calibration. The observer was still an intimate part of the measurement process. We have progressed to more complicated instruments, powerful data acquisition systems, and communication channels that allow us to monitor measurements made anywhere in the world. But the user of the data may not even know what the sensing instrument looks like or how it is exposed to the atmosphere. The data presentation may not indicate measurement uncertainty. No wonder we find people using terms like "ground truth." Anyone familiar with the measurement process knows that there is no such thing as measurement truth. We can only estimate the variable in question and attempt to quantify the uncertainty.

1.4.6 Human Judgment

Users of the data must ultimately decide to accept, reject, or question the data with or without adequate knowledge for such a judgment. They can be assisted by better instrumentation education, by improved overall measurement system design, and by insisting upon availability of information about system specifications.

1.4.6.1 Instrumentation Education

The objective of a basic instrumentation course is to teach the physics of instruments, the meaning of commonly used instrument performance characteristics and specifications, and the kind of error, especially exposure error, characteristic of each type of instrument. Instrument physics must be the foundation of an instrument course in order to understand the instrument, its fundamental limitations, and how it might be improved.

1.4.6.2 Improved Measurement System Design

The essential first step of any measurement system design is a clear statement of goals. Following this, one must recognize the applicable standards and adhere to them to the extent compatible with the goals. There are standards for instrument exposure, calibration, performance specification, and data acquisition procedures available from ASTM.

If the measurement is meant to be representative of a fairly large area, as in a mesoscale or synoptic observing system, then, for example, the wind measurements should be made at a height of 10 m above level terrain. There should be a clear fetch in all directions with no obstruction closer than 10 times the height of the obstruction. If the purpose of the wind measurement is to characterize flow around a building or in complex terrain, then this standard is not compatible with the goal and there is good reason for not following it. In this case, the goal and the deviation from the standard should be clearly stated so that any user of the data will know what was done and why. This will reduce the chance of misusing the data.

Traditionally, automation of measurement systems has been limited by the technology available at the time of design. Limits have been imposed by the cost of local computer processing power and by communication constraints. As one might expect in times of rapid growth of computer and communication technology, currently available measurement system designs lag far behind technological limits. This growth has been so rapid that even design concepts are lagging. As noted above, the introduction of measurement technology has not been without real tradeoffs: the loss of involvement of the end user with the measurement process and the increased difficulty of maintaining quality assurance. However, technological developments can be, and should be, channeled to help alleviate the problems originally caused by the introduction of technology.

Measurement system design can include the local computation and transmission of derived quantities (standard deviation, and minimum and maximum values in the averaging period, to mention some simple examples) to facilitate improved data quality assurance. In some cases, redundant sensors can be used effectively.

1.5 Quality Assurance

Quality assurance methods and/or final data format may depend on the primary end-user of the data; for instance, will it be used for research or by the general public? For example, relative humidity (RH) sensor inaccuracy specifications allow for the RH reported by the sensor to be in excess of 100% (as high as 103%). A sensor reporting an RH of 102% may be operating correctly but could cause problems for under-informed users. For example, modelers ingesting RH may need to be aware that values greater than 100% are possible and do not necessarily indicate super-saturation conditions. In addition, those unaware of instrument inaccuracy specifications may think the data are incorrect if RH is above 100%.

A more complete data quality assurance program can be developed for a network of stations than for a single measurement station and so this discussion will address the issue of quality assurance for a network. A single station assurance program would be a subset of this.

In designing a measurement system, it is useful to consider the impact of automation on data quality.

Schwartz and Doswell (1991) claim that automation has been accompanied by a decrease in data quality but that this decrease is not inevitable. When a measurement system is automated, some sources of human error are eliminated, such as personal bias and transcription mistakes incurred while reading an instrument, to name only two. However, another, more serious form of error is introduced: the isolation of the observer from the measurement process.

One tends to let computers handle the data under the mistaken assumption that errors are not being made or that they are being controlled. Unless programs are specifically designed to test the data, the computer will process, store, transmit, and display erroneous data just as efficiently as valid data. Automatic transmission of data tends to isolate the end user from the people who understand the instrumentation system. Utilization of computers in a measurement system allows data to be collected with finer time and space resolution. Even if the system is designed to let observers monitor the data, they can be overwhelmed by the sheer volume and unable to effectively determine data quality. Automation, or the use of computers in a measurement system, can have a beneficial impact on data quality if the system is properly designed. Inclusion of data monitoring programs that run in real time with effective graphic displays allows the observer to focus on suspect data and to direct the attention of technicians.

The objective of the data quality assurance (QA) system is to maintain the highest possible data quality in the network. To achieve this goal, data faults must be detected rapidly, corrective action must be initiated in a timely manner, and questionable data must be flagged. The data archive must be designed to include provision for status bits associated with each datum. *The QA system should never alter the data but only set status bits to indicate the probable data quality.* It is probably inevitable that a QA system will flag data that are actually valid but represent unusual or unexpected meteorological conditions. Flagged data are available to qualified users but may not be available for routine operational use.

The major components of a QA program are the design of the measurement system, laboratory calibrations, field intercomparisons, real-time monitoring of the data, documentation, independent reviews, and publication of data quality assessment. Laboratory calibrations are required to screen sensors as they are purchased and to evaluate and recalibrate sensors returned from the field at routine intervals or when a problem has been detected. Field intercomparisons are used to verify performance of sensors in the field since laboratory calibrations, while essential, are not always reliable predictors of field performance. QA software is used to monitor data in real time to detect data faults and unusual events. A variety of tests can be used, from simple range and rate tests to long-term comparisons of data from adjacent stations. Documentation of site characteristics and sensor calibration coefficients and repair history is needed to answer questions about possible data problems.

Independent reviews and periodic publication of data quality are needed since people close to the project tend to develop selective awareness. These aspects of the overall QA program must be established early and enforced by project leaders. The whole QA program should, ideally, be designed before the project starts to collect data.

1.5.1 Laboratory Calibrations

Laboratory calibration facilities are required to verify the calibration of suspect instruments and to obtain a new calibration for instruments that have drifted out of calibration or have been repaired. However, a laboratory calibration is not necessarily a good predictor of an instrument's performance in the field. This is because laboratory calibrations never replicate all field conditions. For example, a laboratory calibration of a temperature sensor would never include the effects of solar and earth radiation, nor would it be subject to poor coupling with the atmosphere due to low wind speeds.

The ultimate calibration tool for wind sensors is a wind tunnel. A good wind tunnel with adequate reference instrumentation to determine the flow speed is an essential tool for establishing a complete calibration of a wind sensor. A far less expensive alternative would be to use simple motor calibration sets provided by the wind sensor vendors. These sets test the anemometer transducer but not the cup wheel or propeller.

Since it is difficult to measure air temperature in the field to $0.1°C$, due to radiation error, the laboratory calibration is relatively easy to implement. All that is needed is a temperature transfer standard with errors less than about $0.03°C$, a temperature chamber, a bath, and a stirrer. With the temperature standard, it is not necessary to set the chamber to precise temperatures. Nor is it necessary to hold the chamber temperature constant; it is only necessary to control the rate of change of temperature. The rate must be low enough that errors induced by spatial gradients between the temperature standard and the test sensors and the errors induced by the time response of the sensors are small compared to the acceptable error in the sensors.

Rate errors are fairly easy to detect; if the test sensors lag the reference sensor during increasing temperature and lead it when the temperature if falling, the rate of change is too high. The response of the sensors will be a function of the kind of bath and the amount of stirring. The bath also affects spatial gradients. These can be detected by correlating errors with sensor position.

1.5.2 Field Intercomparisons

Two types of field intercomparisons should be performed to help maintain data quality. First, a field intercomparison station should be established and, second, when technicians visit a station, they should carry portable transfer standards and make routine comparison checks.

The field intercomparison station should comprise of operational sensors and a set of reference sensors (higher quality sensors). Both should report data to the base station but the reference station data should be permanently flagged (marked as suspect or otherwise not suitable for operational use).

Portable transfer sensors can be used to make reference measurements each time a technician visits a station. This method can detect drift or other sensor failures that could otherwise go undetected. These sensors can include a barometer and an Assmann psychrometer. In addition, technicians can carry a lap-top computer to read current data, make adjustments to calibration coefficients when sensors are changed, set the data logger clock (if it cannot be set remotely), and reload the data logger program after a power interruption.

1.5.3 Data Monitoring

Neither laboratory calibration nor routine field intercomparisons will provide a real-time indicator of problems in the field. In a system that collects and reports data in real time, bad data will be transmitted to users until detected and flagged. The volume of data flow will likely be far too great to allow human observers to effectively monitor the data quality. Therefore, a real-time, automatic monitoring system is required.

The monitor program should have two major components: scanning algorithms and diagnostic algorithms. The function of the scanning algorithms is to detect outliers while the diagnostic algorithms are used to infer their probable cause. The monitor program can analyze the incoming data stream using statistical techniques adapted from exploratory data analysis, and knowledge of the atmosphere, knowledge of the measurement system, and using objective analysis of groups of stations during suitable atmospheric conditions.

Exploratory data analysis techniques are resistant to outliers and robust, that is, insensitive to the shape of the data's probability density function. Knowledge of the atmosphere allows us to place constraints on the range of some variables such as relative humidity that would be flagged if it were reported greater than approximately 103% (due to sensor inaccuracy specifications). Knowledge of the measurement sys-

18 Meteorological Measurement Systems

tem places absolute bounds on the range of each variable. If a variable exceeds these limits, a hardware failure may have occurred.

The monitor program must be tailored for the system and should be developed incrementally. Initially, it could employ simple range tests, while more sophisticated tests can be added as they are developed. The QA monitoring program will never be perfect; it will fail to detect some faults and it will label some valid data as potentially faulty. Therefore, *the monitor program must not delete or in any way change data* but set a flag associated with each datum to indicate probable quality. The monitor program should have a mechanism to alert an operator whenever it detects a probable failure. Some of these alerts will be false alarms, that is, not resulting from hardware failure, but may indicate interesting meteorological events.

1.5.4 Documentation

There are several kinds of documentation needed: documentation of individual station characteristics, a station descriptor file, and a sensor database.

Station characteristics can be documented by providing an article describing the station and its instrumentation. In addition, there should be a file of panoramic photographs showing the fetch in all directions and the nature of the land. Aerial photographs can also provide valuable information about a site, as can high-resolution topographical maps.

As part of the system database, a system descriptor file can include the location and elevation of each station and the station type (e.g., standard meteorological station, special purpose agricultural station, or sensor research station).

It is necessary to maintain a central database of sensors and other major components of the system including component serial number, current location, and status. Some sensors have individual calibration coefficients so there must be a method of accounting for sensors to insure that the correct calibration coefficients have been entered into the appropriate data logger. It is also necessary to keep records of how long a component has been in service and where it was used, so that components that suffer frequent failures can be identified. This would help to determine if the component was seriously flawed or if the defect was characteristic of the component design.

This kind of accounting cannot be left to chance; if it is, sensors will inevitably be matched with the wrong calibration coefficients or put back into service without having been repaired or recalibrated. Some sensors require periodic recalibration but it is not feasible to recalibrate them all at once. Therefore, a formal database system should be set up. All technicians should be required to report maintenance activity, including swapping components, and this information should be entered into the database. The database system should be able to generate reports indicating the serial number of every component at a station, the number of components awaiting repair at any given time, the number of spares available, the history of any given sensor or sensor type, etc.

1.5.5 Independent Review

For much the same reason that scientific proposals and papers are reviewed, periodic independent reviews of a network's performance should be invited. It is always possible for people in constant close proximity to a project to become blind to problems and this would help alleviate this.

1.5.6 Publication of Data Quality Assessment

There will be frequent data faults in any network, even with the data quality assurance program outlined above. To assist critics in making a realistic assessment, it would be desirable to publish, periodically, an honest appraisal of the network performance including all data faults, causes when known, and action taken.

1.6 Scope of This Text

In situ or immersion sensors, the subject of this text, are those in direct contact with the atmosphere being measured. Examples are thermometers, anemometers, hygrometers, pressure sensors, rain gauges, etc. Remote sensors monitor the state of the atmosphere at distances great enough to eliminate interaction between the sensor and the parcel of air being sensed. Remote sensors include radars, lidars, sodars, and radiometers.

Sensors discussed in this text include the sensors commonly used in surface networks. Cloud height sensors are included, even though they are remote sensors by the above definition, because they are used in surface synoptic scale networks. There is a brief discussion of radar as used to estimate rainfall in order to contrast this measurement with ordinary rain gauges.

QUESTIONS

1. Define the following terms:

    analog signal conditioning
    analog-to-digital converter
    atmospheric coupling
    calibration standards
    data quality assurance
    digital signal processing
    drift error
    dynamic characteristics
    dynamic error
    exposure error
    exposure standards
    functional model
    in-situ or immersion sensors
    instrument
    measurand
    performance standards
    primary input
    procedural standards
    remote sensors
    secondary input
    sensor

signal
static characteristics
static error
static sensitivity
transducer

BIBLIOGRAPHY

American Society for Testing and Materials, 1987: *1987 Annual Book of ASTM Standards, Atmospheric Analysis: Occupational Health and Safety.* Section 11, Vol 11.03. American Society for Testing and Materials, Philadelphia, PA.

Bradley, J.T., K. Kraus, and T. Townsend, 1991: Federal siting criteria for automated weather observations. *Preprints 7th Symp. on Meteorological Observations and Instrumentation, New Orleans,* LA. American Meteorological Society, Boston, MA, pp. 207–210.

Brock, F.V., and P.K. Govind, 1977: Portable Automated Mesonet in operation. *J. Appl. Meteor.,* 16(3), 299–310.

Brock, F.V., G.H. Saum, and S.R. Semmer, 1986: Portable Automated Mesonet II. *J. Atmos. Oceanic. Technol.,* 3(4), 573–582.

Brock, F.V., K.C. Crawford, R.L. Elliott, G.W. Cuperus, S.J. Stadler, H.L. Johnson, and M.D. Eilts, 1993: The Oklahoma Mesonet: a technical overview. *J. Atmos. Oceanic Technol.,* 12, 5–19.

Elliott, R.L., F.V. Brock, M.L. Stone, and S.L. Harp, 1994: Configuration decisions for an automated weather station network. *Appl. Eng. Agric.,* 10, 45–51.

Huffman, G.J., and J.N. Cooper, 1989: Design issues in nearly real-time meteorological-data systems and sites. *J. Atmos. Oceanic Technol.,* 6, 353–358.

Meyer, S.J., and K.G. Hubbard, 1992: Nonfederal automated weather stations and networks in the United States and Canada: A preliminary survey. *Bull. Am. Meteor. Soc.,* 73(4), 449–457.

Militzer, J.M., S.R. Semmer, K.S. Norris, T.W. Horst, S.P. Oncley, A.C. Delany, and F.V. Brock, 1995: Development of the prototype PAM III/Flux-PAM surface meteorological station. *Preprints 9th Symp. on Meteorological Observations and Instrumentation, Charlotte, NC.* American Meteorological Society, Boston, MA, pp. 490–494.

Norment, H.G., 1992: Calculation of Wyngaard turbulence distortion coefficients and turbulence ratios; and influence of instrument-induced wakes on accuracy. *J. Atmos. Oceanic Technol.,* 9, 505–519.

Office of the Federal Coordinator (OFCM), 1987: *Federal Standard for Siting Meteorological Sensors at Airports.* FCM-S41987, Washington, DC, 17 pp.

Oost, W.A., 1991: Flow distortion by an ellipsoid and its application to the analysis of atmospheric measurements. *J. Atmos. Oceanic Technol.,* 8, 331–340.

Schwartz, B.E., and C.A. Doswell III, 1991: North American rawinsonde observations: problems, concerns and a call to action. *Bull. AMS* 72(12), 1885–1896.

Stokes, G.M., and S.E. Schwartz, 1994: The Atmospheric Radiation Measurement (ARM) Program: Programmatic background and design of the cloud and radiation test bed. *Bull. Am. Meteor. Soc.,* 75, 1201–1221.

Wolfson, M.M., 1989: The FLOWS automatic weather station network. *J. Atmos. Oceanic Technol.,* 6, 307–326.

World Meteorological Organization, 1983: *Guide to Meteorological Instruments and Methods of Observation.* WMO No. 8, 5th ed., Geneva, Switzerland.

GENERAL INSTRUMENTATION REFERENCES

The volumes marked with an asterisk are included even though they do not specifically address meterological applications. They are good, general instrumentation books that treat many sensors and measurement systems used in meteorology.

Dally, J.W., W.F. Riley, and K.G. McConnell, 1984: *Instrumentation for Engineering Measurements*, 2nd ed. New York, John Wiley, 584 pp.

Dobson, F., L. Hasse and R. Davis, (eds.), 1980: *Air–Sea Interaction: Instruments and Methods*. Plenum Press, New York, 801 pp.

Doebelin, E.O., 1983: *Measurement Systems: Application and Design**, 3rd ed. McGraw-Hill, New York, 876 pp.

Fraden, J., 1993: *AIP Handbook of Modern Sensors**., American Institute of Physics, New York, 552 pp.

Fritschen, L.J. and L.W. Gay, 1979: *Environmental Instrumentation*. Springer-Verlag, New York, 216 pp.

Lenschow, D.H. (Ed.), 1986: *Probing the Atmospheric Boundary Layer*. American Meteorological Society, Boston, MA, 269 pp.

Middleton, W.E.K., and A.F. Spilhaus, 1953: *Meteorological Instruments*. University of Toronto Press, Canada, 286 pp.

Nachtigal, C.L., 1990: *Instrumentation and Control: Fundamentals and Applications**. John Wiley, New York, 890 pp.

Simid, D.A., 1986: *Compendium of Lecture Notes on Meteorological Instruments for Training Class III and Class IV Meteorological Personnel*, Vols I and II. WMO-622. World Meteorological Organization, Geneva, 361 pp.

Sydenham, P.H. (Ed.), 1982: *Handbook of Measurement Science**, Vol. 1, *Theoretical Fundamentals*. John Wiley, Chichester, 654 pp.

Wang, J.Y., and C.M.M. Felton, 1983: *Instruments of Physical Environmental Measurements*. Vol. I, 2nd ed. Kendall/Hunt, Dubuque, 378 pp.

2

Barometry

The objective of barometry is to measure the static pressure exerted by the atmosphere. Static pressure is the force per unit area that would be exerted against any surface in the absence of air motion. It is an isotropic, scalar quantity. Dynamic pressure is the force per unit area due to air motion. It is a vector quantity, following the wind vector. This chapter is concerned with determining the static air pressure and doing so in the presence of air motion (wind) that requires special measurement techniques.

2.1 Atmospheric Pressure

The Earth's atmosphere exerts a pressure on the surface of the Earth equal to the weight of a vertical column of air of unit cross-section. Since air is a fluid, this pressure, or force, is exerted equally in all directions. The static pressure at the surface is given by

$$p(O) = \int_0^\infty g(z)\rho(z)\mathrm{d}z \qquad (2.1)$$

where $g(z)$ = acceleration due to gravity at height z above sea level in m s^{-2}, and ρ = density as a function of height, kg m^{-3}. The SI unit of pressure is the pascal, abbreviated as Pa. In meteorology, the preferred unit of pressure is the mb or the hPa (equivalent magnitude). Table 2-1 lists some conversion factors for units currently in use in pressure measurement and also for some units no longer favored. Standard sea level pressure in various units is shown in table 2-2.

Table 2-1 Pressure unit conversion factors.

1 mb	= 1.000000 hPa
1 in. Hg @ 273.15 K	= 33.86390 hPa
1 mm Hg @ 273.15 K	= 1.333240 hPa
1 lbf in^{-2} (psia)	= 68.94790 hPa
1 Std. atmos.	= 1013.250 hPa

The last line of table 2-2 refers to the units of lbf in^{-2}, also called psi (pounds per square inch). Pressure measurements are often called absolute (psia), gauge (psig), or differential (psid). Absolute pressure is simply the total static pressure exerted by the gas (or fluid) and so the barometric pressure is also the absolute pressure. Gauge pressure is the pressure relative to ambient atmospheric pressure. Pressure in an automobile tire is measured relative to atmospheric pressure so it is gauge pressure, not absolute pressure. Differential pressure is the pressure relative to some other pressure. Gauge pressure is a special case of differential pressure.

In addition to the static pressure there is a dynamic pressure exerted by wind flow. When the wind blows on and around the barometer or the structure containing the barometer, the dynamic pressure produces a pressure error given by

$$\Delta p = \tfrac{1}{2} C \rho V^2 \qquad (2.2)$$

where Δp is the dynamic pressure in Pa, ρ is the air density in kg m^{-3}, V the wind speed in m s^{-1} and C is a coefficient whose magnitude is typically close to unity. The coefficient C can be either positive or negative since the pressure error can be of either sign, depending upon the shape of the structure and the wind direction.

Physical principles commonly employed for measuring atmospheric pressure are the *direct* measurement of the pressure, or force per unit area, exerted by the weight of the atmosphere, and *indirect* techniques such as measuring the boiling point of a liquid exposed to atmospheric pressure.

2.2 Direct Measurement of Pressure

Some common direct methods of measuring the atmospheric pressure are to balance the force due to atmospheric pressure against the weight of a column of mercury, or to balance the atmospheric force against a spring force in the aneroid barometer.

Table 2-2 Standard atmospheric pressure in various units.

1 std. atmos. =	101325 Pa
($g = 9.80665$ m s^{-2})	1013.25 HpA
	101.325 kPa
	1013.25 mb
	760.000 mm Hg @ 273.15 K
	760.000 torr[1] @ 273.15 K
	29.9213 in. Hg @ 273.15 K
	14.6959 lbf in^{-2} (psi)

[1] A torr = 1 mm Hg at standard gravity and 273.15 K. It is named after Torricelli.

2.2.1 Mercury Barometers

A manometer is the simplest form of mercury pressure sensor or barometer. Two forms are shown in fig. 2-1. With both ends open, it measures differential or gauge pressure, whereas, with one end sealed and enclosing a vacuum, it measures absolute pressure. A manometer is a bit awkward to use, as one must measure the height in both arms of the tube and take the difference to get the raw output, h, in m.

An improved version of the mercury barometer offers high accuracy and easy calibration and, in the Fortin design, is somewhat portable. It has excellent long-term stability. However, it is difficult to automate the reading and the mercury barometer is not suitable for field use. In addition, mercury presents a health risk. A simple mercury barometer, shown in fig. 2-2, is a column of mercury enclosed in a glass tube which is sealed at the top, and with a reservoir of mercury at the bottom. There is almost a vacuum at the top of the tube. Pressure at the top of the tube will be equal to the vapor pressure of mercury plus pressure due to any residual gas left there. In a well-maintained barometer, this pressure is negligible (< 10 Pa) so we can consider it to be a vacuum. Note that the mercury surface in fig. 2-2 is shown curved because mercury does not wet glass. This curved surface is called the meniscus.

The weight of the mercury in the column is balanced against the force of the atmosphere. The height of the column, relative to the surface of the mercury in the reservoir, is determined using an attached scale. To read the height, the level of mercury in the reservoir is adjusted to the fiducial point (i.e., reference point). Then the movable index is manually set to the top of the mercury column and, to facilitate reading the scale, a vernier scale provides for interpretation between the main scale divisions as shown in fig. 2-3.

In the example shown in fig. 2-3, both main scales are marked in units of hPa with the smallest division 1 hPa. The scale on the left has a 10-division vernier that allows interpolation to 0.1 hPa. The reading is 911.00 hPa from the main scale plus 0.80 hPa taken from the vernier division that is most nearly aligned with a main scale division. The scale on the right uses a 20-division vernier that, in the example, is read as 915.00 hPa from the main scale plus 0.65 hPa from the vernier.

Mercury is used as a barometric fluid because:

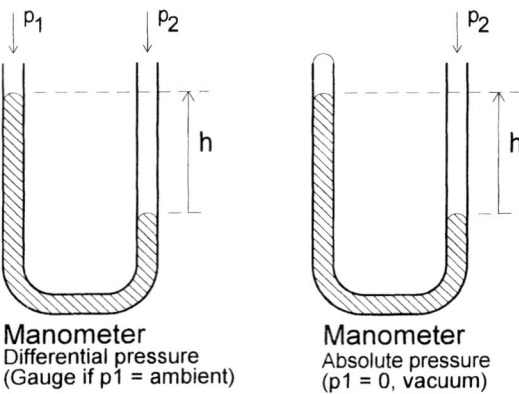

Fig. 2-1 Differential and absolute manometer.

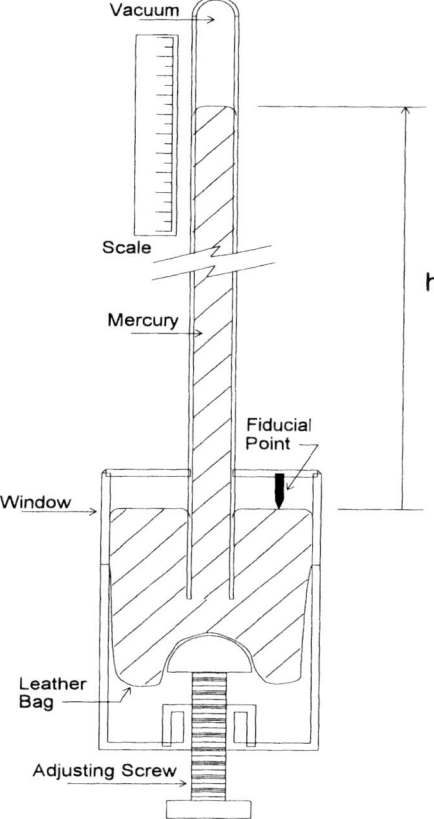

Fig. 2-2 A simple mercury barometer.

(a) it has a high density (13 595.1 kg m^{-3} at 0°C) thus the column can be of reasonable length (about 0.760 m);

(b) it has a low vapor pressure (0.021 Pa at 0°C) that has little effect on the vacuum at the top of the mercury column;

(c) it is easily purified and is chemically quite stable, but the vapor is toxic so extreme care in handling is required; and

(d) it is a liquid for a wide range of temperatures including room temperature (−38.87°C to 356.58°C).

The calibration equation for a mercury barometer is $p_1 = \rho_m g h$, where p_1 is the estimated pressure determined from the measured height (h), ρ_m is the density of mercury, g is the acceleration due to gravity, and h is the height of the mercury column relative to the level of mercury in the reservoir. This equation is obtained from $F = ma$ since $\rho_m h A_c$ = mass of mercury in the column when A_c cross-sectional

26 Meteorological Measurement Systems

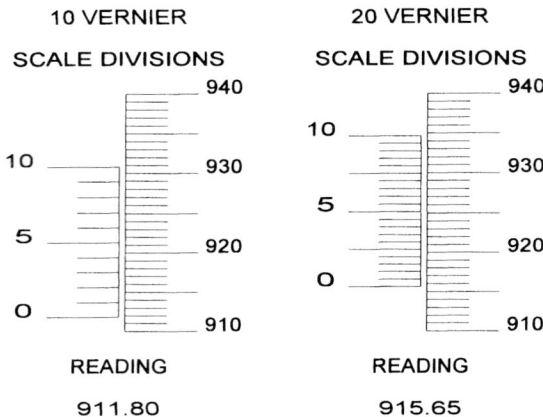

Fig. 2-3 Illustration of two different implementations of vernier scales in barometers.

area of the column (recall pressure is force per unit area so $F/A_c = p_1$). The transfer equation (which shows the sensor raw output, h in this case, as a function of the measurand, p in this case) is $h = p/\rho_m g$. To increase the sensitivity of a barometer it would be necessary to find another barometric fluid with the desirable properties of mercury but with less density. That would also make the barometer longer and more cumbersome.

The main sources of error for mercury barometers are:

(a) Dynamic wind pressure, defined in eq. 2.2, superimposed on the static pressure can produce significant (several millibars) positive or negative error. Mercury barometers are usually mounted inside a building where the pressure effect of wind may be a function of door and window openings and of the building ventilation system. Sometimes it is necessary to provide a special pressure vent to the outside called a static port that is designed to reduce sensitivity to wind speed and direction (see sect. 2.5).

(b) The density of mercury is a function of temperature and, in addition, the attached scale will have some linear coefficient of expansion. These temperature effects must be compensated by measuring the temperature of the barometer and computing a temperature correction, C_T.

(c) Since the force of the atmosphere is balanced against the weight of a column of mercury, local gravity must be known accurately and a gravity correction computed, C_G.

(d) The presence of gas in the tube above the mercury will cause an error. Mercury vapor does not seriously degrade the assumed vacuum because of its low vapor pressure as noted above. However, the presence of air or water vapor is a source of error.

(e) In small-bore barometer tubes, the surface tension of mercury will cause a depression of the mercury column. In a 5 mm internal diameter tube, the error will be about 200 Pa, and this error decreases to 27 Pa for a 13 mm tube. The correction for this error is usually incorporated into the index correction.

(f) The barometer must be kept vertical.

(g) Impurities of all kinds affect the density of mercury and, therefore, the reading. Oxidized mercury changes the surface tension effect. Contaminated mercury appears to have a dull surface. Clean mercury has a bright mirror-like surface that is helpful in setting the fiducial point.

A functional view of a mercury barometer is shown, somewhat idealistically, in fig. 2-4(a) and more realistically in fig. 2-4(b) which shows that dynamic pressure due to wind flow and the temperature of the barometer affect the output reading. The measurand (the primary input we want to measure) is static pressure but, unfortunately, dynamic pressure and temperature effects are also inputs and result in errors.

A raw barometer reading, p_1, is converted to station pressure, p_s, by applying the index correction, the temperature correction, and the gravity correction. (Define error $\equiv$ (observed value) $-$ (acual value). Then correction $\equiv$ $-$error.)

$$p_2 = p_1 + C_X + C_T \tag{2.3a}$$
$$p_s = p_2 + C_G \tag{2.3b}$$

The index correction C_X, is obtained by comparison with a reference barometer. If no index correction is given, assume that it is zero. The temperature correction C_T is developed from the known thermal expansion coefficients for mercury and for the scale. For Fortin barometers, this becomes

$$C_T = -p_1(\beta - \alpha)T \tag{2.4a}$$

where β = volume coefficient of expansion of mercury, α = linear coefficient of expansion of the scale, T = barometer temperature in Celsius, and p_1 = observed barometer reading, hPa. This equation was obtained using the expression for the volume expansion, $\Delta V = \beta V \Delta T$, and the expression for linear expansion of the

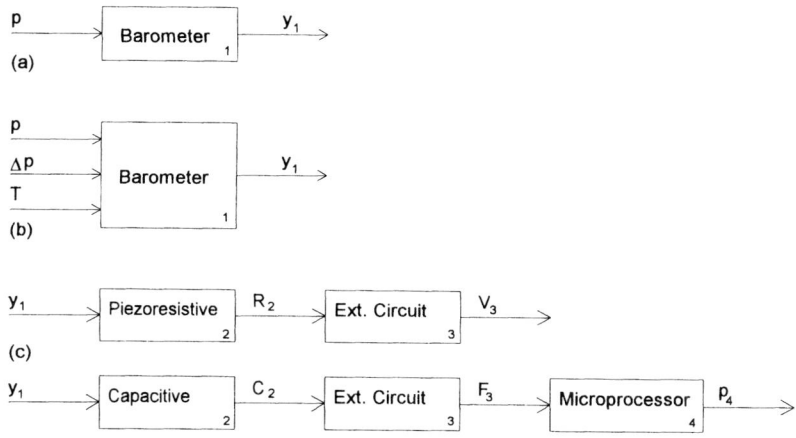

Fig. 2-4 Functional diagram of a barometer.

scale, $\Delta L = \alpha L \Delta T$. Since the reference temperature for thermal expansion is 0°C, $\Delta T = T$ when expressed in Celsius. Assuming a brass scale, and substituting the appropriate coefficients for mercury and brass ($\alpha = 1.84 \times 10^{-5}$ K^{-1}, $\beta = 1.818 \times 10^{-4}$ K^{-1}), the correction becomes, to a close approximation,

$$C_T = -1.63 \times 10^{-4} p_1 \Delta T \qquad (2.4b)$$

The correction for local gravity, C_G, is given by

$$C_G = \frac{g_L - g_0}{g_0} p_2 \qquad (2.5a)$$

where p_2 = barometer reading with temperature and index corrections, g_0 = standard gravity = 9.80665 m s^{-2} and g_L = local gravity. Note that standard gravity is the reference value used in barometry; it is not the value of gravity at latitude 45° at sea level. Compute local gravity (WMO, 1983) by first computing gravity at sea level at the barometer latitude ϕ,

$$g_\phi = 9.80616(1 - 2.6373 \times 10^{-3} \cos(2\phi) + 5.9 \times 10^{-6} \cos^2(2\phi)) \qquad (2.5b)$$

and then the elevation effect using

$$g_L = g_\phi - 3.086 \times 10^{-6} z + 1.118 \times 10^{-6}(z - z') \qquad (2.5c)$$

where z = barometer elevation in meters and z' = mean elevation within a 150 km radius. These expressions are valid for stations located on continents and away from the seashore. In eqs. 2.3–2.5, the pressure can be in any units as long as the same units are used throughout the calculation. The height must be in m and temperature must be in degrees C.

EXAMPLE

A mercury barometer reads: $p_1 = 941.23$ hPa; the temperature $T = 21.2$°C. The index correction is zero (unknown), so $C_X = 0$. The latitude is 40.00° and the elevation is 652 m. Calculate the station pressure.

SOLUTION

$$C_T = -1.63 \times 10^{-4} p_1 T = -1.63 \times 10^{-4} \times 941.23 \times 21.2 = -3.25 \text{ hPa}$$

$$p_2 = 941.23 + 0 - 3.25 = 937.98 \text{ hPa}$$

$$g_\varphi = 9.80616(1 - 2.6373 \times 10^3 \cos(80) + 5.9 \times 10^{-6} \cos^2(80)) = 9.80167 \text{ m s}^{-2}$$

$$g_L = 9.80167 - 3.086 \times 10^{-6} \times 652 = 9.79966 \text{ m s}^{-2}$$

$$C_G = p_2(g_L - g_0)/g_0 = 937.98(9.79966 - 9.80665)/9.80665 = -0.67 \text{ hPa}$$

$$p_s = p_2 + C_G = 937.98 - 0.67 = 937.31 \text{ hPa}$$

It is reasonable to ask how accurately must the latitude and elevation be determined. Do we need to measure the exact height of the barometer in the building? We can determine this by calculating the derivative of C_G with respect to z. $\partial G_L/\partial z = (p_2/g_0)\partial g_L/\partial z = (p_2/g_0)(-3.086 \times 10^{-6} \text{ m s}^{-2}/\text{m})$ if $z' = z$. In the above example, the derivative of C_G with respect to z becomes -3×10^{-4} hPa/m. If we are willing to tolerate an error of 0.05 hPa due to an error in gravity correction, we need to determine the height of the barometer to ±169 m.

Note that this is height tolerance for the gravity correction only. It is often necessary to correct the pressure to a common reference level, sea level, for synoptic observations. For this purpose, the actual height of the barometer must be determined with much greater accuracy since, near sea level, an 8 m height error is equivalent to a pressure error of 1 hPa.

The effect of an error in determining the latitude of the barometer is given by

$$\frac{\partial g_L}{\partial \phi} = -2a_0 \sin(2\phi)(a_1 + 2a_2 \cos(2\phi)) \tag{2.6a}$$

where $a_0 = 9.80616$, $a_1 = -2.6373 \times 10^{-3}$, and $a_2 = 5.9 \times 10^{-6}$. The units of the above equation are m s^{-2}/rad, so that

$$\frac{\partial C_G}{\partial \phi} = \frac{p_2}{g_0} \frac{\partial g_L}{\partial \phi} \tag{2.6b}$$

where the units are (pressure units of p_2)/rad, usually hPa/rad. Again, if we are willing to tolerate an error of 0.05 hPa, and using the above example, $\partial g_L/\partial \phi = 8.88 \times 10^{-4}$ m s^{-2}/deg and $\partial C_G/\partial \phi = 0.0850$ hPa/deg. This means we would need to determine the latitude of the barometer to ±0.6 deg or ±36 min.

2.2.2 Aneroid Barometers

An aneroid barometer consists of an evacuated chamber with a flexible diaphragm that moves in response to applied pressure. The word "aneroid" means without fluid. The restoring force is a spring that may be part of the diaphragm itself. Two common types of aneroid barometers are generally available: the welded or soldered metallic-diaphragm sensor, and the integrated circuit, silicon-diaphragm sensor.

The metallic-diaphragm sensor is a circular capsule made of a metal selected for excellent elastic properties. Consider a simple, evacuated, aneroid chamber with a flat metallic diaphragm on one side as shown in fig. 2-5(a).

The calibration equation has the form (Doebelin, 1983)

$$p = \frac{16Et^4}{3R^4(1-\nu^2)}\left[\frac{y}{t} + 0.488\left(\frac{y}{t}\right)^3\right] \tag{2.7}$$

where p = pressure in Pa, E = modulus of elasticity, N m^{-2}, y = deflection of the diaphragm center, m, t = diaphragm thickness, m, R = diaphragm radius, m, and ν = Poisson's ratio. Poisson's ratio is related to the ratio of lateral and axial strain, which is approximately 1/3 for metals.

EXAMPLE

For a metal diaphragm, let $E = 1.0 \times 10^{11}$ N m^{-2}, diaphragm thickness $= 8.0 \times 10^{-4}$ m and radius $= 5.0 \times 10^{-2}$ m. Poisson's ratio is 1/3. Calculate the pressure when the deflection ratio $y_r = y/t = 1.1, 1.2,$ and 1.3.

SOLUTION

Substitution of these values into eq. 2.7 yields $p = 68\,794, 80\,344,$ and $93\,276$ Pa.

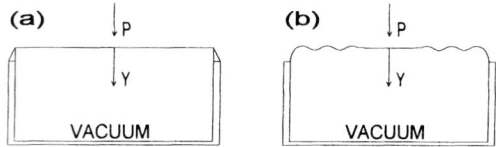

Fig. 2-5 Cross-section of a simple aneroid chamber with a flat diaphragm in (a) and with a corrugated diaphragm (b).

To find the static sensitivity, first write the calibration equation as

$$p = c_0[y_r + c_1 y_r^3] \qquad (2.8)$$

plotted in fig. 2-6, using the numeric values given in the above example. The coefficients c_0 and c_1 in eq. 2.8 may be evaluated by comparison with eq. 2.7. Then, since the static sensitivity is the derivative of y_r with respect to p,

$$\frac{dy_r}{dp} = \frac{1}{c_0 + 3c_0 c_1 y_r^2} \qquad (2.9)$$

EXAMPLE

Continuing the analysis of the previous example, the static sensitivity of the flat-plate diaphragm when $y_r = 1.1$, 1.2, and 1.3 is 9.18×10^{-4}, 8.18×10^{-4}, and 7.32×10^{-4} hPa^{-1}.

The static sensitivity decreases as the pressure increases and the diaphragm is forced to greater deflections. Note that the deflections shown are very small, on the order of the diaphragm thickness, and that the response is nonlinear (because eq. 2.9 is nonlinear in y_r). The linearity could be improved by making the diaphragm stiffer but that would reduce the sensitivity.

It is difficult to work with such small deflections. That is why aneroid diaphragms usually have corrugations that increase the sensitivity and improve the linearity. Wildhack et al. (1957) reported the properties of several corrugated diaphragms including a design referred to as NBS Shape 1. (NBS refers to the National Bureau of Standards, the predecessor of NIST, the National Institute of Standards and Technology.) The normalized transfer equation for this shape is

$$y_r = \frac{y}{t} = \frac{2.25 \times 10^5 \, D(1 - v^2)}{tE} \left(1000 \frac{t}{D}\right)^{-1.52} p \qquad (2.10)$$

where $D = 2R$ and, as above, pressure is in Pa. This equation is plotted in fig. 2-6 using the values listed in the example above. The static sensitivity is 1.06×10^{-3} hPa^{-1}.

The typical aneroid capsule consists of two corrugated diaphragms welded together. These capsules are usually evacuated to about 1 Pa although, in some designs, an inert gas at about 70 hPa is left in the capsule since the gas provides some temperature compensation. A calibration equation should be a function of the diaphragm shape and amount of residual gas. Since diaphragm designs can vary so much, an empirical calibration equation is often used:

$$p = c_0 + c_1 y + c_2 y^2 + T(c_3 + c_4 y + c_5 y^2) + T^2(c_6 + c_7 y + c_8 y^2) \qquad (2.11)$$

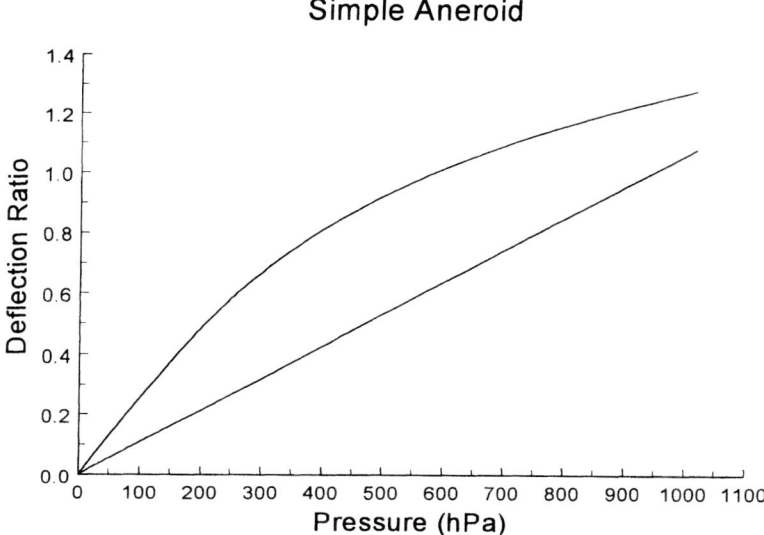

Fig. 2-6 Deflection ratio y/t, versus pressure for a flat-plate diaphragm (curved line) and for a corrugated diaphragm (straight line).

where y is the raw sensor output, T is the sensor temperature, and the c_i are the coefficients to be determined in the calibration.

A capsule may drive a dial display, as in fig. 2-7, or a recording arm through a mechanical linkage, or provide an electrical output by mechanically driving a variable resistor or transformer. In another version, the electrical signal is generated by a variable capacitor as shown in fig. 2-8. As the diaphragm deflects, it changes the distance between the capacitor plates and thereby the capacitance. When this capacitor is part of an oscillator circuit, the capacitance change produces a frequency change related to the atmospheric pressure. Figure 2-4(c) shows the functional diagram for an aneroid with piezoresistive or capacitive transducers. A piezoresistor changes resistance in response to an applied force that deforms the resistor. In this case, the resistor is fastened to or part of the diaphragm and the applied force is due to the movement of the diaphragm.

A more recent development is the silicon-diaphragm sensor made using integrated circuit (IC) technology. As with other aneroids, IC pressure sensors may be designed to be either absolute or differential sensors. An absolute pressure sensor has a reference chamber evacuated to about 3×10^{-3} Pa. In one version, bridge resistors are etched onto the diaphragm and, with a reference voltage, the output is a voltage signal from the bridge. In another version, the diaphragm and the bottom plate are electrically insulated thus forming a capacitor. Integrated circuit sensors were developed to meet other pressure sensor needs and, in the early versions, were not suitable for barometry. Significant design improvements have been made (see fig. 2-9), and integrated circuit sensors now have accuracy comparable with other aneroid barometers.

The irregular shape of the top plate is designed to make the output more linear. Temperature compensation must be provided. A silicon-diaphragm sensor is similar

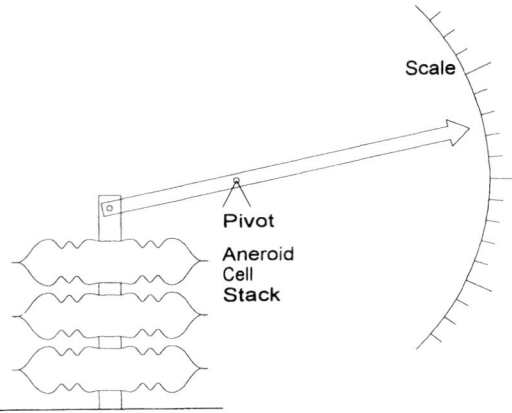

Fig. 2-7 Stack of aneroid cells linked to a dial indicator.

to the metal-diaphragm device as it is subject to many of the same error sources: temperature sensitivity, hysteresis, nonlinearity, and drift.

Major sources of error in aneroid barometers are listed below:

(a) Aneroid barometers are subject to the same exposure errors as mercury barometers and therefore must be equipped with some sort of static port.

(b) Virtually all barometers are subject to temperature-induced errors. Aneroid barometers exhibit a temperature error that is nonlinear, a function of ambient pressure, and which changes functional form from one sensor to the next, even among those of the same type and manufacture. Kim and Wise (1983) have identified six sources of temperature error in integrated-circuit pressure sensors: temperature dependence of the piezoresistive components, residual gas expansion in the reference cavity, resistor tracking errors, junction leakage currents, thermally induced stress, and packaging effects. In addition, the electronic circuits required to condition (amplify, filter, oscillate, etc.) the sensor output signal are typically sensitive to temperature. Pressure sensors usually incorporate some form of temperature compensation but, due to the many sources of temperature effects, there is always some residual temperature error. The sensors and signal conditioning circuits may also be sensitive to other interfering inputs such as supply voltage.

(c) Some aneroid sensors have a most troublesome characteristic: a hysteresis effect whose magnitude is significant in meteorological applications. It arises from defects or irregularities in the diaphragm material and/or shape. This is a poorly documented effect because it appears as a time-dependent or transient offset that may go unnoticed or be discarded as an outlier.

(d) The nonlinearity of aneroid sensors is easily handled with an on-board microprocessor. It typically requires a second- or third-order calibration polynomial. This source of error is not nearly as troublesome as the temperature, hysteresis, and drift errors.

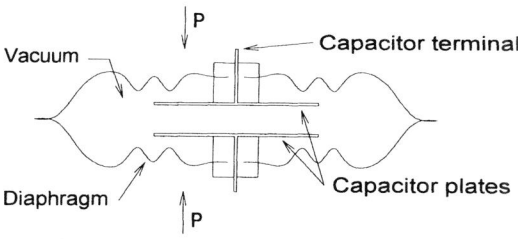

Aneroid capsule with capacitive transducer

Fig. 2-8 Aneroid capsule with built-in capacitive transducer.

(e) Drift is a long-term change in the sensor sensitivity or offset due to diaphragm creep. To compensate for drift, both Garratt et al. (1986) and Richner (1987) made periodic checking and offset correction a key feature of operational mesoscale barometer networks. This is the only way to minimize drift errors because, in general, it is impossible to know how drift will occur as a function of time. Frequent checks will not eliminate drift errors, nor allow one to correct the data, but they will greatly reduce the amount of bad data when drift errors occur.

A barometer can be designed using an integrated circuit aneroid as the sensor, with circuits to convert the small diaphragm motions to a voltage, with a built-in temperature sensor, with an analog-to-digital converter, and with a microprocessor. The microprocessor controls the measurement process by sampling the voltage output of the aneroid and the output of the temperature sensor and, typically, some reference quantities. The microprocessor applies a calibration equation, with coefficients stored in on-board ROM (Read-Only Memory), corrects for temperature effects, and reports pressure in convenient units (Pa, hPa, etc.) to the user (often another computer.) Typical specifications of an integrated-circuit barometer and of two other aneroid barometers are listed in table 2.3.

Another variation of the aneroid barometer is the Bourdon tube, shown in fig. 2-10. In this configuration, the inside of the tube is at ambient pressure and the tube is enclosed in an evacuated chamber. The tube cross-section is non-circular, flat with rounded ends. In this case, as the pressure increases, the tube tries to assume a circular cross-section; this causes its shape to become straighter and the raw output, the distance y, increases. A variety of transducers can be used to measure the raw output. In one case, a quartz element is used whose resonant frequency is a function of the force applied. In this case, the distance moved, or the force applied by the Bourdon tube, is converted to a frequency. Another arrangement is to have the tube evacuated and enclosed in a chamber open to atmospheric pressure.

2.3 Indirect Measurement of Pressure

A pressure measurement technique is called indirect if it does not respond directly to the force due to atmospheric pressure but, instead, responds to some other variable that is a function of pressure.

34 Meteorological Measurement Systems

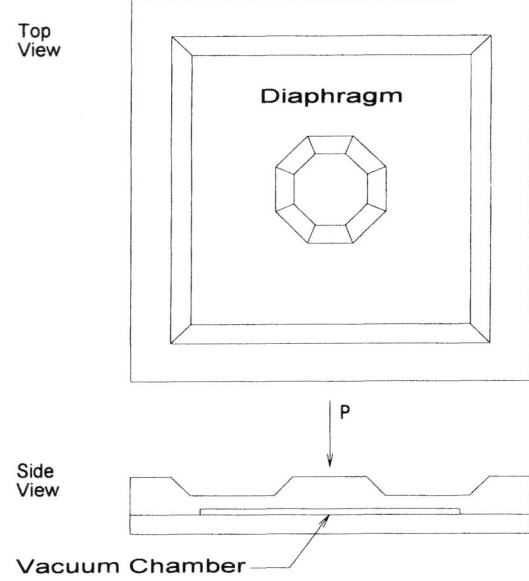

Fig. 2-9 Top view and cross-section of an integrated circuit.

2.3.1 Boiling Point of a Liquid

The boiling temperature of a liquid depends upon the liquid itself and the atmospheric pressure. For example, the boiling point of pure water at standard sea-level pressure is 373.15 K and decreases with decreasing pressure. A pressure sensor utilizing this property is called a hypsometer (literally a height meter). It comprises a flask of the hypsometric fluid, heated to maintain continuous boiling, and a temperature sensor to measure the boiling point temperature. One possible sensor con-

Table 2-3 Comparison of three aneroid barometers.

Parameter	IC Barometer	Aneroid #1	Aneroid #2
Press range	800–1060 hPa	800–1060 hPa	800–1060 hPa
Temp. range	−25 to 50°C	−25 to 50°C	−25 to 50°C
Resolution	0.01 hPa	0.01 hPa	0.01 hPa
Power	0.3 W	0.07 W	0.8 W
Errors			
Linearity	0.25 hPa	0.25 hPa	0.03 hPa
Hysteresis	0.03 hPa	0.10 hPa	0.026 hPa
Repeatability	0.03 hPa	0.10 hPa	0.026 hPa
Temperature (over 80°C)	≤ 0.20 hPa	0.40 hPa	0.43 hPa
Drift (per year)	0.2 hPa	0.30 hPa	0.13 hPa
Cal. uncertainty	0.20 hPa	0.07 hPa	0.10 hPa
Total error (root mean square)	0.43 hPa	0.58 hPa	0.46 hPa

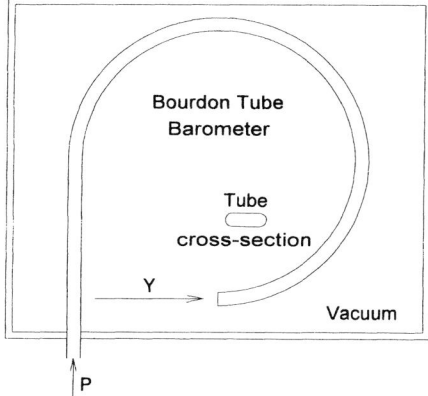

Fig. 2-10 Bourdon tube type of aneroid barometer.

figuration is shown in fig. 2-11. If the boiling point of the fluid is below ambient temperature, a heater is not required. Freon (a fluorochlorohydrocarbon) has been used in radiosondes for this reason. The temperature of the sensor housed inside the radiosonde is much higher than the air temperature at upper levels, due to the time required for the radiosonde body to lose heat. Richner et al. (1996) showed that it is possible to use a water hypsometer, with a heater, in a radiosonde. A condensing column is sometimes used to recapture the fluid; it condenses on the cooler walls and drains back into the boiling chamber. The temperature sensor must be positioned carefully to obtain a representative measure of the boiling temperature.

The relation between the vapor pressure and temperature is known as the Clausius–Clapeyron equation

$$\frac{\mathrm{d}\ln(p/p_0)}{\mathrm{d}T} = \frac{L}{RT^2} \qquad (2.12)$$

where L is the latent heat of vaporization, R is the gas constant, and T is the temperature of the vapor, assumed to be the same as that of the fluid. Table 2-4 gives the values of these parameters for several possible hypsometric fluids.

Equation 2.12 can be integrated to yield the calibration equation

Table 2-4 Parameters for several hypsometric fluids. Boiling temperature is given at standard pressure (101 325 Pa).

Fluid/Units	Boiling Temp. T_0 (K)	Latent Heat L (J kg^{-1})	Gas Constant R (J K^{-1} kg^{-1})
Water	373.15	2.5×10^6	461.5
Freon-11	296.97	1.8×10^5	60.52
Freon-13	191.75	1.5×10^5	79.59
Carbon disulfide, CS_2	319.45	3.6×10^5	109.21
Carbon tetrachloride, CCl_4	349.45	21×10^5	54.05

36 Meteorological Measurement Systems

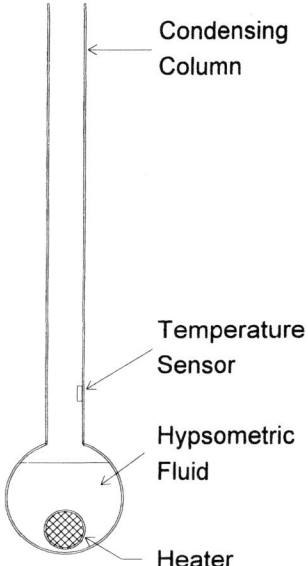

Fig. 2-11 Hypsometer implementation.

$$p = p_0 \exp\left[\frac{L}{R}\left(\frac{1}{T_0} - \frac{1}{T}\right)\right] \quad (2.13)$$

where $p = p_0$ when $T = T_0$; $T_0 =$ boiling point when $p_o =$ sea-level pressure. Equation 2.13 is not exact for water because the latent heat of water is a function of temperature and because water vapor is not a perfect gas. However, the error is small and acceptable in this context. This equation can be inverted to give the transfer equation

$$T = \frac{T_0}{1 - \frac{RT_0}{L}\ln\left(\frac{p}{p_0}\right)} \quad (2.14)$$

Water could, in principle, be used as a hypsometric fluid for surface pressure measurement. As illustrated in fig. 2-12, the static sensitivity of a hypsometer at sea-level pressure is very poor; at sea level, a large change in pressure produces a small change in boiling temperature. To work at sea level, high performance temperature sensors are required. Since the static sensitivity improves with decreasing pressure, a reasonable application of a hypsometer is in a radiosonde, especially as the accuracy of an aneroid sensor deteriorates as the pressure declines toward the lower limit of the aneroid range. In the past, radiosonde hypsometers used Freon, carbon disulfide, or carbon tetrachloride as a working fluid and the hypsometer was designed to supplement the aneroid sensor. This configuration did not require very precise temperature measurement in the hypsometer since it was used only at lower pressures where the hypsometer sensitivity improved sharply and where the typical aneroid errors became excessive. Richner et al. (1996) reported a radiosonde hypsometer which measured temperature to within 10 mK with sufficient accuracy to provide the pres-

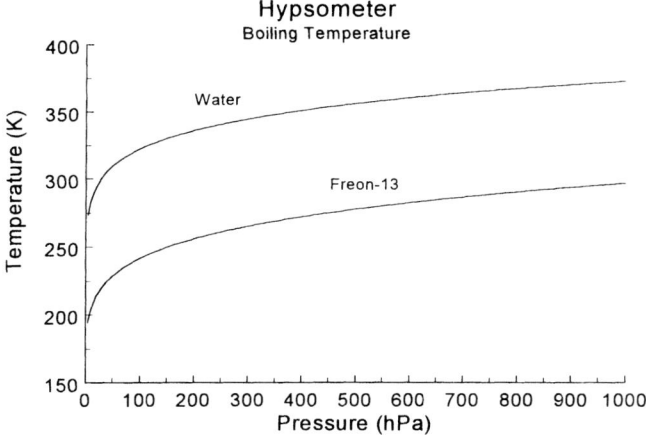

Fig. 2-12 Transfer plot for a hypsometer.

sure measurement for the whole flight, thus eliminating the need for an aneroid sensor.

EXAMPLE
Calculate the pressure error due to an error of 10 mK in a water barometer at sea level.

SOLUTION
First, find dp/dT.

$$\frac{dp}{dT} = \frac{Lp_0}{RT^2} \exp\left[\frac{L}{R}\left(\frac{1}{T_0} - \frac{1}{T}\right)\right]$$

but at sea level, where $T = T_0$, the exponential term disappears and

$$\Delta p = \frac{Lp_0}{RT^2} \Delta T = \frac{2.5 \times 10^6 \times 101\,325}{461.5 \times (373.15)^2} 0.01 = 39\,\text{Pa}$$

This is the worst-case error since the error decreases with decreasing pressure.

2.4 Comparison of Barometer Types

Each type of barometer described above can be designed in many different ways. The best of each type may be comparable in performance but it is useful to list, in a general way, the advantages and disadvantages of each type. Cost is not listed as a factor because it changes rapidly as technology changes. For example, at one time it would have been prohibitively expensive to build a barometer with a digital computer built in to apply the calibration equation. This is now commonplace.

2.4.1 Mercury Barometers

Mercury barometers of the Fortin type have the following characteristics:

- Simple physical concept. It is easy to see and understand the way such barometers work. With careful attention to design, they require no calibration (assuming one has a calibrated scale.)
- Difficult to automate. Requires manual reading.
- Difficult to transport.
- Must be kept vertical for operation.
- Requires temperature correction for expansion of the mercury and of the scale.
- Requires a gravity correction because atmospheric pressure is balanced against the weight of the mercury column.
- Mercury vapor is toxic.
- Improper handling may introduce air bubbles into the vacuum chamber, causing errors.
- Mercury contamination may affect accuracy.
- Some hysteresis due to mercury sticking to glass.
- The height of the barometer cannot be reduced.

2.4.2 Aneroid Barometers

Consider only the best aneroid barometers, whether large diaphragm, silicon diaphragm, or Bourdon tube style.

- Very small size.
- Readily automated.
- Insensitive to orientation, motion, and shock (within reasonable ranges) thus very portable.
- No gravity correction required.
- Users are not exposed to toxic materials.
- While the fundamental concept is simple, implementation is sufficiently complex that calibration is always required.
- Temperature sensitivity is high and does not have a simple, predictable correction as with mercury barometers.
- Subject to unpredictable drift, thus requiring frequent monitoring and recalibration.

2.4.3 Hypsometer

A hypsometer is still the most specialized type of barometer, having a very limited range of applications.

- Small size.
- Can be automated.
- Reasonably portable but is sensitive to orientation.

- No gravity or temperature correction required.
- Physical concept is simple and does not require calibration (assuming a calibrated temperature sensor is available) but does require careful implementation.
- No drift or hysteresis.
- Extreme nonlinearity makes implementation at sea-level pressure very difficult (may require highly accurate temperature measurement).

2.5 Exposure Error

By definition, a barometer is designed to measure the static pressure and must, therefore, be isolated from dynamic wind effects caused by air flow and from other pressure perturbations such as those caused by building air conditioning or ventilation. It is impractical to have a barometer inside a building unless it is equipped with a good static port placed in a suitable location. Dynamically induced error in a building will be a function of wind speed, direction, and number and location of open doors and windows. A static port is designed to be insensitive to horizontal wind effects but will have some sensitivity to the vertical component of airflow. Thus, if a barometer is located inside a building, its static port must be located outside of the significant pressure field of the building. A rule of thumb is that this field can extend to 2.5 times the building height and to a horizontal distance up to 10 times the height of the building.

Essentially the same problem occurs when the barometer is exposed inside a box mounted on a tower. In both cases, the barometer should be connected to a static port by a tube. A typical static port is shown in fig. 2-13. The support tube causes a small asymmetry so that when the tilt angle (the angle of the flat plate with respect to the wind vector) is 0°, the dynamic pressure error is not quite zero. The coefficient C (Akyuz et al., 1991) in eq. 2.2 is listed in table 2-5 for various tilt angles. This static port performs quite well for tilt angles less than 10°. This does not seem to be a very demanding requirement, but note that the angle is with respect to the wind vector, which is not necessarily horizontal. The wind vector may deviate from the horizontal when the static port is located on or near a building, near flow obstructions, or in complex terrain. Even over flat terrain free of obstructions, the higher the port and the more turbulent the wind, the greater the possible fluctuations in the tilt angle. Finally, while it is relatively easy to install the static port with a tilt angle of less than ±10° with respect to the horizontal (assuming the wind vector is horizontal), it could be accidentally misaligned later and cause large dynamic errors.

When $C = 0.2$, the dynamic error, at a wind speed of $20\,\mathrm{m\,s^{-1}}$, is 40 Pa (0.4 mb). This is a fairly large error when combined with the other sources of error in the

Table 2-5 Coefficient C (eqn. 2.2) for a flat-plate port vs. tilt angle.

Tilt angle	+10°	0°	−10°	−15°	−30°
Coefficient C	+0.10	−0.08	−0.20	−0.80	−1.70

40 Meteorological Measurement Systems

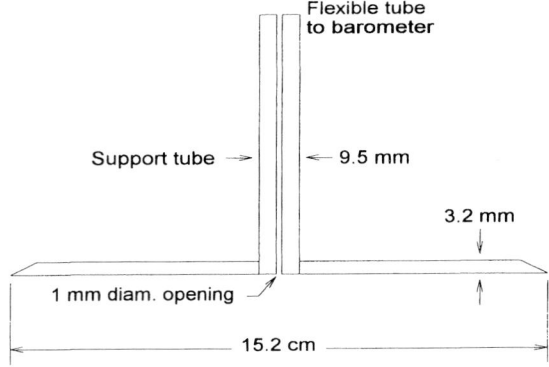

Fig. 2-13 Cross-section of a flat-plate static port.

barometer. This type of static port is unsuitable for use on a buoy, where the tilt angles will often exceed 10°. As with most other meteorological sensors, exposure errors can readily exceed all other sources of error in a well-designed measurement system.

An improved static port, reported by Nishiyama and Bedard (1991), is shown in fig. 2-14. They claim smaller errors and less sensitivity to the tilt angle.

2.6 Laboratory Experiment

The objective of this experiment is to become familiar with the Fortin barometer and to calibrate other pressure sensors with respect to this barometer.

Read the Fortin barometer, a microbarograph, an aneroid, and one other if available, on five separate occasions separated by at least four hours. Record the raw reading in the table (see p. 42) in units native to the instrument in question and convert to SI units (use hPa for pressure and Celsius for temperature).

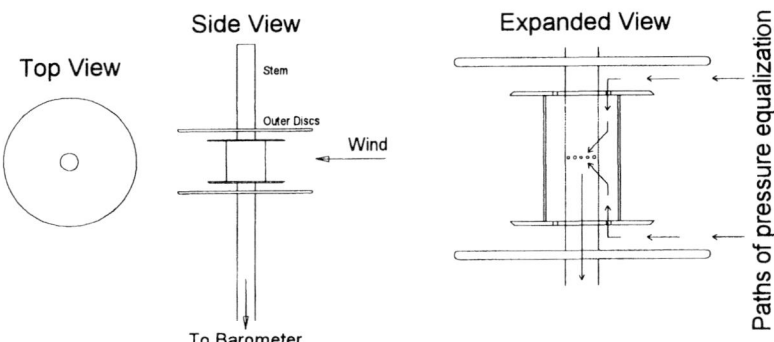

Fig. 2-14 Static pressure port designed by Nishiyama and Bedard.

Procedure

Read the Fortin barometer

(a) Open the barometer case and read the thermometer immediately to the nearest 0.1°C. The desired temperature is that of the barometer; after you open the case, convection currents may affect the thermometer reading.

(b) Use the bottom screw to adjust the level of mercury in the cistern until the mercury just touches the fiducial point.

(c) Gently tap the cistern to move any mercury sticking to the glass. Check the level and repeat (b) as necessary.

(d) Adjust the scale index to the top of the mercury column. Keep your eye level with the mercury meniscus in the tube. The zero line should be tangent to the uppermost point of the mercury.

(e) Read the pressure p_1 using the vernier, to 0.1 hPa.

(f) Lower the level of mercury in the cistern until the mercury is no longer touching the fiducial point, turn off the light, and close the barometer case.

Apply correction to the Fortin barometer reading

(a) Index correction. The index correction C_I is stated in the barometer certificate. It was obtained by comparison with a standard barometer.

(b) Temperature correction. Use eq. 2.4a or 2.4b or look up the correction in tables. Apply the index and temperature correction: $p_2 = p_1 + C_x + C_T$.

(c) Gravity correction. Determine the height and latitude of the barometer and compute C_G using eq. 2.5. Show this calculation. The station pressure is given by $p_s = p_2 + C_G$.

Read other barometers

Read the microbarograph, the aneroid, and other barometer, if available. Convert readings to hPa and compute the correction relative to the Fortin barometer.

Barometric Observations Report Name:

	Observation	1	2	3	4	5
1	Date					
2	Time					
3	Fortin, p_1					
4	Index corr.					
5	Temperature					
6	Temp. corr.					
7	p_2					
8	Gravity corr.					
9	p_s = Station press.					
10	Microbarograph					
11	Corr. = (9)–(10)					
12	Aneroid					
13	Corr. = (9)–(12)					
14	Other barometer					
15	Corr. = (9)–(14)					

Average microbarograph correction = _____
Average aneroid correction = _____
Average other barometer correction = _____

2.7 Calibration of Barometers

Calibration of barometers is a difficult and demanding task requiring measurement errors on the order of a few pascals. This is often done using a dead-weight tester that balances the weight of known metallic masses against the pressure to be measured.

It is possible to develop a barometer check station that can evaluate barometers without reaching standards laboratory performance. Such a scheme is illustrated in fig. 2-15. It is a computer-controlled system that can expose a group of barometers under test to their range of pressures (say 800 to 1100 hPa) and temperatures (−30 to 50°C) and compare them to a reference barometer. The reference barometer is a good barometer, although not of standards quality, that is kept at room temperature and not removed from the laboratory. It must be kept at the same height as the barometers under test or else the differential static column of air will create an offset. The reference barometer should be recalibrated periodically at a standards laboratory. In fig. 2-15, the valves controlled by the computer are designated V1, V2 and V3. Pressure gauges are labeled PG1 and PG2. The pressure gauges are selected to monitor the high and low reservoirs where high accuracy is not needed, but they must be capable of withstanding a large range of pressures, say 0 to 2000 hPa.

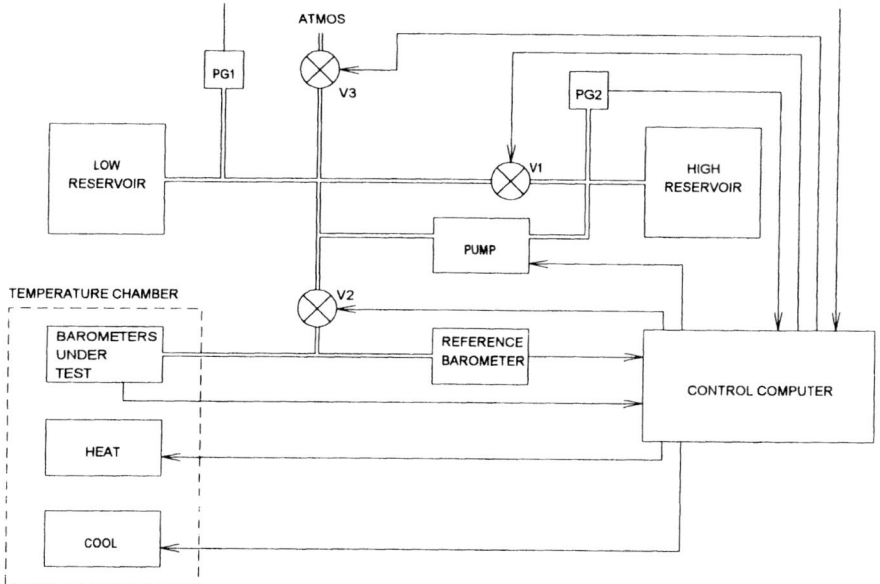

Fig. 2-15 A barometer check station designed to test barometers over their range of pressure and temperature.

QUESTIONS

1. With respect to the mercury barometer, how much inaccuracy can we tolerate in the measurement of temperature if we want the pressure error ≤ 0.05 hPa? Assume the pressure is 960 hPa.
2. Name three corrections commonly used with a mercury barometer.
3. Calculate the static sensitivity of a hypsometer using water at a pressure of 1000 hPa, 100 hPa, 10 hPa, 100 Pa, and 10 Pa. What temperature sensor resolution is required to resolve pressure to 10 Pa at each level? Why is it difficult to use hypsometers for general-purpose barometry?
4. Why does the mercury barometer require a gravity correction while the aneroid does not? Does a hypsometer have a gravity correction? Name a barometer that (a) balances atmospheric pressure with a spring; (b) balances atmospheric pressure with weight; (c) does not balance atmospheric pressure.
5. Name three factors that limit the resolution of a Fortin barometer.
6. Why is it likely that a "temperature compensated" aneroid barometer will have significant residual temperature errors?
7. What is the physical reason for hysteresis error in an aneroid? Is it possible for a mercury barometer to exhibit hysteresis? What about the effect of mercury sticking to the glass tube? Is there a hysteretic mechanism in a hypsometer?
8. Calculate the static sensitivity of a mercury barometer. How could you improve the static sensitivity of a mercury barometer?
9. What is the dynamic wind error for $C = 1$ when the wind speed is 20 m s^{-1}? If the dynamic pressure error is d_0 Pa when $V = V_0$, what is it when $V = V_1$?
10. A certain manometer has an inside diameter of 5 mm. The length of the bend in the tube, below y_1 in fig. 2-16, is 0.15 m. The maximum height of the manometer, h_m,

is the distance from the top of the barometric fluid in one column to the top of the fluid in the other column when the maximum expected atmospheric pressure $(p = p_m)$ is applied. A brass scale (not shown) is provided to measure h. Assume the brass scale reads true at 0°C (a dangerous assumption). The density of mercury (Hg) is 13595 kg m^{-3} at 0°C.

(a) If the temperature of the barometric fluid, mercury, is $T_f = 0°C$, and $p_m = 1050$ hPa, what is h_m? What is the volume of the mercury in the manometer, and its mass?

(b) If the barometric fluid were water, what would h_m and the mass of the water be?

(c) The coefficient of volume expansion of mercury is $\beta = 1.818 \times 10^{-4}$ K^{-1}. If $T_f = 25°C$, $p = 1050$ hPa, what is the change in h? Express this error in hPa.

(d) The coefficient of linear expansion of the brass scale is $\alpha = 1.84 \times 10^{-5}$ K^{-1}. When the scale temperature $T_s = 25°C$, what will a true distance of 0.7600 m appear to be? Express this error in hPa.

(e) Given another manometer with identical dimensions but with double the tube diameter: for the same pressure as in the original manometer, how does h change? How about the mass of H_g?

(f) When the pressure decreases, h decreases. Under what conditions would $h = 0$? Can h go negative? Explain.

(g) Why is the mass of mercury computed in part (a) not equal to the mass of water computed in part (b)?

11. The STATIC SENSITIVITY of a sensor is the rate of change of the raw output with respect to the input, when the input is changing very slowly. If the sensor is linear, the static sensitivity is constant over the range, but changes for a nonlinear sensor. What is the static sensitivity of a flat-plate (radius = 50 mm) diaphragm aneroid barometer at a deflection ratio of 1.35? How does this compare with the static sensitivity of the corrugated diaphragm aneroid discussed in the text? What would be its static sensitivity at the same pressure?

12. What is the raw output for the following sensors: a hypsometer, a mercury barometer, and an aneroid barometer?

13. Write the hypsometer transfer equation and identify all of the terms.

14. List two sources of error for an aneroid barometer that cannot be corrected.

15. Show that $\Delta h = \beta h \Delta T$, where β is the volume coefficient of expansion of the barometric fluid in a barometer or manometer.

16. (Advanced topic) Is there a limit to the height, i.e. a minimum pressure, to which a water hypsometer can be flown and continue to operate?
   Hint: Consider the triple point of water.

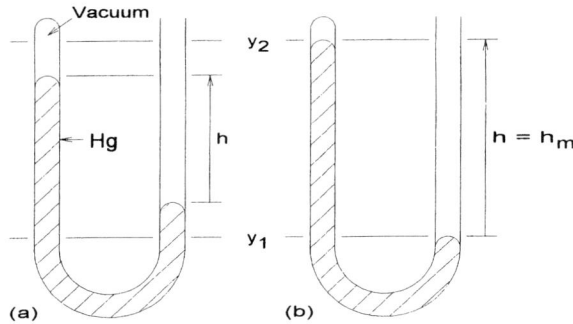

Fig. 2-16 Manometer problem.

17. Your group is going to build a water barometer to be installed in a building. Assume the temperature in the building is 22°C. Your task is to develop specifications for the scale. Do you need to know the tube diameter? If the average station pressure is 970 hPa the scale should cover the range of 970 hPa ± 50 hPa. How high will the scale mid-point be above the free surface of water outside of the tube, i.e. above the fiducial point? Length of the scale? Distance between 1 hPa scale markers?

18. You are leading the group building the water barometer. The others come to you, with some proposed design changes. One student suggests the height of the barometer tube could be reduced by increasing the diameter of a 1 m section of the tube by a factor of 10. Another student proposes to reduce the barometer height by keeping the tube diameter constant but bending the tube into a corkscrew pattern. You consult the fluid mechanics chapter of your physics book, and tell them why they are wrong.

19. Why have aneroid barometers replaced mercury barometers?

BIBLIOGRAPHY

Akyuz, F.A., H. Liu, and T. Horst, 1991: Wind tunnel evaluation of PAM II pressure ports. *J. Atmos. Oceanic Technol.*, 8, 323–330.

Brock, F.V,. and S. Fredrickson, 1995: The Oklahoma Mesonet barometer calibration system: high quality, practical and cost-effective. *Preprints 9th Symposium on Meteorological Observations and Instrumentation, Charlotte, NC.* American Meterological Society, Boston, MA, pp. 459–464.

Chan, C.K., and M.E. Simpson, 1979: Thermal-transient effects on pressure measurements. *Exper. Mech.*, 19, 324–330.

Chiswell, S.M., and R. Lukas, 1989: The low-frequency drift of Paroscientific pressure transducers. *J. Atmos. Oceanic Technol.*, 6, 389–395.

Dobson, F.W., 1980: Air pressure measurement techniques. In *Air–Sea Interaction: Instruments and Methods.* F. Dobson, ed. L. Hasse, and R. Davis. Plenum Press, New York, 801 pp.

Doebelin, E.O., 1983: *Measurement Systems: Application and Design.* McGraw-Hill, New York, 876 pp.

Garratt, J.R., I.G. Bird, and J. Stevenson, 1986: An electrical-readout, oven-controlled, aneroid barometer for meteorological application. *J. Atmos. Oceanic Technol.*, 3, 605–613.

Jones, B.W., 1992: The elimination of temperature effects in microbarometers. *J. Atmos. Oceanic Technol.*, 9, 796–800.

Kim, S.-C., and K.D. Wise, 1983: Temperature sensitivity in silicon piezoresistive pressure transducers. *IEEE Trans. Electron Dev.*, ED-30, 802–810.

Kodama, M., and Y. Ishida, 1967: Development of a barometric sensor insensitive to high winds. *J. Meteor. Soc. Japan*, 45, 191–195.

Lee, Y.S., and K.D. Wise, 1982: A batch-fabricated silicon capacitive pressure transducer with low temperature sensitivity. *IEEE Trans. Electron Dev.*, ED-29, 42–48.

Liu, H., and G.L. Darkow, 1989: Wind effect on measured atmospheric pressure. *J. Atmos. Oceanic Technol.*, 6, 5–12.

Meteorological Office, 1981: *Measurement of Atmospheric Pressure,* Vol. 1 of *Handbook of Meteorological Instruments,* 2nd ed. Meteorology Office 919a, Her Majesty's Stationery Office, London, 29 pp.

Nishiyama, R.T., and A.J. Bedard, Jr., 1991: A "quad-disc" static pressure probe for measurement in adverse atmospheres: with a comparative review of static pressure probe designs. *Rev. Sci. Instrum.*, 62, 2193–2204.

Pike, J.M., 1984: Realistic uncertainties in pressure and temperature calibration reference values. *J. Atmos. Oceanic Technol.*, 1, 115–119.

Richner, H., 1987: The design and operation of a microbarograph array to measure pressure drag on the mesoscale. *J. Atmos. Oceanic Technol.*, 4, 105–112.

Richner, H., J. Joss, and P. Ruppert, 1996: A water hypsometer utilizing high-precision thermocouples. *J. Atmos. Oceanic Technol.*, 13, 175–182.

Snow, J.T., M.E. Akridge, and S.B. Harley, 1992: Basic meteorological observations for schools: atmospheric pressure. *Bull. Am. Meteor. Soc.*, 73, 781–794.

Voorthuyzen, J.A., and P. Bergveld, 1984: The influence of tensile forces on the deflection of circular diaphragms in pressure sensors. *Sensors and Actuators*, 6, 201–213.

Wildhack, W.A., R.F. Dressler, and E.C. Lloyd, 1957: Investigations of the properties of corrugated diaphragms. *Trans. ASME*, 79, 65–82.

World Meteorological Organization, 1983: *Guide to Meteorological Instruments and Methods of Observation*, 5th ed. WMO No. 8. Geneva, Switzerland.

3

Static Performance Characteristics

Sensor performance characteristics are generally divided into at least two categories: static and dynamic. Additional categories sometimes used include drift and exposure errors. The performance of sensors in conditions where the measurand is constant or very slowly changing can be characterized by static parameters. Dynamic performance modeling requires the use of differential equations to account for the relation between sensor input and output when the input is rapidly varying. Static characteristics due to friction or other nonlinear effects would vastly complicate the differential equations so, even when the input is not steady, static and dynamic characteristics are considered separately. Static characteristics are determined by carefully excluding dynamic effects. Dynamic characteristics are assessed by assuming that all static effects have been excluded or compensated.

3.1 Some Definitions

Many of these terms have been encountered in chaps. 1 and 2, although without formal definitions.

Analog signal. A signal whose information content is continuously proportional to the measurand. If an electrical temperature sensor has a voltage output, that voltage signal fluctuates with the sensor temperature. Voltage output would be continuously proportional to the measurand (temperature) and is analogous to it, hence we refer to the sensor output as an analog signal.

Data display. Any mechanism for displaying data to the user. The stem of a mercury-in-glass thermometer with attached scale is a data display.

Data storage. A memory element or mechanism for holding data and later recovering them such as a disk or magnetic tape. Again, this could be as simple as a piece of paper.

Data transmission. The process of sending a signal from one place to another. The data transmission medium could be a piece of paper, a magnetic tape, radio or light waves, or telephone wires.

Digital signal. A signal whose information content varies in discrete steps. The step size can be made arbitrarily small such that a plot of a digitized signal could also resemble the analog signal. However, the granularity of a digital signal will be revealed if it is examined in sufficient detail.

Drift. Drift is sometimes considered to be a static characteristic but is really a separate category since it is a change in the output caused by a physical change in the instrument. An instrument that has just been calibrated has no drift error. But subsequent physical changes, such as a leak in an aneroid diaphragm, may cause an apparent bias error at a later time. Drift error may occur gradually or abruptly, depending on the nature of the physical change. Drift is generally unpredictable and can be corrected only by recalibration.

Instrument. An instrument is a sensor plus other transducers as required, and a data display element. A mercury-in-glass thermometer is an instrument since it incorporates a data display element: the visible column of mercury and attached scale.

Measurand. A measured quantity, such as temperature, pressure, or wind speed, which is the sensor input. The measurand can never be determined exactly – there is always some error associated with measurement.

Primary and secondary inputs. The primary input to a sensor is the desired input, the measurand. Secondary inputs are usually undesired but unavoidable, such as the temperature sensitivity of a pressure sensor. If the desired input of a mercury-in-glass thermometer were air temperature, then unwanted or secondary inputs would be the temperature of the observer's hand holding the thermometer or solar radiation that heats the thermometer.

Sensor. An element that receives energy from the measurand and produces an output signal related in some way to the measurand. Typical output signals are mechanical deflection or position, rotation rate, resistance, voltage, frequency, etc. The sensor always extracts some energy from the measured medium and always adds some noise to the signal. This makes a perfect measurement impossible. The measurand is an analog variable (e.g., varying continuously) and the sensor output is always an analog signal. Note that, in some instruments, an analog-to-digital conversion element is closely integrated with the sensor. Then the output may be perceived as digital but, when one examines the instrument carefully, the sensor and the converter can be distinguished.

Signal conditioning. Operations performed on a signal to convert it from one form to another (e.g., from resistance to voltage), to increase the amplitude (amplification or gain), to reduce noise (filtering), or to compensate for undesirable side effects such as temperature sensitivity of a pressure sensor (see chap. 2).

Signal. An information-bearing quantity. A voltage proportional to temperature is a signal that conveys information about the temperature. If the frequency of a time-varying voltage is proportional to the measurand, then the frequency contains the information. In this case, the voltage or current, while perhaps critical for other reasons, contains no information.

Transducer. A transducer is a device that converts energy from one form to another. All sensors are transducers but not all transducers are sensors. A sensor is the primary transducer that interacts with the measurand. An instrument usually contains other transducers to convert the energy into a useful form. Transducers can be active or passive, depending on whether they do or do not require an external source of power.

EXAMPLE
The mercury-in-glass thermometer converts heat energy into a change in volume of the mercury in the bulb and then into height of the mercury column. A mercury-in-glass thermometer is a passive transducer as it does not have an external source of power.

EXAMPLE
The primary sensor in an aneroid instrument is the diaphragm which converts pressure into a deflection. A secondary transducer coverts the deflection into an electrical signal.

3.2 Static Calibration

Static calibration is performed by varying one input, usually in a stepwise fashion, over the possible range of values while holding other inputs, if any, constant. At each measurement step, the output is observed in steady-state conditions, that is, when the input has been held steady long enough for the output to stabilize, as shown in fig. 3-1. After the output is constant, input and output measurements are recorded and the input is advanced to the next level and the process repeated. In this way we develop an input–output or transfer relation that is assumed to be valid under the stated conditions of the other possible inputs (e.g., we might calibrate a temperature sensor at 10 degree increments and then assume that the relation we obtained holds true for all temperatures in between the steps).

The vertical lines in fig. 3-1 show one period during which the input and output signals are stable and calibration measurements could be taken. The ultimate objective of calibration is to define instrument accuracy. To perform a static calibration, we must use reference instruments to measure input and output that are an order of magnitude more accurate than the instrument under test. The sensor transfer or input-to-output relation is plotted in fig. 3-2, the abscissa is the input with appropriate units: °C, K, etc.; the ordinate is the actual physical sensor output, or raw output, in volts, meters, etc. For a mercury-in-glass thermometer, the abscissa of the transfer plot would be temperature, °C, while the ordinate would be the height of the mercury column, perhaps in mm.

In fig. 3-2, line (a) shows a linear sensor with a straight line fit, line (b) shows a sensor with some nonlinearity where a straight line model has been fitted to the data, and line (c) shows the same data as in (b) but with a curve fitted to the data. The data in line (c) have been displaced by 5 units above that in line (b) for clarity. Random errors are the same in (b) and (c) but the residual nonlinearity error is smaller in (c). The scatter in the measured values is caused by random (non-repetitive) errors that cannot be individually eliminated because the errors are not predictable. This causes uncertainty in the measurement.

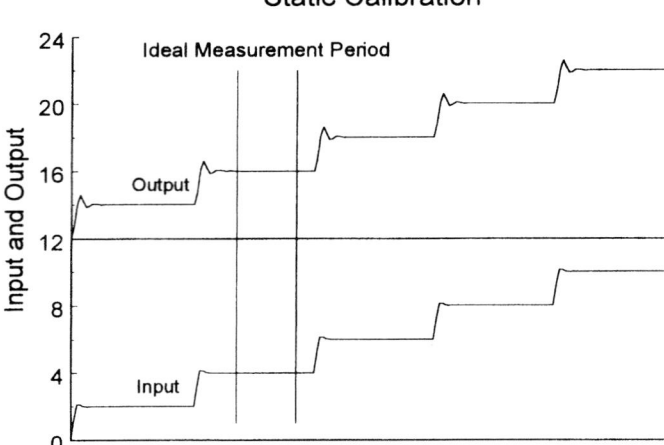

Fig. 3-1 Static calibration process. Input is shown in the lower panel and output in the upper panel.

3.2.1 Definition of Terms Related to the Transfer Plot

The transfer plot is, as mentioned above, a plot of the sensor raw output versus the measurand over the design range of the sensor. Data for a transfer plot are always obtained under static conditions.

Range. The measurand interval over which a sensor is designed to respond is called the range. For a pressure sensor, this might be 700 hPa to 1100 hPa.

Span. The algebraic difference between the upper and lower range values is called the span. If the range is 700 hPa to 1100 hPa, the span is 400 hPa.

Static sensitivity. The static sensitivity is the slope of the transfer curve, $S_S \equiv d(\text{raw output})/d(\text{input})$. If this curve is a straight line, the sensitivity is constant over the range of the instrument, $S_S = \text{constant}$. If the transfer curve is not a straight line, the sensitivity varies over the range and the sensor is said to be nonlinear. An ideal sensor would have large static sensitivity that is constant, or nearly so, over the entire desired range. Sensor design or data logging systems often include a microprocessor and this makes it easy to correct a small nonlinearity. A sensor with zero sensitivity is a useless sensor. An example would be to use a brick for a pressure sensor. There is no detectable physical change in the brick with changes in pressure and therefore it has zero sensitivity and is useless as a pressure sensor.

Resolution. The smallest change in the primary input that produces a detectable change in the output is called the sensor resolution. The resolution is limited by friction that inhibits response and by noise that masks it. The best possible resolution for an instrument is zero. Yet the resolution of a sensor is often claimed to be infinite! What is meant is that the resolution is infinitesimal which is an unattainable ideal state.

Linearity. If a straight line can be fitted to the data of the transfer plot such that the residual errors, that is, the differences between the data points and the straight-line fit,

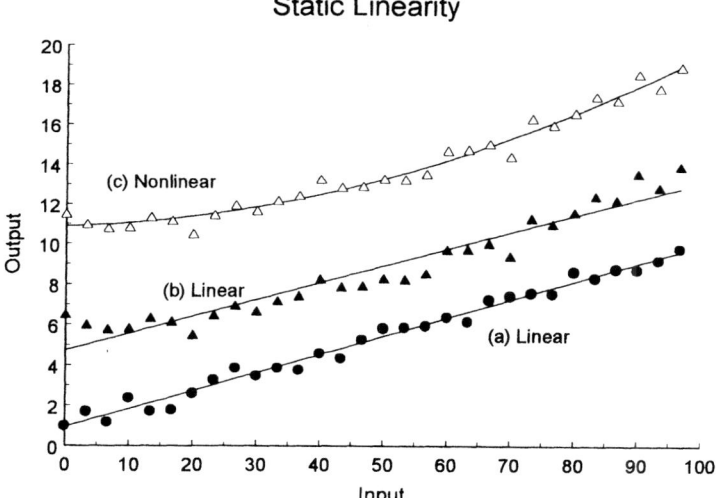

Fig. 3-2 Static linearity concepts. The abscissa is the input and the ordinate is the output.

are randomly scattered about the line, as shown in fig. 3-2(a), then the sensor is said to be linear. Small nonlinearities that produce systematic deviations from a straight line that are within the accuracy specification of the instrument are often ignored, at the expense of decreased accuracy. Large systematic deviations that cannot be ignored indicate the presence of a significant nonlinearity. The straight line is usually obtained by a least-squares fit. An instrument is said to be linear if the errors due to nonlinearity are small enough to be accepted within the required accuracy specification.

Hysteresis. Hysteresis is present when the sensor output for a given input depends upon whether the input was increasing or decreasing. This can be a nonlinear effect caused by mechanical friction in some sensors. Aneroid barometers and many humidity sensors exhibit hysteresis. Figure 3-3 is a transfer plot for a humidity sensor (not a typical sensor) that follows a different path for increasing input than for decreasing input even though the input is changed very slowly.

Threshold. Threshold is a special case of hysteresis that occurs in some instruments when the input is at or near zero. If the output remains zero while the input slowly increases, there is a threshold effect and the threshold value is the value of the input when the output starts to change. As with hysteresis, this is usually caused by static friction. See the discussion of anemometers (chap. 7) for an example of the threshold effect.

Stability. An instrument is said to be stable and free from drift if repeated calibrations over some period of time reproduce the transfer curve. The time period can vary from days to months to years, depending on the sensor in question.

Random error or noise. After a model has been fitted to the data (a straight line in the simplest case) the remaining error is called residual error. If this error is nonsystematic, it is called random error or noise. Error due to noise can be predicted only in the statistical sense and cannot be corrected. From the instrumentation perspective, noise is the part of the output signal that did not originate from the input signal.

Humidity Sensor Calibration

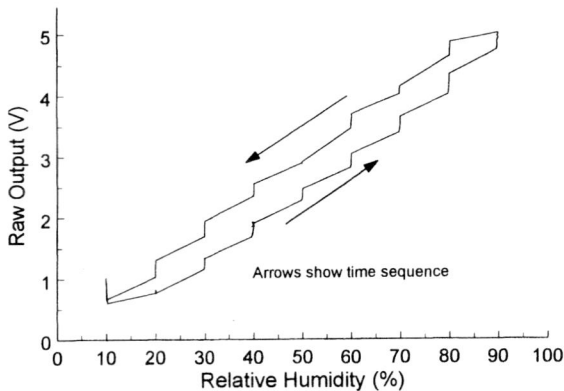

Fig. 3-3 An example of hysteresis error in a humidity sensor.

Noise originates from secondary inputs, from interaction of the sensor with the measurand, or from the sensor itself. As an example of the latter, consider an ordinary resistor with current flowing through it. Random variations in the voltage across the resistor (noise voltage) will appear, caused by the motion or vibration of molecules in the resistor. Power lines near a field site can also cause noise, or 60 Hz noise can result from room lighting and electrical equipment.

3.2.2 Calibration Procedure

As noted above, the calibration procedure starts with the accurate measurement of X_i, the primary input, and Y_1, the raw sensor output, at N points over the design range of the sensor. The units of X_i are those of the measurand, while the units of Y_1 will be those appropriate to the raw sensor output. These data are then plotted and form the transfer plot. The next step is to fit a straight line, or curve if necessary, to the data, preferably using the least-square procedure. There are other techniques, but the least-square is well known and generally appropriate to this objective. The objective is to develop a transfer equation that can be used to convert the observed output Y_1 to an estimate of the known input X_i. Ideally, the transfer equation and the calibration equation would be obtained from knowledge of sensor physics, as was the case with the mercury barometer and the hypsometer. In other cases, as with the aneroid barometer, the transfer model will be a polynomial of the form

$$\hat{Y}_1 = a_0 + a_1 X_i + a_2 X_i^2 + a_3 X_i^3 + \cdots \qquad (3.1)$$

which reduces to

$$\hat{Y}_1 = a_0 + a_1 X_i \qquad (3.2)$$

when the sensor is linear. In all cases $\hat{Y}_1$ is an approximation to Y_1 because of the random errors noted above. It would be possible to obtain an exact fit if there were no measurement uncertainty, but random errors are always present. The least-squares procedure is a method of determining the coefficients a_0, a_1, etc. by minimizing the square of the differences between the line and the observed data. If the transfer model were linear, we would find the coefficients a_0 and a_1 that would minimize the function

$$E = \sum_{n=1}^{N} (a_0 + a_1 X_{in} - Y_{1n})^2$$

where the summation is over N data values (from $n = 1$ to $n = N$).

This can be done by setting the following equations to zero:

$$\frac{\partial E}{\partial a_0} = 2 \sum (a_0 + a_1 X_{in} - Y_{1n}) = 0$$

$$\frac{\partial E}{\partial a_1} = 2 \sum (a_0 + a_1 X_{in} - Y_{1n}) X_{in} = 0$$

which produces the following equations, called the normal equations:

$$\left. \begin{array}{l} a_0 N + a_1 \sum X_{in} = \sum Y_{1n} \\ a_0 \sum X_{in} + a_1 \sum X_{in}^2 = \sum X_{in} Y_{1n} \end{array} \right\} \quad (3.3)$$

The solution for the coefficients is given below:

$$\left. \begin{array}{l} D = N \sum X_{in}^2 - \left(\sum X_{in} \right)^2 \\ a_0 = \left(\sum Y_{1n} \sum X_{in}^2 - \sum X_{in} \sum X_{in} Y_{1n} \right) / D \\ a_1 = \left(N \sum X_{in} Y_{1n} - \sum X_{in} \sum Y_{1n} \right) / D \end{array} \right\} \quad (3.4)$$

Static sensitivity is $S =$ slope of the line $= a_1$, and a_0 is the offset. Figure 3-2(a) shows an example of fitting a linear model, eqn. 3.2, to a set of data. The data are apparently scattered randomly about the fitted line, indicating that the sensor being modeled is indeed linear (or very close to it). The vertical distance between a plotted datum, Y_{1n}, and the line is $\hat{Y}_{1n} - Y_{1n}$.

When the sensor is nonlinear, as shown in fig. 3-2(b) and (c), one still has the option of fitting a linear model, line (b), or a nonlinear model, line (c). A nonlinear model can be more troublesome, especially when the measurement system does not include a computer, but the advantage is that the resulting errors or deviations from the model are smaller in line (c) than in line (b).

The next step is to determine the calibration equation, another polynomial,

$$X_1 = c_0 + c_1 Y_1 + c_2 Y_1^2 + c_3 Y_1^3 + \cdots \quad (3.5)$$

and, as before, the linear form is

$$X_1 = c_0 + c_1 Y_1 \quad (3.6)$$

where the quantity X_1 is the best estimate of the true but unknowable input X_i. The c coefficients can be obtained from the a coefficients by

$$c_0 = -a_0/a_1$$
$$c_1 = 1/a_1$$

although it would be better to find these coefficients using a least-squares procedure similar to that given above. The units of X_1 are the same as for X_i. We can refer to the sensor output either as Y_1 if we are thinking of the raw output, or as X_1 if we wish to consider the output after applying the calibration equation (3.6). The subscript 1 is used to designate the output of the sensor; the output of subsequent elements in the measurement system chain, described in chap. 1, would be designated X_2, X_3, etc.

3.2.2.1 Bias, Imprecision, and Inaccuracy

The calibrated output for each measurement, $X_1 = c_0 + c_1 Y_1$, can be calculated and the instrument error can be defined as

$$\varepsilon_n \equiv X_{1n} - X_{in} \tag{3.7}$$

shown in the left-hand graph in fig. 3-4. A long vertical line in the right-hand graph marks the mean. The horizontal line shows the spread of the distribution, called the standard deviation. A Gaussian distribution is given by

$$w(x) = \frac{1}{\sigma\sqrt{2\pi}} \exp\left[-\frac{(x-\mu)^2}{2\sigma^2}\right] \tag{3.8}$$

where μ = population mean, σ = population standard deviation, and $w(x)$ is the probability of observing a given value of x. The population of x is all of the values that could be realized whereas a sample of x comprises only the values actually measured.

If this model is appropriate, the sample mean and standard deviation of the errors are given by

$$\bar{\varepsilon} = \frac{1}{N}\sum_{n=1}^{N} \varepsilon_n \quad \text{and} \quad S_\varepsilon = \sqrt{\frac{1}{N-1}\sum_{n=1}^{N}(\varepsilon_n - \bar{\varepsilon})^2}. \tag{3.9}$$

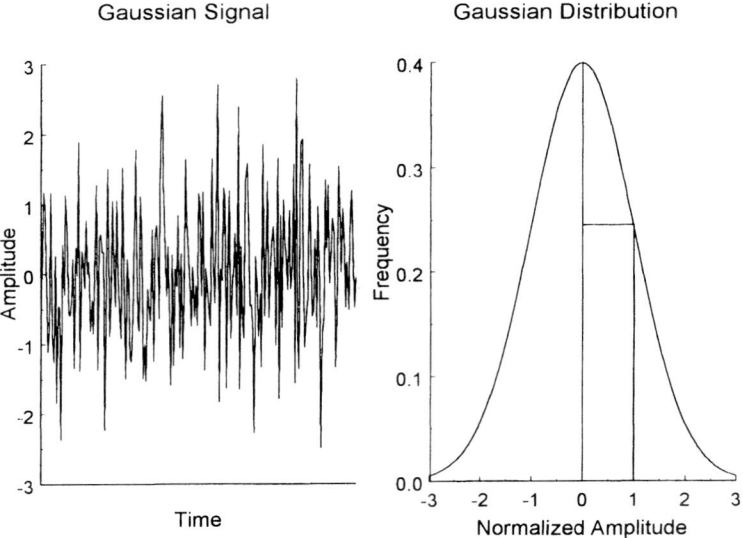

Fig. 3-4 Sample of a Gaussian distributed signal in the left-hand panel and the corresponding Gaussian distribution in the right-hand panel.

After calibration, the average error should be zero. The standard deviation is a measure of dispersion and, if the underlying distribution of errors is a Gaussian distribution, then 95% of the errors will lie within two standard deviations from the mean (66% of the errors are within one standard deviation of the mean). Thus, we can define:

$$\left.\begin{array}{ll} \text{Bias} & \equiv \bar{\varepsilon} \\ \text{Imprecision} & \equiv 2S_\varepsilon \\ \text{Inaccuracy} & \equiv \text{Bias} \pm \text{Imprecision} = \bar{\varepsilon} \pm 2S_\varepsilon \end{array}\right\} \quad (3.10)$$

Imprecision or uncertainty is a measure of the noise or scatter in the measurement and the bias is the systematic error. *The purpose of static calibration is to remove the bias and to numerically define the imprecision.* However, bias may still be present in an instrument if the instrument has drifted since the last calibration. (Recall that drift is the slow change in instrument parameters with time and that most sensors are subject to some drift.) Another source of inaccuracy is secondary inputs when adequate compensation is not provided (e.g., when a aneroid barometer is sensitive to changes in air temperature and this is not accounted for in a calibration).

Root-mean-square (RMS) error is commonly used in addition to the above terms:

$$\text{RMS} \equiv \sqrt{\frac{1}{N}\sum_{n=1}^{N}\varepsilon_n^2} \quad (3.11)$$

which is very similar to the definition for S_ε, the standard deviation of the errors. The difference is that the RMS combines the average and standard deviation of the error.

Figure 3-5 shows several examples of sensors with bias and imprecision. Each panel shows a time series of data where the input is the constant value shown as the solid line in the middle of each plot. In panel (a), the bias is evidently zero but the imprecision is large. In panel (b), the imprecision is much less, but there is significant

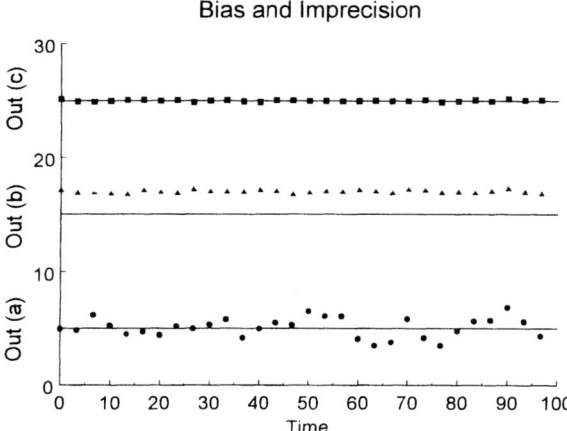

Fig. 3-5 Bias and imprecision in a sensor. The sensor in panel (a) has no bias but is imprecise; the sensor in (b) is more precise but has a large bias, while the sensor in (c) has no bias and very little imprecision.

bias. Panel (c) shows a better sensor; it has no bias and there is little imprecision. The sensor in panel (c) has the least inaccuracy.

Note that the above definitions are not universally applied. There is considerable imprecision in the common usage of the terms, especially of accuracy and resolution. It is common to state the accuracy of an instrument as a small quantity when clearly what is meant is that the inaccuracy is small. Frequently, the precise definition of inaccuracy, or accuracy, is not indicated.

3.3 Example of a Static Calibration

A pressure sensor is to be calibrated over the range 80 to 110 kPa. To simplify matters, we assume that it has no temperature sensitivity. The first two columns in table 3-1 show respectively the input X_i in kPa, determined using a much more accurate sensor, and the raw output Y_1 in volts. The data in the remaining two columns are derived as a result of subsequent calculations, as shown below.

Step 1. There are $N = 13$ data pairs $(X_{i1}, Y_{11}), (X_{i2}, Y_{12}), \ldots (X_{in}, Y_{1n}), \ldots (X_{iN}, Y_{1N})$ give in the first two columns of the table. Accumulate the following sums: $\sum X_{in}$, $\sum X_{in}^2$, $\sum Y_{1n}$, $\sum Y_{1n}^2$, $\sum X_{in} Y_{1n}$, where all of the summations are over $n = 1$ to N. In the example, the above sums are 1235, 118 462.5, 77.9618, 513.087, and 7633.982, respectively. This calculation can be done on a computer or on a hand calculator. The normal computing procedure would be to store the sums and all other intermediate results internally at full computer precision (at least six significant digits). If it is done on a hand calculator, it is important to store these sums in internal registers or, if this is not possible, to record them to the full calculator precision. Equation 3.4 calls for taking the difference between two large numbers: $N \sum X^2 - (\sum X)^2$. In this case, with just 13 numbers, $N \sum X^2 - (\sum X)^2 = 13^*118\,462.5 - (1235)^2 = 1\,540\,012.50 - 1\,525\,225.000 = 14\,787.50$. As N becomes larger it becomes more difficult to take the difference between large numbers without error being induced by finite computer

Table 3-1 Calibration data for a pressure sensor. Note the different units for each column.

Input X_i (kPa)	Raw Output Y_1 (V)	Calibrated Output X_1 (kPa)	Error ε (Pa)
80.000	3.0033	80.038	38
82.500	3.4949	82.495	−5
85.000	3.9882	84.961	−39
87.500	4.4937	87.487	−13
90.000	5.0049	90.042	42
92.500	5.5004	92.518	18
95.000	6.0055	95.042	42
97.500	6.4839	97.433	−67
100.000	6.9801	99.913	−87
102.500	7.5008	102.515	15
105.000	7.9971	104.995	−5
107.500	8.5061	107.539	39
110.000	9.0029	110.022	22

precision. It frequently becomes necessary to double the precision of the arithmetic (usually to 15 decimal digits).

Step 2. Use the linear model shown in eqn. 3.2 and solve for the coefficients using the least squares equation, eqn. 3.4. The result is

$$D = 14\,787.5 \qquad a_0 = -13.012\,21 \text{ V} \qquad a_1 = 0.200\,097\,6 \text{ V/kPa}.$$

The static sensitivity $= a_1$. Not all of the sums obtained in step 1 were used here; the remaining sums will be used in the next step. If the static sensitivity is not needed, this step can be skipped.

Step 3. Find the coefficients c_0 and c_1 in the calibration eqn. 3.6 using

$$D = N \sum Y^2 - \left(\sum Y^2\right) = 592.0885$$

$$c_0 = \left(\sum X \sum Y^2 - \sum Y \sum XY\right)/D = 65.029\,88 \text{ kPa}$$

$$c_1 = \left(N \sum XY - \sum X \sum Y\right)/D = 4.997\,467 \text{ kPa/V}$$

Step 4. Use the calibration equation $X_1 = c_0 + c_1 Y_1$ to calculate the values in column 3 of table 3-1 and then find the residual errors $\varepsilon = X_1 - X_i$ in column 4.

Step 5. Accumulate the sums $\sum \varepsilon$ and $\sum \varepsilon^2$ ($-8.803\,147 \times 10^{-6}$ and $4.223\,338 \times 10^{-2}$).

Step 6. Compute the average error and the standard deviation of the errors in column 4 using eqn. 3.8.

The bias $= -8.803\,147 \times 10^{-6}$ kPa and the imprecision $= 8.446\,676 \times 10^{-2}$ kPa. If the sensor output indicates 104.544 kPa, we can estimate that the 'true' value of the measurand lies between 104.460 and 104.628 kPa. There is no justification for expressing the output to six significant digits since there is uncertainty in the fourth digit. It would be appropriate to write the output, in this case, as 104.5 kPa. With a reasonable number of significant digits, bias $= 0.0$ Pa and imprecision $= 84$ Pa. We can use the static sensitivity, $0.200\,097\,6$ V/kPa, to convert the imprecision of 84 Pa with respect to the calibrated output to 17 mV with respect to the raw output. This suggests that we need only to measure the raw output voltage to about 10 mV, and shows that the last two digits in column two of table 3-1 are not significant.

3.4 Multiple Sources of Error

In the above treatment, the instrument errors were treated as if they originated from a single source. In reality, there may be several independent sources of error in a single instrument. If the error due to each source is independent (e.g., they do not depend on each other), then these errors can be combined using root-sum-square (RSS), given as

$$\text{RSS} = \sqrt{\varepsilon_1^2 + \varepsilon_2^2 + \varepsilon_3^2 + \cdots} \qquad (3.12)$$

Note the similarity to the RMS definition.

As an example, the errors occurring in a certain barometer are listed in table 3-2. Imprecision due to linearity is 0.15 hPa and imprecision due to hysteresis is 0.03 hPa. These error sources are independent, meaning that the occurrence of hysteresis error

Table 3-2 Errors for a certain barometer.

Error Source	Imprecision
Linearity	0.15 hPa
Hysteresis	0.03 hPa
Repeatability	0.03 hPa
Calibration uncertainty	0.20 hPa
Temperature dependence	0.10 hPa
Long-term stability	+0.20 hPa/year
Combined error	0.3 hPa + 0.2 hPa/year

is independent of the magnitude of the linearity error, as are the others listed in the table. With the exception of long-term stability, we can determine the overall imprecision by taking the square root of the sum of the squares of the individual error components. In this case, the combined imprecision is 0.3 hPa, much less than the simple sum of the error components. If long-term stability, or drift, is not a randomly fluctuating error (as is usually the case), it should be accounted for separately, as in the above table.

Calibration uncertainty indicates that the manufacturer does not and cannot have a perfect standard to use in calibrating barometers. The calibration uncertainty (or imprecision) is calculated in much the same way as above. The error due to each component of the pressure standard can be evaluated and combined using the RSS technique.

3.5 Significant Figures

If a thermometer has an imprecision equal to 0.12 K then an observation or indicated reading of 273.781 K could be stated as 273.781 ± 0.12 K. But the last two digits of the reading are not significant since the correct value is estimated to lie between 273.661 K and 273.901 K. The result should be rounded to four significant digits and reported as 273.8 ± 0.1 K. If the precision of the sensor is unknown, one may be able to estimate the precision, to use in recording the results, from some knowledge of the sensor. For example, if a thermometer has scale markings to 0.5 degrees, one might interpolate to 0.1 degree but not to 0.01 degree. The results would then be recorded to 0.1 degree. Note that the average of many readings would still be recorded to just 0.1 degree.

All measurements have some degree of uncertainty. We have expressed this as the inaccuracy (or imprecision) which should be attached to every measurement. Another way of expressing uncertainty is by the number of significant digits (or significant figures) used to report a datum. Instead of reporting the pressure as 967.23 ± 0.12 hPa, we could write it as 967.2 hPa with the understanding that there is some uncertainty in the last digit. All of the digits, including the uncertain one, are called significant digits.

Some rules for applying significant digits:

(1) All nonzero digits are significant: 457 cm (three significant digits), 0.25 kg (two digits).

(2) Zeros between nonzero digits are significant: 1001.3 hPa (5 digits).
(3) Zeros to the left of the first nonzero digit in a number are not significant; they indicate the position of the decimal point: 0.03 mV (one significant digit).
(4) Zeros to the right of the decimal point are significant: 980.00 hPa (5 digits).
(5) When a number ends in zeros that are not to the right of the decimal point, the zeros are not necessarily significant: 900 hPa (1, 2, or 3 significant digits). This ambiguity can be removed by writing the numbers in exponential notation.

1.01×10^5 Pa   three significant digits
1.013×10^5 Pa   four significant digits
1.0130×10^5 Pa   five significant digits
1.01300×10^6 Pa   six significant digits

Sometimes we can remove the ambiguity by the context. For example, if data from a certain barometer are reported as 900, 910, 909, ...we can safely assume that there are three significant digits.

When performing mathematical operations on measured quantities, the result is limited by the least accurate measurement. If we measured the density of mercury as 13 595.1 kg m^{-3}, the acceleration due to gravity as 9.806 65 and the height of the column of mercury as 0.7590 m then the pressure is

$$p = 13\,595.1 \times 9.806\,65 \times 0.7590 = 101\,191.6920$$

However, the height was only measured to four significant digits so the pressure must be written as 1.012×10^5 Pa.

Consider another example. The radius of the mercury column is measured as 5.00×10^{-3} m. The circumference of the column is

$$2\pi r = 2 \times 3.141\,592\,654 \times 5.00 \times 10^{-3} = 0.031\,415\,926\,54 \text{ m}$$

which should be written as 0.0314 because the radius was measured to 3 significant digits. What about the first number? The number 2 is an exact, integer quantity. There are exactly two radius lengths in one diameter, not 1.9999 or 2.0001. Pi is a special number that can be calculated to whatever precision is desired. The number of digits written for π should be such that it never limits the precision of the result (as in the case shown here).

In addition or subtraction, the result should be reported to the same number of decimal places as that of the term with the least number of decimal places (assuming all of the quantities are reported in consistent units). Thus the difference between 980.123 71 hPa and 979.2 hPa is 0.9 hPa.

These rules are difficult to apply when many operations are performed, involving both multiplication/division and addition/subtraction. When these operations are performed in a computer and computer precision is carried throughout the operations (as it should be) the user must set the precision of the final output according to the above rules. One can usually apply the multiplication/division rule unless the result is an error value; then use the addition/subtraction rule. Errors are the difference between the observed value and the reference value.

When using hand calculators or digital computers, intermediate results are stored to the precision of the computer but the user has control over the number of digits displayed in the output. *Don't confuse the precision of the computer with the pre-*

cision of the instrument! In general, mathematical operations do not improve the precision of the result but careless handling of the intermediate results can decrease the precision. For example, in computing the least-squares coefficients, eqn. 3.4, or the mean and standard deviation in eqn. 3.7, it is essential to avoid rounding or truncating the sums. Most computers carry six significant digits in real, or floating point, arithmetic that is usually sufficient. There are cases where six significant digits are inadequate; when large quantities of data, say more than 1000 points, are to be processed, it may be necessary to use double precision arithmetic. When using a hand calculator, always save intermediate products, especially sums, to the full machine precision. Then round the final result to the appropriate precision.

QUESTIONS AND PROBLEMS

1. What is the formal name for the sensor input?
2. Is the sensor output always an analog signal? If not, can you think of any exceptions?
3. If the sensor output is a frequency, is this signal analog or digital?
4. A certain sensor is designed to measure pressure between 700 and 1100 hPa. After fitting a straight-line equation, the transfer coefficients are $a_0 = -7.00$ V and $a_1 = 0.0100$ V hPa^{-1}. Average and standard deviation of the residual errors are 0.00 hPa and 0.25 hPa respectively. Evaluate: (a) the bias, (b) the imprecision, (c) the inaccuracy, (d) the span, (e) the static sensitivity. When the output = 3.000 V, what is the input?
5. If a certain barometer has a linearity error of 0.1 hPa, hysteresis error of 0.02 hPa, repeatability error of 0.2 hPa, and the vendor's calibration uncertainty is 0.2 hPa, what is the overall error?
6. Indicate the number of significant digits in the following:

Significant digits	Expression
	$3.42 + 273.15$
	9.00×10^2
	$1013.25 \text{ hPa} - 1012.5 \text{ hPa}$
	1.2×97.12

7. Calculate the static sensitivity of an aneroid barometer in mid-range:

p (hPa)	800	900	1000	1100
V (Volts)	1.00	2.35	3.70	5.00

8. A calibration check (not a complete recalibration) yields the following results: Average error = 0.7 K, standard deviation of the error = 0.15 K.
   (a) What is the inaccuracy?
   (b) If this sensor were recalibrated, what would you expect the average error to be?
9. Imprecision is a measure of what in an instrument?
10. Discuss the difference between threshold and hysteresis.
11. If static calibration removes the bias, how could a sensor ever exhibit any bias?
12. What causes imprecision in a sensor?
13. Discuss the difference between sensitivity and resolution of a sensor.

14. Given the following set of data resulting from the calibration of a temperature sensor:

Index	Input (K)	Output (V)
1	240	−6.59
2	245	−5.46
3	250	−4.17
4	255	−3.10
5	260	−2.12
6	265	−0.95
7	270	0.41
8	275	1.51
9	280	2.53
10	285	3.54
11	290	4.42
12	295	5.72
13	300	6.79
14	305	8.03
15	310	8.98
16	315	10.16
17	320	11.17

(a) Plot the transfer curve.
(b) Calculate the least-squares best-fit line $Y_1 = a_0 + a_1 X_i$.
(c) What is the static sensitivity?
(d) Find the coefficients in the calibration equation $X_i = c_0 + c_1 Y_1$.
(e) What is the bias?
(f) What is the imprecision?
(g) What is the inaccuracy?

15. Is it possible for a sensor to be too accurate? Why don't we always use the most accurate sensor available?

16. Contrast accuracy and precision.
   (a) Is it possible for a sensor to be very accurate but not precise?
   (b) Is it possible for a sensor to be very precise but not accurate?
   (c) Can you envision circumstances in which you would choose a sensor which is precise but inaccurate?

17. Define the following terms: measurand, sensor, transfer equation, calibration equation, bias, imprecision, span, drift, error, hysteresis, exposure error, raw output, secondary inputs, threshold.

18. Can we develop a correction for drift? Explain.

BIBLIOGRAPHY

Doebelin, E.O., 1983: *Measurement Systems: Application and Design*, 3rd ed. McGraw-Hill, New York, 876 pp.

Hofmann, D., 1982: Measurement errors, probability and information theory. In *Handbook of Measurement Science*, Vol. 1, ed. P.H. Sydenham. John Wiley, Chichester.

Rabinovich, S., 1993: *Measurement Errors: Theory and Practice*. American Institute of Physics, New York, 271 pp.

4

Thermometry

Air temperature is one of the most fundamental of all meteorological measurements and directly effects our everyday lives. It has been measured for centuries, using countless different techniques. One might assume that accurate air temperature measurements are readily made. Indeed, it is possible to make very accurate air temperature measurements but it can be a remarkably hard task, especially when limitations such as power consumption, reliability, and cost are involved.

Errors in the measurement of air temperature in excess of 2 to 3°C are not uncommon in many networks. Errors of this magnitude are generally acceptable for the general public who is most interested in what clothes to wear for the day. However, numerical models at all scales of motion (mesoscale, synoptic scale, or climate models) are greatly affected by errors even as large as 1°C. Errors of just 1°C in a mesoscale model have been shown to be the deciding factor between no storms initiated and intense storms (Crook, 1996). In addition, errors as small as 0.2°C can change the prediction of a global climate model, depending on its dependency on initial conditions (DeFelice, 1998).

Measurement of air temperature near the surface of the earth is facilitated by the vast array of temperature sensors and supporting electronic modules that are readily available. Accuracy is limited not by technology but by our ability to use it and by our ability to avoid exposure error, that is, to provide adequate coupling with the atmosphere.

The preferred temperature scales are Celsius and Kelvin. These scales can be used almost interchangeably (except when absolute temperature is required) because a temperature difference of 1 K is equal to a temperature difference of 1°C. The Fahrenheit scale is still in general use by the U.S. public. Some common temperature reference points are shown in table 4-1. The triple point of water is the temperature and pressure where all three phases, gas, liquid, and solid, can coexist.

Thermometry

Table 4-1 Several near-ambient temperature reference points listed in various scales. A pressure of 1013.25 hPa is assumed for the ice and steam points and 6.106 hPa for the triple point of water.

Scale	Symbol	Ice	Triple	Steam
Thermodynamic	K	273.15	273.16	373.15
Celsius	°C	0.00	0.01	100.00
Fahrenheit	°F	32.00	32.018	212.00

Temperature sensors can be categorized according to the physical principle that they use: thermal expansion, thermoelectric, electrical resistance, electrical capacitance and some other effects (fig. 4-1). Direct indicating instruments (bimetallic strips and liquid-in-glass thermometers) can use displacement directly. Sensors designed to work with data loggers usually convert the output to a voltage signal and may then amplify that signal to obtain a useful voltage. Only a few kinds of temperature sensors (thermal expansion, thermoelectric, and electrical resistance) are considered in this chapter. There are many more that have been used in meteorological applications.

4.1 Thermal Expansion

One way thermal expansion is exploited to sense temperature is through the use of bimetallic strips and liquid-in-glass thermometers. In both cases the difference between expansion coefficients of two materials is used. Some linear and volume expansion coefficients are listed in table 4-2. Linear expansion is given by $\Delta L = \alpha L_0 \Delta T$, where α is the coefficient of linear expansion and L_0 is the length of the material when ΔT is zero, ΔL is the change in length, and ΔT is the temperature change from some arbitrary temperature where L_0 was measured. Volume expansion is given by $\Delta V = \beta V_0 \Delta T$, where β is the volume coefficient of expansion and V_0 is the volume when $\Delta T = 0$.

4.1.1 Bimetallic Strip

A bimetallic strip is a pair of metals with different thermal expansion coefficients that have been bonded together. At the reference temperature, the temperature at which bonding took place, the strip maintains its original shape. When the temperature changes, the strip bends in a circular arc, for small deflections, due to the differential

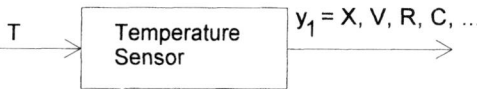

Fig. 4-1 Raw output of a temperature sensor may be displacement, voltage, resistance, capacitance, etc.

Table 4-2 Linear and volume expansion coefficients for various materials.

Material	Linear Expansion α (K^{-1})	Volume expansion β (K^{-1})
Aluminum	2.4×10^{-5}	7.2×10^{-5}
Brass	2.0×10^{-5}	6.0×10^{-5}
Copper	1.7×10^{-5}	5.1×10^{-5}
Glass	0.4 to 0.9×10^{-5}	1.2 to 2.7×10^{-5}
Steel	1.2×10^{-5}	3.6×10^{-5}
Invar	0.09×10^{-5}	0.27×10^{-5}
Quartz	0.04×10^{-5}	0.12×10^{-5}
Ethanol		75×10^{-5}
Carbon disulphide		115×10^{-5}
Glycerin		49×10^{-5}
Mercury		18×10^{-5}

expansion of the two components of the strip, as shown in fig. 4-2 where one end is held in a fixed position.

For small temperature changes, deflection of the free end of the strip is given by

$$y = \frac{K \Delta T L^2}{t} \quad (4.1)$$

where K is a constant that is a function of the characteristics of the two metals, ΔT is the ambient temperature minus the strip formation temperature T_0, L is the length of the strip, and t is the thickness of the strip. A bimetallic strip is used in thermostats to open or close a relay and in thermometers where the strip positions a pointer on a temperature scale.

The static sensitivity of a bimetallic strip is given by

$$S = \frac{\mathrm{d}y}{\mathrm{d}\Delta T} = \frac{\mathrm{d}y}{\mathrm{d}T} = \frac{KL^2}{t} \quad (4.2)$$

For the example given above, a bimetallic strip is linear since the static sensitivity is constant. It is a function only of K, L, and t, which are determined during the sensor design.

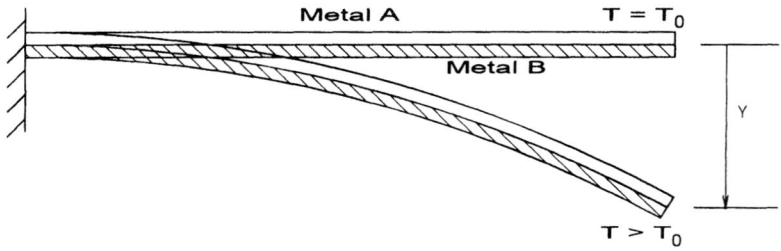

Fig. 4-2 A bimetallic strip anchored at one end while the other end is free to deflect as the temperature changes from the temperature at which the strip was formed.

4.1.2 Liquid-in-Glass Thermometer

A liquid-in-glass thermometer is a glass tube with a bulb at one end filled with the liquid and a scale fastened to or etched on the glass tube as shown in fig. 4-3. The liquid is usually mercury or spirit (alcohol or toluol). Mercury can be used only above $-39°C$, its freezing point, whereas alcohol can be used down to $-62°C$.

One way of classifying these thermometers is by the immersion required. Immersion types are partial, total and complete. A partial immersion thermometer should be placed in the bath liquid (water or oil) until the bulb and a small portion of stem, indicated by an immersion line, are immersed in the liquid to be measured. For a total immersion thermometer the bulb and the portion of the stem containing the thermometric fluid are immersed. Partial and total types of thermometers are used for calibrating other sensors. In a complete immersion thermometer, the bulb and the entire stem are immersed. This type is used for air temperature measurement.

Two special liquid-in-glass thermometers are used to measure the minimum and the maximum temperature. The minimum thermometer uses alcohol with a dumbbell in the stem (see the detail in fig. 4-3); it is called a dumbbell because the original shape of the ends was like a bell). The thermometer is mounted horizontally; the alcohol flows around the dumbbell as the temperature increases and leaves the dumbbell in a fixed position. When the temperature decreases, the meniscus of the alcohol does not let the dumbbell pass but drags it down to indicate the minimum temperature. The force due to friction of the dumbbell along the stem must be less than the force required to break through the surface tension of the meniscus.

A maximum thermometer uses mercury and has a constriction in the stem as illustrated in the detail in fig. 4-3. The bulb is mounted slightly higher than the rest of the column, and as the temperature increases the volume of mercury in the bulb increases and mercury is forced through the constriction. When the temperature decreases, the column breaks at the constriction. The remaining column above the constriction indicates the maximum temperature. This type of thermometer can be reset by swing-

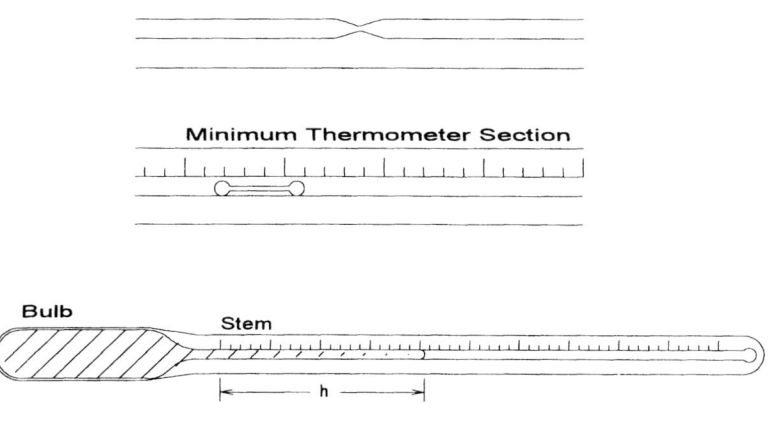

Fig. 4-3 A liquid-in-glass thermometer showing the bulb and stem and indicating the raw output, h. Above, a detail of a minimum thermometer and of a maximum thermometer.

66 Meteorological Measurement Systems

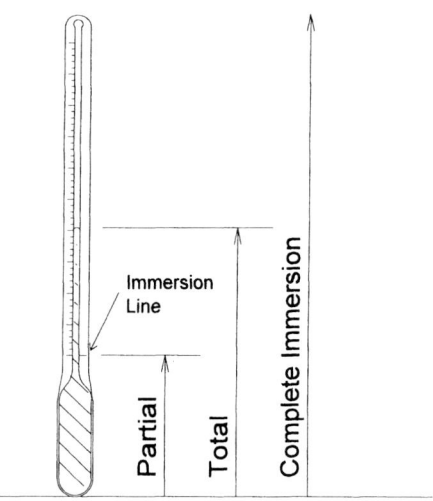

Fig. 4-4 Illustration of partial, total, and complete immersion type thermometers.

ing it briskly, but carefully, to force the mercury below the constriction to fill the void (this type of thermometer is often used in hospitals, although other sensors are becoming more prevalent).

For liquid-in-glass thermometers, as the temperature increases both the thermometric fluid (mercury, alcohol, etc.) and the glass expand:

$$\Delta V_d = V_0 \beta_d \Delta T = \pi r^2 \Delta h \qquad (4.3)$$

where β_d = the differential volume expansion coefficient, $V_d = V_0$ when $\Delta T = 0$, r = the capillary radius, and Δh = change in column height. When the thermometric fluid is mercury, $\beta_d = 1.6 \times 10^{-4}\,°C^{-1}$. Expansion of the fluid in the capillary is often ignored as the volume is small compared to that of the bulb. However, for precision measurements, it cannot be ignored and that is why precision thermometers are designated for partial, complete, or total immersion.

EXAMPLE

A mercury-in-glass thermometer ($\beta_d = 1.6 \times 10^{-4}\,°C^{-1}$) has a bulb volume of 150 mm³ and a capillary tube 15 cm long with a 0.1 mm diameter. Capillary tube volume is 1.178 mm³; thus, even when mercury has expanded to fill the tube, the tube contains only 0.8% of the total volume of 151.178 mm³ while the bulb contains 99.2% of the total volume.

Static sensitivity of a liquid-in-glass thermometer is

$$S = \frac{dh}{dT} = \frac{V_0 \beta_d}{\pi r^2} \qquad (4.4)$$

assuming that the capillary radius r is constant. The sensitivity of a liquid-in-glass thermometer can be increased by increasing the bulb volume and by decreasing the capillary radius. A small capillary also minimizes the error that could be caused by an emergent stem. The emergent stem is that portion of the stem that is above the immersion line and is filled with thermometric fluid. This cannot happen in a complete immersion thermometer.

EXAMPLE

For the thermometer in the above example, the static sensitivity is given by

$$S = \frac{150 \times 1.6 \times 10^{-4}}{\pi (0.05)^2} = 3.06 \text{ mm K}^{-1}$$

so, if the temperature of the bulb increases by 1 K, the mercury column will push the capillary up 3.06 mm. One could make 1 degree marks every 3.06 mm up the stem of the thermometer. Range of the thermometer $= 150/3.06 = 49$ degrees. Allowing some margin at the top and bottom, we could design the thermometer to cover a range of 40°, say from 0°C to 40°C.

4.2 Thermoelectric Sensors

The junction of two dissimilar metals forms a thermocouple. When the two junctions are at different temperatures, a voltage is developed across the junction. By measuring the voltage difference between the two junctions, the difference in temperature between the two junctions can be calculated. Alternatively, if the temperature of one junction is known and the voltage difference is measured, then the temperature of the second junction can be calculated. Thermocouples provide a wide useful temperature range, are inherently differential, are rugged, reliable and inexpensive, and may have fast response. The main disadvantage of thermocouples is the very low output, on the order of 40 µV/°C, a slight nonlinearity, and the need for calibration. A common thermocouple configuration is shown in fig. 4-5 and comprises two junctions and an amplifier. Junction 1 is at temperature T_1, the measured temperature, and junction 2, the reference junction, is held at a constant temperature T_2. Metal "A" is used from junction 1 to the amplifier and from junction 2 to the amplifier. Metal "B" is used between junctions 1 and 2. If metal is copper and metal is constantan, a copper–nickel alloy, the junction pair is called a copper-constantan thermocouple. Some thermocouple types in common use are listed in table 4-3.

In fig. 4-5 an amplifier is used to increase the small voltage difference to a more useful level: $V_3 = G(V_1 - V_2)$. The voltage developed by the two junctions in fig. 4-5, called the "Seebeck effect", is $\Delta V = V_1 - V_2$. It will appear if no current is allowed to flow in the circuit.[1]

There are some observed "laws of thermocouple behavior" used as a rule-of-thumb guide to thermocouple circuit construction (Doebelin, 1983). These laws will be discussed with respect to fig. 4-5.

"*Law*" #1. The voltage across a thermocouple is unaffected by temperatures elsewhere in the circuit, provided the two metals used are each homogeneous. Thus one can use lead wires made of thermocouple metals.

"*Law*" #2. If a third metal is inserted in either wire A or B and if the two new junctions are at the same temperature, no effective voltage is generated by the third metal. This means that a real voltmeter (or amplifier) can be used. The

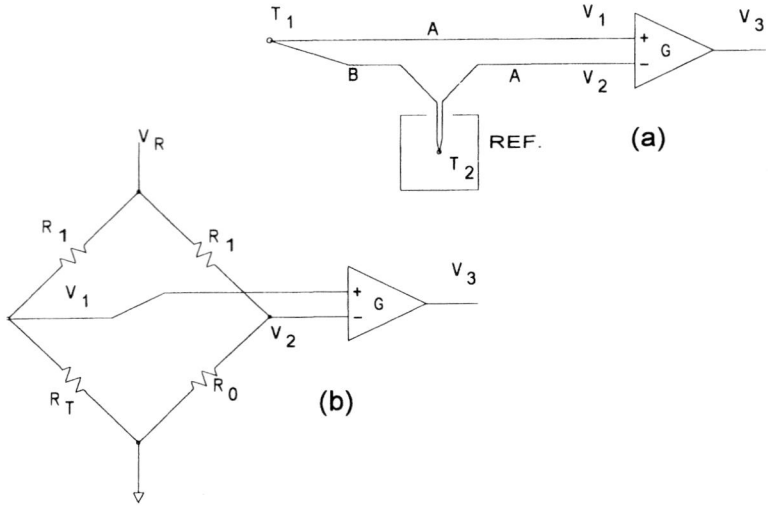

Fig. 4-5 A thermocouple temperature measurement system (a) and an RTD bridge circuit (b). The two thermocouple metals are labeled "A" and "B".

terminals of a voltmeter are usually made of a third metal and can be close together. It is important to make sure the terminals of the voltmeter are at the same temperature.

"Law" #3. If a metal C is inserted in one of the AB junctions, then no net voltage is generated so long as junction AC and BC are at the same temperature. This means that the two wires or a junction can be soldered together and the presence of the third metal, solder, will not affect the voltage if there is no temperature gradient across the solder junction.

"Law" #4. If the thermocouple voltage of metals A and C is ΔV_{AC}, and of junction B and C is ΔV_{BC}, then for metals A and B the voltage will be $\Delta V_{AC} + \Delta V_{BC}$. It is not necessary to calibrate all possible pairs of metals, but only each metal against a standard, to build a table of thermocouple voltages.

"Law" #5. If a thermocouple produces voltage ΔV_a when the junction temperatures are at T_1 and T_2, and V_b when at T_2 and T_3, it will produce $V_a + V_b$ for junction temperatures T_1 and T_3. Standard tables or equations can be used, even if the reference junction is not at 0°C.

Table 4-3 Some common thermocouple types.

Type	Metal
T	Copper and costantan
J	Iron and constantan
E	Nickel (10% chromium) and constantan
K	Nickel (10% chromium) and nickel (5% aluminum and silicon)

Thermometry 69

A thermocouple is inherently a differential temperature sensor; it measures the temperature difference between the two junctions. Absolute temperature measurements can be made only if one of the junctions is held at a known temperature or if an electronic reference junction is used. A block of metal (aluminum, copper, or any highly conductive metal) can be used for the reference temperature. This is done by inserting the reference junction of the thermocouple in the block and simultaneously measuring the temperature of the metal block using a thermistor (or any appropriate temperature sensor). The metal block must be kept in thermal equilibrium, which can be done with minimal effort in most situations. Alternatively, the reference junction can be held at the ice point, using a high quality ice bath, or maintained in an oven.

An electronic reference junction is a device that injects a current to make it appear that a junction is being held at a known temperature. The following discussion assumes that the reference junction is held at 0°C by an ice bath or an electronic reference.

The transfer equation for a thermocouple is

$$\Delta V = (a + b\Delta T)\Delta T \qquad (4.5)$$

where $a = 38.58\ \mu V/°C$ and $b = 0.0428\ \mu V/°C^2$ for copper-constantan.[2] The voltage is given in microvolts (μV).

Let the measurand range be -50 to $50°C$, then the sensor output in μV is given in the second column of table 4-4. It is difficult to compare temperature sensors with different types of output, for example, voltage and resistance, so we could force them to a standard by requiring that the sensor inputs have a common range (-50 to $50°C$). Additionally, we could require the sensor outputs to be matched to the input requirements of an analog-to-digital converter (ADC) that has an input range of -5 to 5 V.

Let us select an amplifier with gain G such that the amplified output will be 5 V when the temperature is $50°C$ and will be 0 V for $0°C$. We can see from table below that when $T = 50°C$, $\Delta V = 2036\ \mu V$, so we need a gain of $5/(2036 \times 10^{-6}) = 2456$. Then the amplifier output will be as given in the last column of table 4-4; the output is plotted over the range in fig. 4-6. The nonlinearity of the thermocouple is also shown in fig. 4-6. Amplifiers capable of higher gain usually cost more than those with lower gain, although this is becoming less true as the electronics become less expensive.

Table 4-4 Copper–constantan thermocouple output when the reference junction is held at 0°C.

T (°C)	$\Delta V = V_1 - V_2$ (μV)	V_3 (V)
-50	-1822	-4.475
-40	-1475	-3.622
-30	-1119	-2.748
-20	-755	-1.853
-10	-382	-0.937
0	0	0.000
10	390	0.958
20	789	1.937
30	1196	2.937
40	1612	3.959
50	2036	5.000

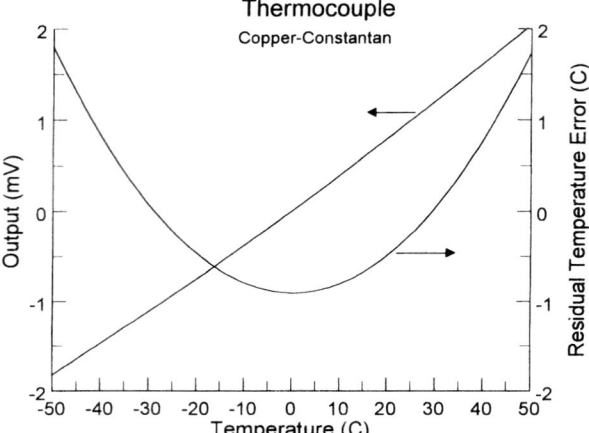

Fig. 4-6 Output of a copper–constantan thermocouple, and the residual nonlinearity.

Another disadvantage to higher gains is that the higher the gain, the more sensitive the amplifier becomes to noise and small offsets.

EXAMPLE

For a copper–constantan thermocouple, $a = 38.58\,\mu\text{V}/°\text{C}$ and $b = 0.0428\,\mu\text{V}/°\text{C}^2$. If the reference junction is held at 0°C and the measuring junction is at 20°C, the voltage developed across the junctions will be $\Delta V = (38.58 + 0.0428 \times 20) \times 20 = 789\,\mu\text{V}$ (see table 4-4, column 2). With an amplifier gain of 2456, the output $V_3 = 789 \times 10^{-6} \times 2456 = 1.937\,\text{V}$ (table 4-4, column 3).

4.3 Electrical Resistance Sensors

An electrical resistance sensor is one whose resistance varies as a function of temperature. Conductive sensors are usually called resistance temperature detectors (RTDs). The other major form of resistance sensors is the class of semiconductors called thermistors.

4.3.1 Resistance Temperature Detectors

Platinum is most commonly used for precision resistance thermometers because it is stable, resists corrosion, is easily workable, has a high melting point, and can be obtained to a high degree of purity. In addition, it has a simple and stable resistance–temperature relationship. However, platinum is sensitive to strain; bending the sensor can change the resistance. The resistance of a platinum sensor is given by

$$R_T = R_0(1 + aT + bT^2) \tag{4.6}$$

with sufficient accuracy for the meteorological temperature range (here taken to be -50 to $50°C$), where R_0 = resistance at $0°C$, R_T = resistance of the sensor at temperature T in $°C$. The values of the coefficients depend upon the purity of the platinum; the coefficient a is usually 0.00385 or $0.00392°C^{-1}$. In this example, $a = 0.00385°C^{-1}$ and $b = -5.85 \times 10^{-7}°C^{-2}$. Figure 4-7 shows the transfer plot for an RTD with $R_0 = 500\,\Omega$.

Because the RTD resistance is fairly low and the change with temperature is small, a bridge circuit, shown in fig. 4-5, is often used. This effectively converts resistance to voltage and can be amplified to a reasonable level using an instrumentation amplifier.[3] The other three resistors in the bridge could be set equal to R_0 (let $R_1 = R_0$); that would make the bridge output $V_3 = 0\,V$ when $T = 0°C$ and maximizes the bridge sensitivity (see Problem 6). In this case, the amplifier output is given by

$$V_3 = G V_R \left[\frac{R_T}{R_T + R_0} - \frac{1}{2} \right] \qquad (4.7)$$

where the amplifier gain G and the reference voltage V_R are constants to be determined.

The sensitivity S is found by using eqns. 4.6 and 4.7 to compute dV_3/dT:

$$S = \frac{dV_3}{dT} = \frac{GV_R R_0^2 a}{(R_T + R_0)^2} \qquad (4.8)$$

To maximize the sensitivity we need to make G and V_R large because these are the only variables left after choosing the RTD.

Another important consideration in the design of the circuit is the need to control sensor self-heating due to current flow through the RTD. The power dissipated, P_D, in R_T is given by

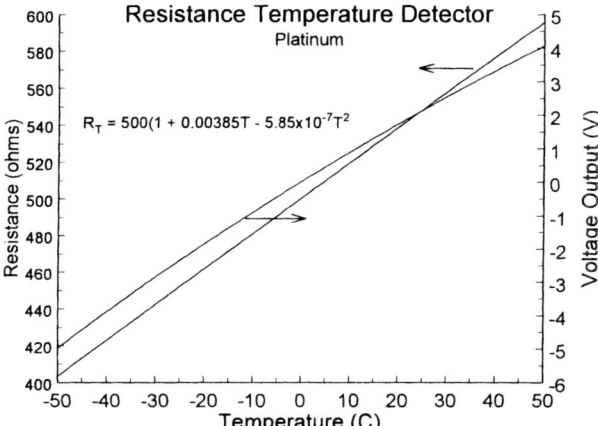

Fig. 4-7 Resistance of 500-ohm platinum RTD and voltage output in a bridge circuit vs. temperature.

$$P_D = I^2 R_T = \frac{V_R^2 R_T}{(R_T + R_0)^2} \tag{4.9}$$

Each RTD has a self-heating specification: the power dissipated P_D divided by the tolerable temperature error T_ε. This sets an upper limit to V_R. Next, the gain G can be selected to provide an adequate signal level to the next block in the measurement system (in the model measurement system shown in fig. 1-1 this would be the analog-to-digital converter which will have some limit on the input voltage range).

If the selected RTD has a self-heating specification of 5.9 mW/°C in air flowing at $1\,\mathrm{m\,s^{-1}}$ and if we are willing to tolerate a maximum error of 0.1°C due to self-heating, then the value of V_R can be determined from eqn. 4.8. To simplify the analysis, take the temperature to be $T = 0°C$ since the power dissipated will not vary much over the temperature range. Then $P_D(0) = V_R^2/(4R_0)$; it must be less than 5.9×10^{-4} W. If $R_0 = 500\,\Omega$ then V_R must be less than 1.09 V. Let us choose $V_R = 1.00$ V.

If the measurand range is -50 to $50°C$ then the absolute maximum amplifier output is obtained at $-50°C$. If the next device in the measurement system after the amplifier is an analog-to-digital converter with an input voltage range of -5.00 to 5.00 V, then we need to set G such that V_3 (at $-50°C$) $= -5.00$ V. This yields $G = 93$. Values of R_T and V_3 are listed in table 4-5 for various T. The amplifier output voltage is shown in fig. 4-7 and the residual nonlinearity (due almost entirely to the bridge circuit) is shown in fig. 4-8. A least-squares first-order fit could be used, with $T_3 = c_0 + c_1 V_3$, but the resulting residual error, due primarily to bridge nonlinearity, shown in fig. 4-5, would be too large. Alternatively, a second-order fit using $T_3 = c_0 + c_1 V_3 + c_2 V_3^2$ can be used. Residual errors from both fits are plotted in fig. 4-8 and the results from using the second-order fit are much more acceptable.

EXAMPLE

Given a platinum RTD in a bridge with an amplifier, find the output voltage when the temperature is $-30°C$.

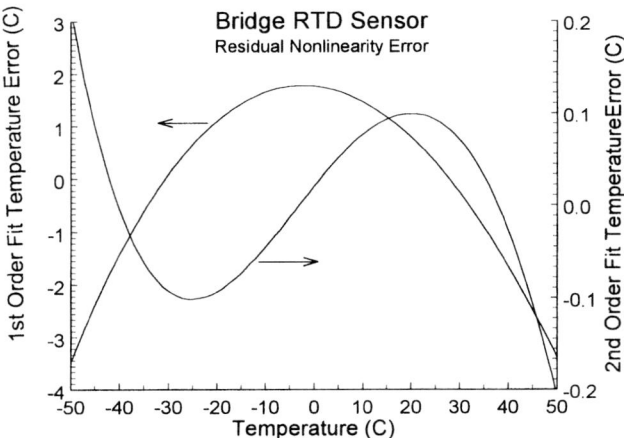

Fig. 4-8 Residual nonlinearity of a first-order fit of an RTD in a bridge circuit. A second-order fit is shown, using a different temperature scale on the right side.

SOLUTION

Using parameters from the above text: $G = 93$, $V_R = 1.000$ V, $a = 0.00385°C^{-1}$, $b = -5.85 \times 10^{-7} °C^{-2}$, $R_0 = R_1 = 500 \, \Omega$.

$$R_T = 500(1 + 0.00385 \times (-30) - 5.85 \times 10^{-7} \times (-30)^2) = 442.0 \, \Omega$$

$$V_3 = 93 \times 1.000 \left[\frac{442.0}{442.0 + 500.0} - \frac{1}{2} \right] = -2.864 \text{ V}$$

Compare these values to those found in table 4-5 columns 2 and 3.

Copper has a coefficient of resistivity ($0.00393°C^{-1}$) about the same as that of platinum, so it is not negligible. Therefore, long lead wires from the sensor to the bridge will affect the accuracy of the circuit. If long lead wires must be used, there are three- and four-wire circuits (from the sensor to the bridge) that reduce or eliminate the effect of the temperature coefficient of the copper wire.

The circuits shown are not necessarily the best circuits but serve to illustrate a possible way to use an RTD. The residual nonlinearity is significant and should not be ignored. When a microprocessor is used in the measuring instrument, the bridge nonlinearity can be easily corrected. Note that the calibration of the RTD circuit is a function of the resistors R_0, the reference voltage V_R, and the amplifier gain G. If any of these change due to drift, temperature sensitivity, or supply voltage sensitivity, the calibration will be affected. A microprocessor in the instrument can be used to control these effects with a little additional circuitry, as shown in fig. 4-9. In this circuit, R_0, R_L, and R_H are special, low temperature coefficient resistors; R_L and R_H are selected to be equal to R_T at the low and high ends of the temperature range, respectively. The microprocessor controls the multiplexer (MUX); in this case, a switch that selects one of three possible inputs. The selected signal is fed to the amplifier and thence to the analog-to-digital (ADC) converter and the converter output is read by the microprocessor. The microprocessor selects, in turn, $N = T$, L, and H, causing the MUX to switch the signal from the bridge legs adjacent to R_T, R_L, and R_H to the amplifier whose output is $V_N = V_T$, V_L, or V_H. Y_N is the digital representation of the analog signal V_N such that $Y_N = G_A V_N$, where G_A is a "gain" that represents the effective multiplier of the ADC. Then

$$Y_N = G_A \, G \, V_R \left[\frac{R_N}{R_N + R_0} - \frac{1}{2} \right] \quad (4.10)$$

Table 4-5 Resistance of a selected platinum RTD and output voltage of a bridge and amplifier combination with $V_R = 1.000$ V and $G = 93$.

T (°C)	R_T (Ω)	V_3 (V)
−50	403.0	−4.994
−30	442.0	−2.864
0	500.0	0
30	557.5	2.528
50	595.5	4.054

74 Meteorological Measurement Systems

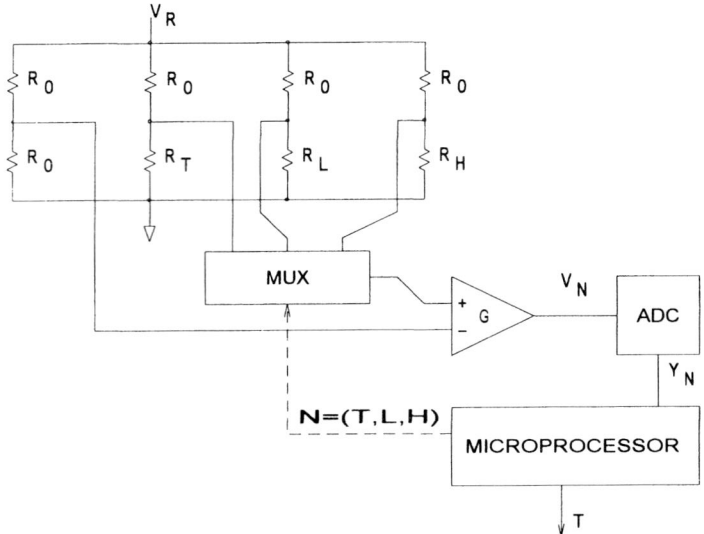

Fig. 4-9 A multiple-leg bridge circuit that, in conjunction with a microprocessor, is insensitive to small changes in the reference voltage or amplifier gain.

and the microprocessor can perform the operation

$$N = \frac{Y_T - Y_L}{Y_H - Y_L} \qquad (4.11)$$

where N is the normalized raw value. This effectively cancels the two gains and the reference voltage. It will also remove any bias or offset inserted by the amplifier or the ADC. The microprocessor can apply the calibration equation

$$T_5 = c_0 + c_1 N + c_2 N^2 + \cdots \qquad (4.12)$$

where T_5 is the estimate of the measurand based on the signal from the fifth element (the microprocessor) of the measurement system. Additional temperature sensors can be incorporated into this scheme by adding more legs to the bridge. See Pike et al. (1983) for an example of a five-leg bridge with two temperature sensors.

4.3.2 Thermistors

Thermistors are temperature-sensitive semiconductors, typically metallic oxides. They are characterized by large and quite nonlinear temperature sensitivity. Most thermistors have a negative temperature slope but a few have a positive slope. One form of the transfer equation is

$$R_T = \exp\left(a_0 + \frac{a_1}{T} + \frac{a_3}{T^3}\right) \qquad (4.13)$$

with T in kelvin. Coefficients for two thermistors are listed in table 4-6 and some resistance values for these thermistors are listed in table 4-7.

Table 4-6 Coefficients for two thermistors.

Thermistor	a_0	a_1	a_3
#1	−5.4019	4356.9	-1.3567×10^7
#2	−4.0103	4493.4	-1.9934×10^7

The resistance of thermistor #1 is plotted in fig. 4-10. The sensitivity decreases with increasing temperature (the slope of the line is becoming less negative with increasing temperature; i.e., it has a positive second derivative).

Despite their nonlinearity, thermistors are used in a wide variety of applications. They are popular because their high resistance makes them less sensitive to the resistance of lead wires and they are available in many configurations. A number of circuits have been devised to linearize thermistors. One of the most popular, using two thermistors and two resistors, is shown in fig. 4-11.

Voltage V_1 is given by

$$V_1 = \frac{R_{T2}(R_2 + R_{T1})V_R}{R_{T2}(R_1 + R_2 + R_{T1}) + R_1(R_2 + R_{T1})} \tag{4.14}$$

When R_{T1} is thermistor #1 from table 4-6, R_{T2} is thermistor #2, $R_1 = 18\,700\,\Omega$, and $R_2 = 35\,250\,\Omega$, then eqn. 4.14 can be approximated by the least-squares fit

$$V_1 \approx V_R(0.651\,07 - 0.006\,796\,6\,T) \tag{4.15}$$

where T is in °C and V_R is the reference voltage. The offset in this approximation, eqn. 4.14, could be countered with a voltage divider and amplifier; see fig. 4-11. In this figure, R_3 and R_4 form a voltage divider such that $V_2 = 0.651\,07\,V_R$. If we set $V_r = 1.00\,\text{V}$ and set the gain $G = 14.7$ such that $V_3 \approx 5.00\,\text{V}$ when $T = 50°\text{C}$, the voltage output will be shown in fig. 4-12 and in table 4-7.

There is a residual nonlinearity that results from the use of eqns. 4.15 as an approximation to eqn. 4.14. This nonlinearity is shown in fig. 4-12 where the nonlinearity (the error relative to a line) is expressed in terms of °C. This circuit is designed to work in the temperature range −30°C to 50°C and there are other circuits appropriate for other ranges. For example, there is a three-thermistor circuit that performs well over the range −50°C to 50°C.

Table 4-7 Resistance vs. temperature for thermistors #1 and #2, and voltages V_1 and V_3 corresponding to fig. 4-11.

Temperature (°C)	#1 Resistance (Ω)	#2 Resistance (Ω)	V_1 (V)	V_2 (V)
−30	106 190	480 648	0.8539	−2.981
−20	58 258	271 225	0.7881	−2.014
−10	33 202	158 122	0.7187	−0.994
0	19 595	95 016	0.6503	0.011
10	11 942	58 738	0.5832	0.997
20	7 495	37 285	0.5157	1.990
30	4 833	24 257	0.4469	3.001
40	3 195	16 147	0.3781	4.012
50	2 162	10 981	0.3122	4.981

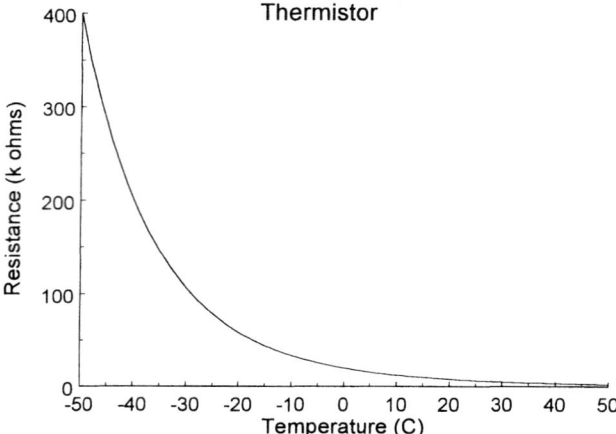

Fig. 4-10 Thermistor #1 resistance vs. temperature.

EXAMPLE

Using coefficients for thermistors #1 and #2 from table 4-6 and a temperature of 25°C, find the resistance of both thermistors. Use eqn. 4.13:

$$R_T = \exp\left(-5.4019 + \frac{4356.9}{298.15} - \frac{1.3567 \times 10^7}{(298.15)^3}\right)$$
$$= 5998.8 \ \Omega \text{ for thermistor \#1.}$$

For thermistor #2, $R_T = 29\,906 \ \Omega$.

4.4 Comparison of Temperature Sensors

If we wish to measure the air temperature, there is a well-established set of sensors available and a large body of knowledge pertaining to their use. We could select a mercury-in-glass thermometer, a bimetallic strip, a thermocouple, a thermistor, or a platinum resistance sensor, and all of these are useful in air temperature measurement. We also have available the elements of a complete measurement system to deliver the temperature information to the user. In the case of a mercury-in-glass thermometer the thermometer is the complete system as it incorporates a means to convert the raw sensor output, the volume expansion of mercury, to a readable display, the mercury column, with a means for incorporating a calibration, the attached scale. Other systems, while using different devices, achieve the same objective. It is never safe to minimize the problems of calibration but it is clearly possible, using established techniques, to obtain a temperature sensor calibrated to a reasonable level of accuracy. And this can be done using any of the temperature sensors described above in addition to many others not mentioned.

We could compare temperature sensors on the basis of cost, reliability, size, and ease of use. Some of these change as manufacturing technology improves. For exam-

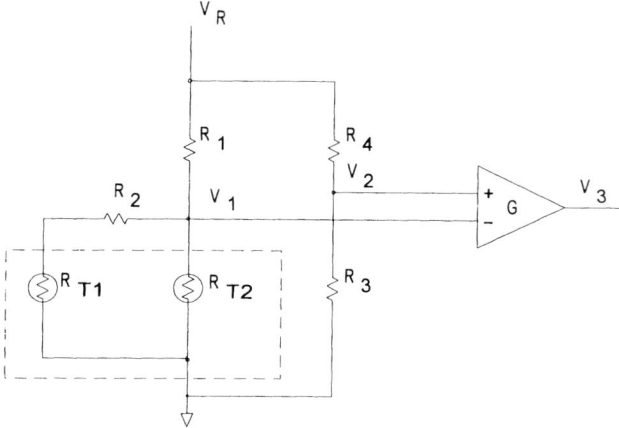

Fig. 4-11 An almost linear circuit with two thermistors. The dashed lines enclose a three-terminal thermistor composite.

ple, RTDs historically were too large and expensive for field use, but now they can be obtained in reasonably small sizes and are cost competitive.

One figure of merit for temperature sensors with various outputs is the amplifier gain required to bring the sensor output into conformance with the input requirements of an ADC. We would like this gain to be as small as possible, given the above requirement, because a high gain amplifier amplifies noise as well as signal and is more expensive. So our figure of merit is the inverse of the log (base 10) of the amplifier gain. We can compare all of the temperature sensors with a common input range of -30 to $50°C$ and require that the amplified output signal be in the range -5 to 5 V. For the copper–constantan thermocouple shown in fig. 4-5, the figure of merit for this thermocouple is $1/\log(2456) = 0.29$. With this measure, the linear thermistor is the best sensor of those discussed (see table 4-8).

4.5 Exposure of Temperature Sensors

If the temperature sensor is exposed to the air, the instrument will indicate a temperature. However, the temperature of the sensor may differ from the air temperature, perhaps by a large amount. Heat flows towards or away from a temperature sensor by conduction, convection, and radiation. Air is very poor conductor of

Table 4-8 Figure of merit for several temperature sensors.

Sensor	Gain	Merit
Thermocouple	2456	0.29
Resistance (RTD)	93	0.51
Linear thermistor	14.7	0.86

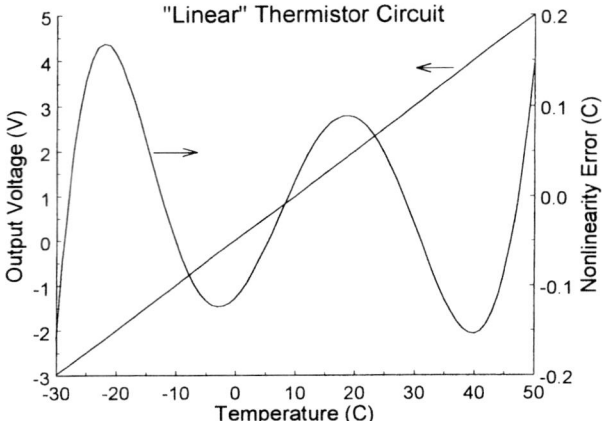

Fig. 4-12 Voltage output (left y-axis) and residual nonlinearity (right y-axis) of a "linear" thermistor circuit.

heat (see table 4-9); common insulation materials trap bubbles of air in a thin matrix of supporting material. The thin matrix exists just to isolate the air bubbles and prevent convection while the air is the primary insulating material. Some mechanical device that is almost certainly a better conductor of heat than still air must support the temperature sensor. The sensor exists in a radiative environment; it attempts to radiate heat to cooler surfaces, including outer space at night. Other, hotter surfaces, including the sun, attempt to radiate heat to the sensor. To accurately measure the air temperature the sensor must be in good thermal contact with the air. That requires air circulation to promote heat transfer by convection while the sensor must be protected from conductive heat flow along the mechanical support, and from radiative heat transfer. The central problem for immersion sensing instruments in meteorology is coupling with the atmosphere.

To illustrate the exposure problem, consider a cylindrical temperature sensor mounted on a wall and exposed to a moving air stream and to solar radiation. There is no shelter of any kind. This is illustrated in fig. 4-13 where R is the solar radiation, T_S is the sensor temperature, T_M is the wall temperature where the mounting bracket is fastened, and T_A and V_A are air temperature and wind speed.

The heat transferred to the sensor by conduction, radiation, and convection is given by

Table 4-9 Thermal conductivity of various substances.

Material	Thermal Conductivity (W m^{-1} °C^{-1})
Silver	406.0
Steel	50.2
Concrete	0.8
Air (calm)	0.024

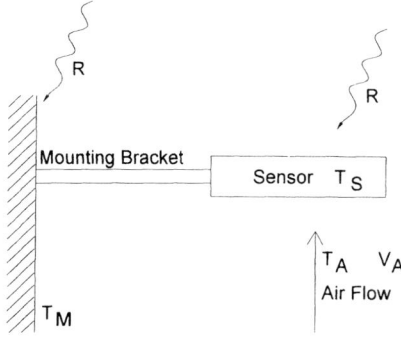

Fig. 4-13 Simple model of heat transfer by conduction, convection, and radiation.

$$H_C = -k_M A_M \frac{\Delta T}{\Delta X} = -k_M A_M \frac{T_S - T_M}{\Delta X} \qquad (4.16a)$$

$$H_R = A_S \alpha_S R = \pi D L \alpha_S R \qquad (4.16b)$$

$$H_V = cL(T_S - T_A)\sqrt{DV_A} \qquad (4.16c)$$

where H_C, H_R, H_V = conductive, radiative, and convective heat flow (W), k_M = mounting bracket thermal conductivity (W m^{-1} K^{-1}), A_M = mounting bracket cross-sectional area (m^2), T_S, T_M = temperatures (K), ΔX = mounting bracket length (m), D, L = sensor diameter and length (m), A_S = sensor surface area (neglecting end areas) = πDL (m^2), α_s = sensor absorptivity (dimensionless), R = global solar radiation (W m^{-2}), $c = 8.011$ W m^{-2} K^{-1} s$^{-1/2}$, an empirical contant, and V_A = air speed (m s^{-1}). The heat conduction equation is the usual expression for heat flow (Fourier law of conduction) across any material, a mounting bracket in this case.

Radiative heat transfer is expressed in terms of the incident solar radiation, a quantity that is usually measured by a pyranometer (see Chap. 10). Solar radiation is received over the surface area of the sensor. Sensor absorptivity α_S is the fraction of incident radiation absorbed by the sensor, therefore $0 \le \alpha_S \le 1$. A value of 0 indicates that the sensor reflects all incident radiation and sensor absorptivity would be 1.0 if it were a black body, that is, if it absorbed all incident radiation.

The convection equation is an empirical equation appropriate for cylinders whose length is large compared to diameter and for light wind speeds, say $0.1 < V_A < 10$ m s^{-1}. This equation is presented to give some feeling for the convective heat flow but should not be used outside of this context. See Richardson and Brock (1995) for more details.

In fig. 4-13, the sun is shining on the sensor, the wall, and on the mounting bracket so all will be warmer than the air temperature. If we ignore heat transfer by conduction (assume the mounting bracket is a good insulator) we can calculate the sensor excess temperature by setting $H_R = H_V$ (this is a steady-state example in which the energy lost by convection must equal the energy gained by absorption of radiation); then

$$T_S - T_A = \frac{\pi}{c}\alpha_S R \sqrt{\frac{D}{V_A}} \qquad (4.17)$$

that we wish to minimize. To do so, we must minimize α_S, R, and D and maximize V_A. But in the simple case posed here, only α_S and D can be controlled through instrument design. This implies that we should make α_S small (highly reflective sensor) and make the sensor diameter as small as possible. It is difficult to make α_S much less than 0.2, especially if the sensor is exposed to the atmosphere for a long time. Dust accretion will gradually increase the absorptivity.

When temperature sensors are mounted on radiosondes, power and weight constraints prohibit the use of a fan to augment wind speed and it is difficult to provide any shelter from solar radiation, so designers usually use a very small diameter sensor that is painted white or silver. For example, Turtiainen et al. (1955) developed a highly reflective radiosonde temperature sensor with a diameter of just 100 µm. Even when it was exposed to direct sunlight at high altitudes, they expected radiation-induced temperature errors to be less than 0.5°C. In their design, the wires supporting the sensing element had a diameter of 100 µm; therefore, heat conduction was minimized since heat conduction is proportional to the cross-sectional area of the supporting wire.

Radiosondes (e.g., Vaisala RS-90) use an unshielded, highly reflective sensor only 0.1 mm in diameter. The radiational heating error is approximately 0.5 K at 10 hPa and, after correction, residual error is on the order of 0.1 K. The time constant of this sensor is approximately 0.5 s.

EXAMPLE

Consider a cylindrical temperature sensor with a diameter of 1 cm and an absorptivity of 0.2. The wind speed is 2 m s^{-1} and the solar radiation is 500 W m^{-2}. The excess temperature error is (using eqn. 4.16)

$$\Delta T = T_S - T_A = \frac{\pi}{8.011} \times 0.2 \times 500 \times \sqrt{\frac{.01}{2}} = 2.7°C.$$

If the sensor diameter were reduced to 1 mm, $\Delta T = 0.9°C$.

In some micrometeorological and agricultural applications temperature measurements near and under leaves are required and radiation shelters of any kind are too cumbersome to use. Therefore, small, highly reflective sensors are used, just as with radiosondes.

In general, the best way to measure air temperature is with an aspirated (fan-forced) radiation shield. The shield protects the sensor from rain, mechanical damage, and some, not all, of the incident radiation. A fan is used to force a steady flow of air over the sensor to maintain relatively high convective heat transfer to offset any remaining radiative or conductive heat transfer. Normally, the shield is designed such that air flows over the sensor, then to the fan, and is exhausted far enough from the sensor that return airflow is unlikely. This is the most expensive kind of shield and requires power, usually from a commercial power line, to operate the fan.

When cost or power constraints prohibit the use of aspirated shields, passive (no fan) shields are used. Two examples are illustrated in fig. 4-14. These are designed to protect the sensor from mechanical damage and incident radiation while impeding natural airflow as little as possible. The latter two design goals are, to some extent, mutually exclusive. A design that blocks all incoming radiation may also completely block natural airflow. Solar radiation that strikes the

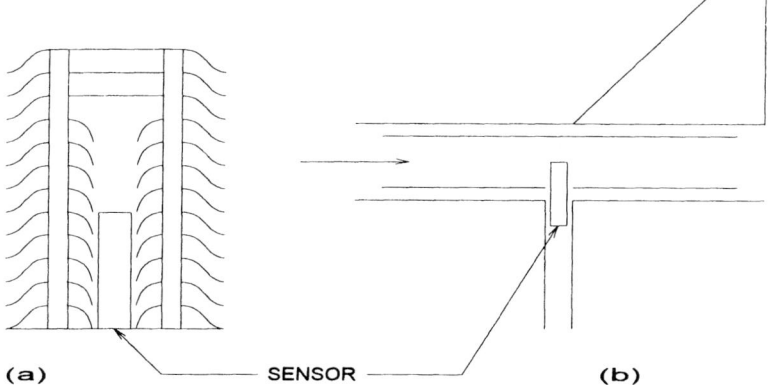

Fig. 4-14 Cross-section of two unaspirated radiation shields, the multiplate shield (a) and the vane shield (b).

shield will either be reflected away, be absorbed by the shield, or will, after some internal reflections, hit the sensor. Radiation that is absorbed by the shield will heat the shield which, in turn, will heat the air flowing over it before the air reaches the sensor. This causes an indirect contribution to excess temperature error. Radiation that reaches the sensor will, if absorbed by the sensor, heat the sensor (direct contribution). Therefore, on a calm, sunny day, the temperature sensor will indicate a temperature in excess of the actual air temperature. This is an example of exposure error.

Since it is not possible to totally protect a sensor from solar radiation and because any shield design will block natural air flow to some extent, it follows that the temperature sensor mounted inside the shield should be selected following the guidelines established for radiosondes: the sensor should have the smallest possible diameter and the lowest possible absorptivity. In addition, the sensor should be short enough to be totally enclosed within the shield. If part of the sensor is exposed to direct or reflected solar radiation, errors due to heat conduction through the sensor will be enhanced. Even a good passive shield design will block only about 80% of the incoming solar radiation and, unfortunately, will block about 80% of the ambient air flow (Richardson et al., 1999).

Radiation errors can reach extremes when there is maximum solar radiation, light winds and a highly reflecting ground surface (snow). These are the conditions simulated in fig. 4-15, which is based on unpublished wind tunnel experiments conducted by G.C. Gill at the University of Michigan in 1983. Actual temperature error due to inadequate exposure will be a function of the temperature sensor design (length, diameter, and absorptivity), shield design (solar radiation absorbed, solar radiation passed to the sensor, ambient air flow blockage), weather conditions (solar radiation, wind speed, and, to some extent, wind direction), and ground reflectivity (amount of solar radiation reflected back to the shield.) None of these factors are included in the manufacturers sensor performance specifications because exposure conditions are not under the manufacturer's control. Exposure errors are always excluded from laboratory calibrations so laboratory calibration performance is not necessarily a good predictor of actual field performance.

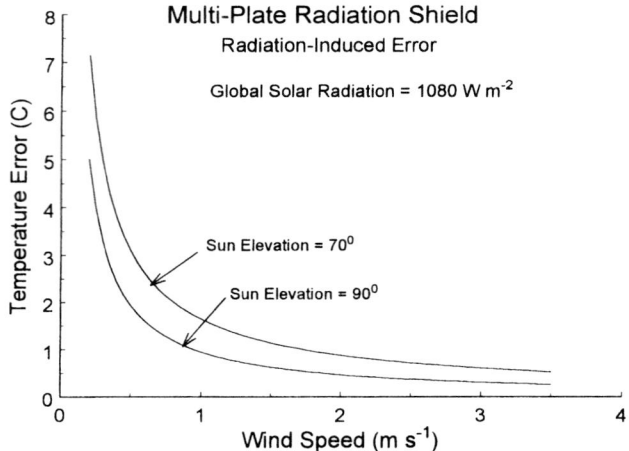

Fig. 4-15 Extreme radiation error in a Gill multiplate shield.

QUESTIONS

(More difficult questions are indicated with an asterisk.)

1. If a bimetallic strip is 5 cm long, 1 mm thick, and has a deflection constant of $5 \times 10^{-5}\,°C^{-1}$, how much will it deflect for a temperature change of 10°C?
    Answer: 1.25 mm.
2. Calculate the static sensitivity of a mercury-in-glass thermometer if the volume of the mercury in the bulb is 200 mm^3, capillary diameter is 0.15 mm, and the difference between the volume coefficients of thermal expansion of mercury and glass is $1.6 \times 10^{-4}\,°C^{-1}$.
3. If the nonlinearity error for a "linear" thermistor, defined by eqn. 4.15, at a certain temperature is 1.1 mV, what is the equivalent temperature error?
    Answer: $-0.16/V_R\,°C$.
4. Given a platinum RTD with $R_0 = 100\,\Omega$, $a = 0.00385\,°C^{-1}$, and a self-heating rating of 0.20 mW/°C at a wind speed of $1\,m\,s^{-1}$, if the maximum tolerable error due to self-heating is 0.05°C and $T = 0°C$, what is the maximum power that can be dissipated in the RTD? What is the static sensitivity at $T = 10°C$?
5. For the RTD in question 4, what is the resistance of the RTD at $T = 30°C$?
    Answer: 111.6 Ω.
6.* Show that the sensitivity of a bridge circuit, as shown in fig. 4-5, is maximized by setting $R_1 = R_0$.
7. Write an expression for the static sensitivity of a bimetallic strip and identify all of the terms.
8. Define:

   thermistor
   linear thermistor
   resistance Temperature Detector
   thermocouple
   self-heating
   bimetallic strip

9. In meteorological applications, what is the primary limitation to the accuracy of air temperature sensors? What can be done about this limitation?
10. Does the density of a thermometric fluid affect the static sensitivity of a liquid-in-glass thermometer?
11. What are the characteristics of an ideal radiation shield?
12. What is the effect of excessive current through a resistance temperature detector?
13. What is the static sensitivity of thermistor #1 (table 4-6) at $0°C$?
14. What is the static sensitivity of a copper–constantan thermocouple when $\Delta T = 10°C$?
15. What is the static sensitivity of the linear thermistor described in the text?
16. Can we ignore the "b" in eqn. 4.5? Let $R_0 = 500\,\Omega$ and $T = 50°C$. What is the temperature error caused by ignoring b?
17.* Derive eqn. 4.13 for the linear thermistor, starting from Ohm's law.
18.* Design a passive (no fan) radiation shield that minimizes solar radiative heat transfer to the sensor, both direct and indirect, while minimizing blockage of natural air flow.
19. What signal conditioning is performed in a mercury-in-glass thermometer?
20. A thermocouple temperature sensor static equation is $V = (a + b\Delta T)\Delta T$ where $a = 38.6\,\mu V\,K^{-1}$ and $b = 0.0413\,\mu V\,K^{-2}$. ΔT is the temperature difference between the two junctions of the thermocouple.
   (a) When $\Delta T = 10\,k$, what is the output V? What are the units?
   (b) When $\Delta T = -10\,K$, what is the static sensitivity?
   (c) When $\Delta T = 40\,K$, what is the static sensitivity?
   (d) Is this sensor linear? Explain.
21. If we wish to decrease the radiation-induced error in a temperature sensor by a factor of two, by how much should we increase or decrease the sensor diameter? What other change could we make to the sensor to reduce this error?
22. If you double the $0°C$ resistance of an RTD, how much does the static sensitivity change?
23. For a thermocouple with $a = 38.63\,\mu V/°C$ (assume $b = 0$) the observed output was $-1250\,\mu V$ when the reference junction temperature was $50°C$. What was the sensing junction temperature?
24. Could a mercury-in-glass thermometer have a hysteresis effect? What about threshold?

Solution of Selected Problems

6. Given the bridge circuit in fig. 4-5 with R_0 equal to R_T when $T = 0°C$, let $R_1 = mR_0$; then

$$V_3 = G V_R \left[\frac{R_T}{R_T + m R_0} - \frac{1}{1+m} \right].$$

The sensitivity, with respect to R_T, is

$$S = \frac{d V_3}{d R_T} = \frac{G V_R m R_0}{(R_T + m R_0)^2}$$

and the optimum value of m can be found by taking the derivative of S with respect to m and setting the result to 0. This yields

$$\frac{dS}{dm} = G V_R R_0 \left[\frac{R_T - m R_0}{(R_T + m R_0)^3} \right] = 0.$$

This will be equal to 0 when $mR_0 = R_T$ and, since $R_T = R_0$ when $T = 0°C$, the optimum value of the constant $m = 1$.

BIBLIOGRAPHY

Brock, F.V., S.J. Richardson, and S.R. Semmer, 1995a: Passive multiplate solar radiation shields. *Preprints 9th Symposium on Meterological Observations and Instrumentation, Charlotte, NC*. American Meterological Society, Boston, MA, pp. 329–334.

Brock, F.V., S.R. Semmer, and C. Jirak, 1995b: Passive solar radiation shields: Wind tunnel testing. *Preprints 9th Symposium on Meterological Observations and Instrumentation, Charlotte, NC*. American Meterological Society, Boston, MA, pp. 179–183.

Cheney, N.R., and J.A. Businger, 1990: An accurate fast response temperature system using thermocouples. *J. Atmos. Oceanic Technol.*, 7, 504–516.

Crook, N.A., 1996: Sensitivity of moist convection forced by boundary layer processes to low-level thermodynamics fields. *Mon. Wea. Rev.*, 124, 1767–1785.

Deacon, E.L., 1980: Slow-response temperature sensors. In *Air–Sea Interaction: Instruments and Methods*, ed. F. Dobson, L. Hasse, and R. Davis. Plenum Press, New York, NY, 801 pp.

DeFelice, T.P., 1998: *An Introduction to Instrumentation and Measurement*. Prentice Hall, Upper Saddle River, NJ, 229 pp.

Doebelin, E.O., 1983: *Measurement Systems: Application and Design*, 3rd ed. McGraw-Hill, New York, 876 pp.

Friehe, C.A., and D. Khelif, 1992: Fast-response aircraft temperature sensors. *J. Atmos. Oceanic Technol.*, 9, 784–795.

Fuchs, M., and C.B. Tanner, 1965: Radiation shields for air temperature thermometers. *J. Appl. Meteor.*, 4, 544–547.

Jacobs, A.F.G., and K.G. McNaughton, 1994: The excess temperature of a rigid fast-response thermometer and its effects on measured heat flux. *J. Atmos. Oceanic Technol.*, 11, 680–686.

Katsaros, K., 1980: Radiative sensing of sea surface temperature. In *Air–Sea Interaction: Instruments and Methods*, ed. F. Dobson, L. Hasse, and R. Davis. Plenum Press, New York, 801 pp.

Kent, E.C., R.J. Tiddy, and P.K. Taylor, 1993: Correction of marine air temperature observations for solar radiation effects. *J. Atmos. Oceanic Technol.*, 10, 900–906.

Krovetz, D.O., M.A. Reiter and J.T. Sigmon, 1988: An inexpensive thermocouple probe-amplifier and its response to rapid temperature fluctuations in a mountain forest., *J. Atmos. Oceanic Technol.*, 5, 870–874.

Larson, S.E., J. Hojstrup, and C.H. Gibson, 1980: Fast-response temperature sensors. In *Air–Sea Interaction: Instruments and Methods*, ed. F. Dobson, L. Hasse, and R. Davis. Plenum Press, New York, NY, 801 pp.

Lawson, R.P., and W.A. Cooper, 1990: Performance of some airborne thermometers in clouds. *J. Atmos. Oceanic Technol.*, 7, 480–494.

Lawson, R.P., and A.R. Rodi, 1992: A new airborne thermometer for atmospheric and cloud physics research. Part I: Design and preliminary flight data. *J. Atmos. Oceanic Technol.*, 9, 556–574.

McGee, T.D., 1988: *Principles and Methods of Temperature Measurement*. John Wiley, New York, 581 pp.

Meteorological Office, 1981: *Measurement of Temperature*, Vol. 2 of *Handbook of Meteorological Instruments*, 2nd ed. Meteorology Office 919b, Her Majesty's Stationery Office, London, 57 pp.

Ney, E.P., R.W. Maas, and W.F. Huch, 1961: The measurement of atmospheric temperature. *J. Meteor.*, 18, 60–80.

Nolan, P.F., and S.G. Jennings, 1987: A new thermocouple thermometer. *J. Atmos. Oceanic Technol.*, 4, 391–400.

Pike, J.M., F.V. Brock, and S.R. Semmer, 1983: Integrated sensors of PAM II. *Preprints 5th Symposium on Meterological Observations and Instrumentation, Toronto, Canada.* American Meteorological Society, Boston, MA, pp. 327–333.

Richard, S. J., 1995: Passive solar radiation shields: Numerical simulation of flow dynamics. *Preprints 9th Symposium on Meteorological Observations and Instrumentation, Charlotte, NC.* American Meteorological Society, Boston, MA, pp. 253–258.

Richards, S.J., and F.V. Brock, 1995: Passive solar radiator shields: Energy budget-optimizing shield design. *Preprints 9th Symposium on Meteorological Observations and Instrumentation, Charlotte, NC.* American Meteorological Society, Boston, MA, pp. 259–264.

Richardson, S.J., F.V. Brokc, S.R. Semmer, and C. Jirak, 1999: Minimizing errors associated with multiple radiation shields. *J. Atmos. Oceanic Technol.*, 16, 1862–1872.

Schmitt, K.F., C.A. Friehe, and C.H. Gibson, 1978: Humidity sensitivity of atmospheric sensors by salt contamination., *J. Physical Ocean.*, 8, 151–161.

Schooley, J.F., 1986: *Thermometry.* CRC Press, Boca Raton, Florida, 245 pp.

Snow, J.T., and S.B. Harley, 1987: Basic meteorological observations for schools: temperature. *Bull. Amer. Meteor. Soc.*, 68, 486–496.

Timko, M.P., 1976: A two-terminal IC temperature transducer. *IEEE J. Solid-State Cir.*, SC-11, 784–788.

Turtiainen, H., S. Tammela, and I. Stuns, 1995: A new radiosonde temperature sensor with fast response time and small radiation error. *Preprints 9th Symposium on Meteorological Observations and Instrumentation, Charlotte, NC.* American Meteorological Society, Boston, MA, pp. 60–64.

NOTES

1. If current were to flow through the circuit, then one junction would be heated and the other cooled. This is called the Peltier effect. This effect is used in some heat pumps where a current is forced through a thermopile (a series of therocouple junctions). For example, the Peltier effect is used in the chilled-mirror hygrometer to heat and cool the mirror.

2. The ASTM has published a standard table of thermocouple voltages versus temperature for various thermocouples, including type T for copper-constantan. One could fit a polynomial to any table for any range of temperatures. That is how the coefficients in the text were generated.

3. An instrumentation amplifier is designed to take the difference between two small voltages, say V_1 and V_2, and to amplify that difference, $V_3 = G(V_1 - V_2)$.

5

Hygrometry

The objective of atmospheric humidity measurements is to determine the amount of water vapor present in the atmosphere by weight, by volume, by partial pressure, or by a fraction (percentage) of the saturation (equilibrium) vapor pressure with respect to a plane surface of pure water. The measurement of atmospheric humidity in the field has been and continues to be troublesome. It is especially difficult for automatic weather stations where low cost, low power consumption, and reliability are common constraints.

5.1 Water Vapor Pressure

Pure water vapor in equilibrium with a plane surface of pure water exerts a pressure designated e'_s. This pressure is a function of the temperature of the vapor and liquid phases and can be obtained by integration of the Clausius–Clapeyron equation, assuming linear dependence of the latent heat of vaporization on temperature, $L = L_0[1 + \alpha(T - T_0)]$,

$$e'_s = e'_{s0} \exp\left[\frac{L_0}{R_v}\left(\frac{T - T_0}{T T_0} + \alpha \ln\left(\frac{T}{T_0}\right) - \frac{\alpha(T - T_0)}{T}\right)\right] \tag{5.1}$$

where $T_0 = 273.15$ K, $L_0 = 2.5008 \times 10^6$ J kg^{-1}, the latent heat of water vapor at T_0, $R_v = 461.51$ J kg^{-1} K^{-1}, the gas constant for water vapor, $e'_{s0} = 611.21$ Pa, the equilibrium water vapor pressure at $T = T_0$, and $\alpha = -9.477 \times 10^{-4}$ K^{-1} = average rate of change coefficient for the latent heat of water vapor with respect to temperature.

Table 5-1 Coefficients for the empirical equation 5.2 for equilibrium vapor pressure over a plane surface of pure water and over ice.

Coefficient	Water	Ice
c_1	−2991.272	0.0
c_2	−6017.0128	−5865.3696
c_3	18.876 438 54	22.241 033
c_4	−0.028 354 721	0.013 749 042
c_5	$0.178\,383\,01 \times 10^{-4}$	$-0.340\,317\,75 \times 10^{-4}$
c_6	$-0.841\,504\,17 \times 10^{-9}$	$0.269\,676\,87 \times 10^{-7}$
c_7	$0.444\,125\,43 \times 10^{-12}$	0.0
c_8	2.858 487	0.691 865 1

Since water vapor is not a perfect gas, the above equation is not an exact fit. The vapor pressure as a function of temperature has been determined by numerous experiments. Wexler (1976, 1977) fitted an empirical equation to the experimental vapor pressure data,

$$e'_s = \exp\left(\frac{c_1}{T^2} + \frac{c_2}{T} + c_3 + c_4 T + c_5 T^2 + c_6 T^3 + c_7 T^4 + c_8 \ln T\right) \quad (5.2)$$

(T in K, e'_s in Pa) and the coefficients for vapor pressure over water and over ice are given in table 5-1.

Both eqns. 5.1 and 5.2 are cumbersome; an equation that is readily inverted but with sufficient accuracy would be preferable. Buck (1981) developed an equation that is easy to use and sufficiently precise over the temperature range −30 to 50°C,

$$e'_s = 6.1121 \exp\left(\frac{17.502\,T}{240.97 + T}\right) \quad (5.3)$$

where T is in degrees Celsius and e'_s is in units of hPa (or mb). Equations 5.1, 5.2, and 5.3 are contrasted in table 5-2 and fig. 5-1. The error in eqn. 5.1 is tolerable but eqn. 5.3 is preferred because it is easier to invert to obtain the dew-point temperature given the ambient vapor pressure.

As noted above, the term equilibrium vapor pressure is more accurate but the term saturation vapor pressure is commonly used. We will use the term saturation vapor pressure.

Contrary to Dalton's[1] law of partial pressures, the total air pressure does have a small affect on the saturation vapor pressure; this is called the enhancement effect (Buck, 1981); see table 5-3. Saturation vapor pressure in a mixture of dry air and water vapor (moist air) is the saturation vapor pressure of pure water vapor times the enhancement factor: $e_s = e'_s(T)f(T,p)$.

There is a pressure effect and a weak temperature dependence. For $p > 800$ hPa, we can use $f = 1.004$. The enhancement factor has been incorporated into the following equations for the vapor pressure, and so eqn. 5.3 becomes

$$e_s = 6.1365 \exp\left(\frac{17.502\,T}{240.97 + T}\right) \quad (5.4)$$

and the equilibrium vapor pressure over an ice surface is

Table 5-2 Comparison of the theoretical equation (5.1) for water vapor pressure with the expression obtained from experimental results (5.2) and the more convenient approximation (5.3).

Temperature (°C)	Experimental results, eqn. 5.2 (hPa)	Error in Buck approx., eqn. 5.3 (hPa)	Error in eqn. 5.1 (hPa)
0	6.1121	0.0	0.0
30	42.4520	−0.0169	−0.0789
50	123.4476	0.2447	−0.5831

Table 5-3 Enhancement factor for various temperatures and pressures.

	Enhancement factor f (dimensionless)		
T(°C)	$p = 1000$ hPa	$p = 500$ hPa	$p = 250$ hPa
−50	1.0058	1.0029	1.0014
−40	1.0052	1.0026	1.0013
−30	1.0047	1.0024	1.0012
−20	1.0044	1.0022	1.0012
−10	1.0041	1.0022	1.0012
0	1.00395	1.00219	1.00132
10	1.00388	1.00229	
20	1.004	1.00251	
30	1.00426	1.00284	
40	1.00467	1.00323	
50	1.00519		

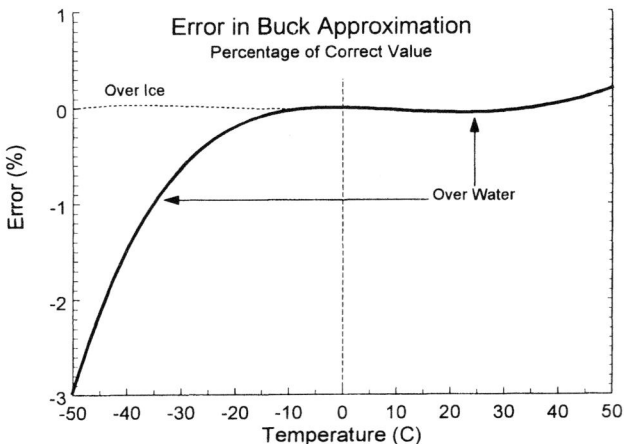

Fig. 5-1 Error in the Buck approximation.

$$e_i = 6.1359 \exp\left(\frac{22.452\,T}{272.55 + T}\right) \tag{5.5}$$

where, as before, T is in units of °C for both of the above equations.

Water vapor saturation pressure varies over two orders of magnitude in the normal temperature range; see fig. 5-2. On the basis of this figure, one would expect the accuracy of almost any humidity instrument to decrease with decreasing temperature.

Figure 5.2 can be used to illustrate several humidity relationships. Let point A represent ambient temperature and vapor pressure. Then the saturation vapor pressure is e_s (point B). If the air parcel were cooled, at constant pressure, until condensation just starts to occur, the new air temperature would be the dew-point temperature T_d and the ambient vapor pressure would be unchanged and would now be equal to the saturation vapor pressure at T_d (point D). Starting at point A again, a thermal bulb covered with water would be cooled by evaporation and the vapor pressure in the immediate vicinity would increase, due to the increased evaporation rate of water molecules, until the temperature of the wet bulb becomes the wet-bulb temperature T_w and the new vapor pressure would be the saturation vapor pressure at T_w, e_{sw} (point C).

5.2 Definitions

There are many variables commonly encountered in the study of humidity.

Absolute humidity, d_v, is the ratio of the mass of water vapor m_v to the total volume of moist air V in units of $kg\,m^{-3}$.

Dew-point temperature, T_d, is the temperature at which ambient water vapor condenses. The frost-point temperature, T_f, is the temperature at which ambient water

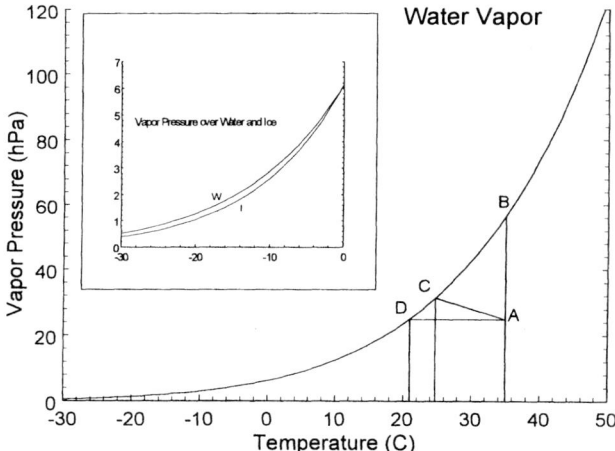

Fig. 5-2 Saturation vapor pressure as a function of temperature. Inset shows saturation vapor pressure with respect to water (top curve) and with respect to ice (bottom curve) for $T < 0°C$.

vapor freezes. The dew- and frost-point temperatures can be obtained from the ambient vapor pressure by inverting eqns. 5.4 and 5.5:

$$\left.\begin{array}{l} T_d = 240.97 \ln(e/6.1365)/(17.502 - \ln(e/6.1365)) \\ T_f = 272.55 \ln(e/6.1359)/(22.452 - \ln(e/6.1359)) \end{array}\right\} \quad (5.6)$$

Mixing ratio, w, is the ratio of the mass of water vapor m_v to the mass of dry air m_d.
Relative humidity, U, is defined as the ratio, expressed as a percentage, of the actual vapor pressure e to the saturation vapor pressure e_s at the air temperature T:

$$U = 100 e/e_s \quad (5.7)$$

This definition always uses saturation vapor pressure with respect to a plane surface of pure water, even for temperatures below freezing. Some of the earliest humidity sensors, and still the most common, are the class of sorption sensors which, as will be shown later, generate an output proportional to relative humidity.

Specific humidity, q, also known as the mass concentration, is the ratio of the mass of water vapor m_v to the mass of moist air, $m_v + m_d$.

Temperature or *dry-bulb temperature* is the ambient air temperature T as measured, for example, by the dry-bulb thermometer of a psychrometer.

Vapor pressure, e, is the partial pressure of water vapor expressed in hPa.

Virtual temperature, T_v, is the temperature that dry air would have if the dry air had the same density as moist air at the same pressure. $T_v \geq T$:

$$T_v = \frac{T}{1 - \dfrac{e}{p}(1 - 0.622)}$$

Wet-bulb temperature, T_w, is the temperature indicated by the wet bulb of a psychrometer, that is, the temperature of a sensor covered with pure water that is evaporating freely into an ambient air stream.

The following relations are useful approximations that are sufficiently accurate for most meteorological applications. The temperature is in degrees Celsius.

$$w = 0.622 e/(p - e)$$
$$e = wp/(0.622 + w)$$
$$q = w/(1 + w); \text{ when } e \ll p, w \approx q \approx 0.622 \, e/p$$
$$d_v = 0.2167 e/(t + 273.15) \, \text{kg m}^{-3}.$$

The formulae for mixing ratio w and specific humidity q are dimensionless; from the definition of these variables, the units are kg/kg. Frequently, w and q are multiplied by 1000 because it is easier to write 15.2 than 0.0152, and then the assigned units are g/kg. The constant 0.622 in the expression for w is the ratio of the gas constant for dry air to the gas constant for water vapor.

EXAMPLE

Given $p = 1000$ hPa, $T = 35.00°C$ and $e = 24.85$ hPa, find the relative humidity and the dew-point temperature. Compute the saturation vapor pressure using eqn. 5.4:

$$e_s = 6.1365 \times \exp\left[\frac{17.502 \times 35.00}{240.97 + 35.00}\right] = 56.48 \text{ hPa}.$$

Use eqn. 5.7 to obtain relative humidity:

$$U = 100 \times 24.85/56.48 = 44.00\%$$

Then we can obtain the dew-point temperature by inverting eqn. 5.4 (to obtain 5.6):

$$T_d = \frac{240.97 \times \ln\left(\dfrac{24.85}{6.1365}\right)}{17.502 - \ln\left(\dfrac{24.85}{6.1365}\right)} = 20.93°\text{C}$$

Instruments that respond directly to relative humidity and those that indicate the dew-point temperature are prevalent. Conversion of error expressed in relative humidity to error in dew-point temperature is a nonlinear process, as shown in figs. 5.3 and 5.4.

In fig. 5-3, an error of $\Delta T_d = 0.20°\text{C}$ is converted to an equivalent error in relative humidity. This situation would arise if the user wished to calibrate a relative humidity sensor using an instrument that measured the dew-point temperature. The dew-point instrument error would be expressed in terms of the dew-point temperature and the user would need to know the equivalent error in percent RH. This is a function of relative humidity and of temperature, so there is a family of curves for temperatures from $-30°\text{C}$ to $50°\text{C}$. Examine the relationships shown in fig. 5-2 to understand the role of temperature in the error conversion.

EXAMPLE

Given air temperature $T = 20°\text{C}$, $e_s = 23.47$ hPa (using eqn. 5.4). If relative humidity $U = 80\%$, then $e = 18.77$ hPa and $T_d = 16.44°\text{C}$. If the dew-point sensor has an error of $+0.20°\text{C}$, the dew-point temperature indicated by the sensor

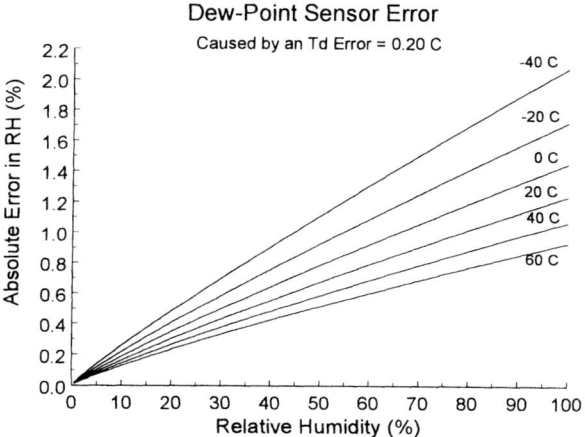

Fig. 5-3 Conversion of a 0.2°C error in dew-point to relative humidity.

92 Meteorological Measurement Systems

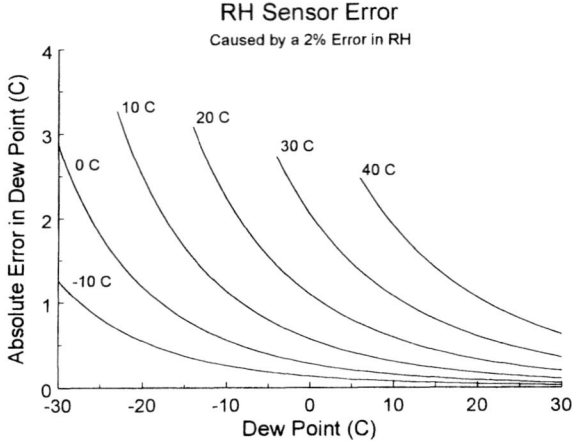

Fig. 5-4 Conversion of a 2% error in relative humidity to the equivalent error in dew-point temperature for $T = -10, 0, 10, 20, 30,$ and $40°C$.

will be $T_d = 16.64°C$. If a line were drawn, on fig. 5-3, vertically from the x-axis value of 80% to the curve for $T = 20°C$ and then horizontally to the left, it would intersect the y-axis at a relative humidity error of about 1% (absolute value). Verify this using eqn. 5.4 to convert $T_d = 16.64°C$ to $e = 18.53\,\text{hPa}$; thus the calculated $U = 78.95\%$, which means a relative humidity error of -1.05%.

If the sensor error had been $-20°C$, the indicated $T_d = 16.24°C$; then $e = 19.01\,\text{hPa}$ and the calculated $U = 80.99\%$ for a relative humidity error of $+0.99\%$. The y-axis of fig. 5-3 represents absolute error, implying symmetry for positive and negative dew-point temperature errors. However, the error is not exactly symmetrical because the slope of the vapor pressure curve, fig. 5-2, is not constant.

A sorption sensor, as will be seen later, generates an output proportional to relative humidity. If such a sensor made an error in relative humidity, that error could be converted to the equivalent dew-point temperature error using the curves in fig. 5-4, which are plotted for air temperatures from $-10°C$ to $40°C$. Again, the error conversion is a function of temperature.

EXAMPLE

Given air temperature $T = 20°C$, then the saturation vapor pressure is $e_s = 23.47$ hPa. If the actual relative humididy is $U = 80\%$, then vapor pressure $e = 18.77$ hPa and dew-point temperature $T_d = 16.44°C$. But if the sensor has a $+2\%$ error in RH, it reports $U = 82\%$. This would be equivalent to $e = 19.24\,\text{hPa}$ and $T_d = 16.83°C$. If we drew a vertical line on fig. 5-4 from the x-axis value $T_d = 16.83°C$ to the $20°C$ isotherm and from there horizontally to the y-axis, the value found there would be $+0.39°C$. In this case, a 2% error in RH is equivalent to a dew-point error of $0.39°C$.

5.3 Methods for Measuring Humidity

Wexler (1970) defined six classes of hygrometric methods based on physical principles: removal of water vapor from moist air, addition of water vapor to moist air, equilibrium sorption of water vapor, attainment of vapor–liquid or vapor–solid equilibrium, measurement of physical properties of moist air, and by chemical reactions.

5.3.1 Removal of Water Vapor from Moist Air

Separation or removal of water vapor from moist air can be accomplished by using a desiccant to absorb water vapor, by freezing out water vapor, or by separation of moist air constituents using a semipermeable membrane. These are standard laboratory techniques that operate on a sample of moist air. After removal of the water vapor the mass of the water vapor and the remaining air sample are determined in a variety of ways and then the humidity can be calculated.

5.3.2 Addition of Water Vapor to Air

Humidity can be determined by measuring the amount of water vapor that must be added to a sample of moist air to achieve complete saturation. This is a laboratory technique, but there is a variation of this method that is suitable for field measurements.

Psychrometry is a method of adding water vapor to moist air where complete saturation is not achieved. The humidity is determined from the cooling of a wet bulb relative to the ambient air temperature. The psychrometer comprises two temperature sensors exposed to the ambient air flow. One sensor, called the dry bulb, measures the ambient air temperature. The other sensor, called a wet bulb, is covered with a wick moistened with water and measures a lower temperature, caused by evaporation of water into the ambient air stream. The wick can be moistened intermittently by dipping into water or continuously by capillary flow through the wick material. Forced ventilation is normally required for optimum performance; natural ventilation may be adequate only when the temperature sensor and wick are very small and/or the ambient wind speed is sufficiently high.

A functional model of a psychrometer is illustrated in fig. 5-5 which shows two separate sensors: a dry-bulb thermometer and a wet-bulb thermometer. The dry-bulb function is simple but the wet-bulb function is more complex. Two primary inputs are shown, temperature (dry-bulb) and vapor pressure, and two secondary inputs, pressure and wind speed. Wet-bulb temperature is only weakly dependent upon pressure and wind speed.

The sources of error in a psychrometer have been well documented and are readily controlled, as noted below.

- *Sensitivity, accuracy, and matching of the temperature sensors.* A psychrometer is less sensitive to the absolute error in the temperature sensors than to the relative error, or matching error, between wet- and dry-bulb sensors.
- *Ventilation rate.* Typically, the ventilation rate should be at least 3 m/s to maximize the heat transfer by convection and evaporation and to minimize

94 Meteorological Measurement Systems

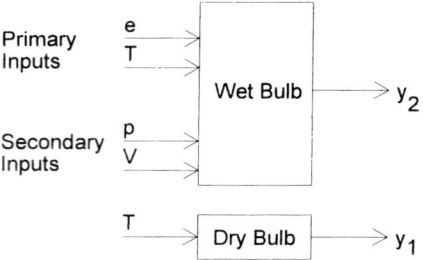

Fig. 5-5 Functional model of a psychrometer. Raw outputs y_1 and y_2 become, after calibration, estimates of T and T_w, respectively.

heat transfer by conduction and radiation. The minimum ventilation rate needed is a function of the sensor's thermal mass. Sensors made from small-diameter thermocouple wire with a fine cloth wick have been successfully used without forced ventilation (Stigter and Welgraven, 1976.)
- *Radiation incident on the temperature sensors.* The sensors must be shielded from direct and reflected solar radiation and from long-wave or earth radiation. This is a major source of error in the field that is not usually a factor in the laboratory.
- *Size, shape, material, and wetting of the wick.* Specially prepared psychrometer wick, available from instrument vendors, should always be used and not ordinary cotton cloth. Most commercial cotton cloth contains hydrophobic or anti-wetting chemicals that will eventually impede wetting of the wick. When used in continuously operating psychrometers, even standard psychrometer wicking should be boiled in a solution of lye and detergent, then boiled in distilled water and flushed with distilled water before use.
- *Relative positions of the wet- and dry-bulb sensors.* The air must not flow from the cooled wet bulb to the dry bulb.
- *Purity of the water used to moisten the wick.* Only distilled or deionized water should be used.

A psychrometer formula, which can be derived from thermodynamic principles, is used to convert the wet- and dry-bulb temperatures to ambient vapor pressure.

$$e = e_{sw} - Ap(T - T_w) \tag{5.8}$$

Temperatures are in °C and pressures may be in any consistent units, typically mb. A theoretical value[2] of the psychrometer coefficient A can be obtained, but usually with some simplifying assumptions; thus Wylie and Lalas (1985) experimentally determined $A = 0.00062°C^{-1}$ for water covered wet bulbs and $A = 0.00054°C^{-1}$ for ice-covered wet bulbs. The psychrometer coefficient is a function of probe diameter, wind speed, and pressure. For pressures above 800 hPa and flow speeds in excess of $2\,\text{m s}^{-1}$ for a 2 mm diameter probe and in excess of $4\,\text{m s}^{-1}$ for an 8 mm diameter probe, the above values can be used. The equilibrium vapor pressure at the wet-bulb temperature is called e_{sw} and is obtained by using the saturation vapor pressure equation. 5.4, with T_w substituted for T.

EXAMPLE

A psychrometer is used to obtain $T = 35.00°C$ and $T_w = 24.70°C$. If the pressure is 1000 hPa, find the relative humidity.

SOLUTION

Find the saturation vapor pressure at the wet-bulb temperature,

$$e_{sw} = 6.1365 \times \exp\left(\frac{17.502 \times 24.70}{240.97 + 24.70}\right) = 31.23 \, \text{hPa}$$

then the ambient vapor pressure is

$$e = 31.23 - 0.00062 \times 1000 \times (35.00 - 24.70) = 24.84 \, \text{hPa}$$

To obtain relative humidity, use the saturation vapor pressure at the ambient temperature calculated in the previous example:

$$U = 100 \times 24.84/56.48 = 44\%$$

With a well-designed instrument, properly used, it is possible to achieve an inaccuracy of less than 1% RH over the dry-bulb range of 5 to 65°C. Figure 5.6 is the input–output or transfer plot of a psychrometer. Each curve shows the wet-bulb depression as a function of relative humidity for various ambient air temperatures. The static sensitivity is the slope of the curve. The sensitivity increases markedly as the temperature increases and slightly as the relative humidity decreases. The error in relative humidity caused by an error of 0.1°C in the wet-bulb depression is shown in fig. 5-7. It would be extremely difficult to achieve an inaccuracy of less than 1% RH for air temperatures below 10°C.

The special advantage of a psychrometer is that the sources of error are documented and are readily checked. The Assmann psychrometer is an excellent example of a relatively inexpensive, hand-held instrument that can be used to check other humid-

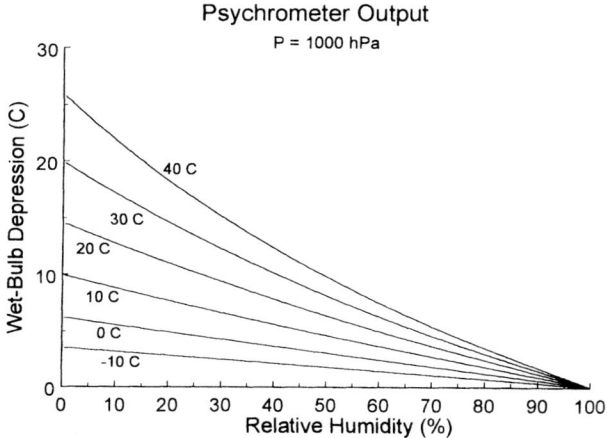

Fig. 5-6 Psychrometer output: wet-bulb depression versus relative humidity. Each curve shows the wet-bulb depression for ambient temperatures from −10 to 40°C.

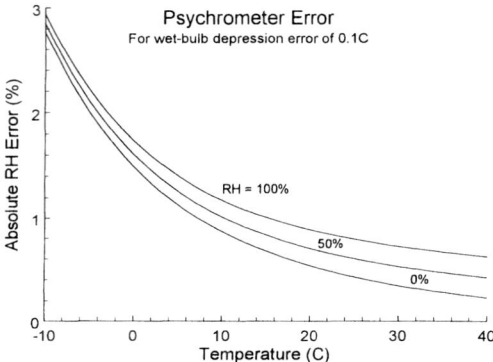

Fig. 5-7 Error induced in a psychrometer due to a 0.1°C error in wet-bulb depression.

ity sensors in the field. The user should periodically check the temperature sensors and verify that the wick is clean and saturated with distilled water and that the instrument is properly exposed. The air intake should not be pointed at the sun or any hot surface, such as the ground on a sunny day, and the instrument must be held upwind of the observer.

Low-power, continuous operation psychrometers have been used by Munro (1980) and Pike et al. (1983). The latter type of psychrometer is shown in fig. 5-8. However, this type of psychrometer fails at low temperatures (below 0°C) and in the presence of salts.

5.3.3 Equilibrium Sorption of Water Vapor

Water vapor interacts with almost every substance, through the process of sorption (absorption and/or adsorption) and sometimes by chemical reaction. This causes the material to expand or contract, or alters electrical resistance or capacitance. When a material exhibits a change that is sufficiently reversible and reproducible and detectable, it can be used as a humidity sensor. It has been observed that, for a selected sensor, dimensions, weight, resistance or capacitance may be proportional to relative humidity.

Some observed characteristics of sorption sensors are listed below.

- Sensor input cannot be relative humidity as there is no physical characteristic of the water vapor molecules in a parcel of air that corresponds to relative humidity. Recall the definition of relative humidity: the ratio of ambient vapor pressure e to the vapor pressure in equilibrium with a plane surface of pure water e_s. It is exceedingly rare to encounter a plane surface of pure water in nature, and even rarer to find water vapor in equilibrium with it. Sensor input could be e and T but not e_s. Therefore, the sensor must somehow compute e_s given T, as we do using a calculator, or it must maintain internally some water substance in liquid form with a free surface, although not necessarily a plane surface.

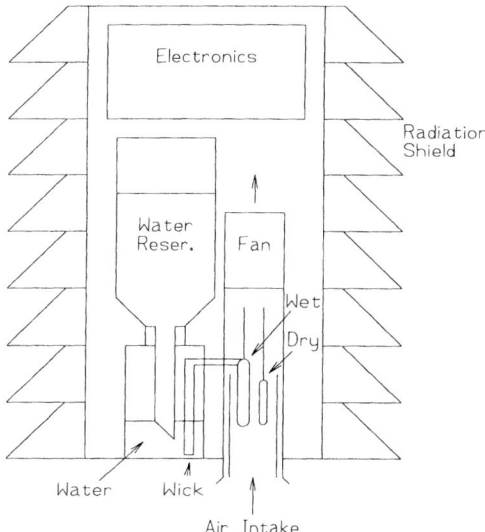

Fig. 5-8 Schematic diagram of a psychrometer designed for use in an unattended automatic weather station.

- Nicholas de Cusa, 1401–1464, observed that the mass of wood was a function of humidity. It was the first hygrometer. Denton et al. (1985) experimentally determined that the mass of a sorption sensor is proportional to ambient relative humidity, demonstrating that the sensor takes up water substance in liquid or solid form.
- It has been determined, from dielectric properties of the sensor, that the water substance in the sensor is in liquid form. Evidently, it forms a thin film of liquid water, just a few molecules thick, on the sensor surface or on the surface of pores in the sensor. We know that it does not change state even at temperatures down to −50°C. There is a large change in the dielectric constant of water when it changes state, and no such change has been observed. It would be especially evident in capacitive sensors. Anderson (1995) showed that the water film thickness, and thus the water mass, is related almost linearly to relative humidity over most of the range although it becomes quite nonlinear as the RH exceeds 90%.
- A sorption sensor must be, at least, slightly hygroscopic (attracting water vapor) as it continues to function in low relative humidity conditions up to 100°C.

From these observations we conclude that sorption sensors must be hygroscopic and porous and that the pores are continuously lined with a thin film of liquid water (unless baked dry). The mass of liquid water is proportional to relative humidity and this water creates secondary effects (changes in capacitance, resistance, etc) in the sensor that we observe. Since the water film may not be pure and certainly is not a plane surface (it may be highly curved), the sensor output may not be linear and most certainly requires calibration. The standard definition of relative humidity is 100 e/e_s where e_s is taken with respect to water at all temperatures, at least down to $-50°C$. Evidently it was recognized that the sensor responded as if liquid water were present long before capacitance sensors came into use.

5.3.3.1 Electric Hygrometers

Electric hygrometers are sorption sensors that take up water which causes a change in an electrical parameter such as resistance or capacitance. The sensor input could be ambient vapor pressure e and temperature T.

A capacitive sensor comprises an optional glass substrate for mechanical strength, a thin gold layer (one plate of the capacitor), a thin layer of polymer (the sorption layer), and a very thin layer of gold (the other capacitor plate). This layer must be thin enough to readily permit water vapor transport, or it must be laid down in strips. The polymer (exactly which polymer is proprietary information) has a low dielectric constant, about 4 (dimensionless). Water has a dielectric constant of about 80 (which is temperature sensitive) so sorption of small amounts of water substance between the plates will affect the capacitance.

Anderson (1995) also showed that the temperature sensitivity of the water film thickness is very low, so the observed temperature sensitivity of capacitive sensors is most likely due to the temperature sensitivity of the dielectric constant of water.

A functional model of the capacitive sensor is shown in the top row of blocks in fig. 5-9. Sensor raw output is the mass of water sorbed, but that mass is sensed indirectly as it affects sensor capacitance. Probe capacitance, shown in fig. 5-10, is in turn converted first to a frequency and then to a voltage by electronics in the sensor probe. There is some nonlinearity evident in fig. 5-10 as the slope of the line is not constant.

Drift can affect the sensor in various ways. Dust can accumulate on the sensor surface and may sorb some water, thus affecting the capacitance. This error is not permanent, cleaning in pure water should restore the calibration. A fine-pore filter is

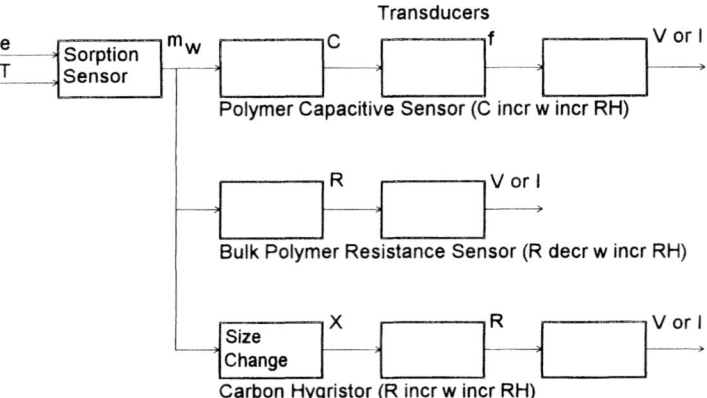

Fig. 5-9 Input–output models for some electric hygrometers.

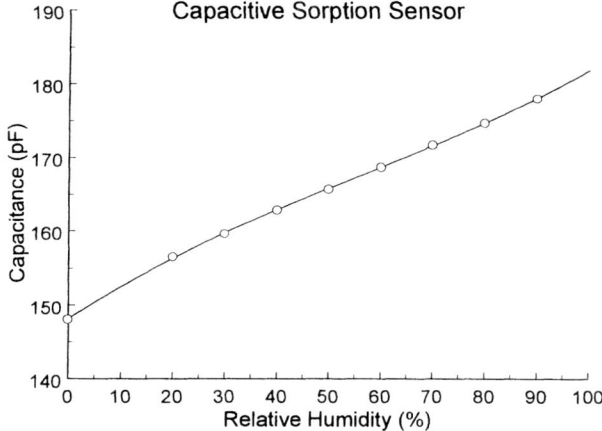

Fig. 5-10 Capacitance of a sorption sensor as a function of relative humidity.

commonly used to prevent dust accumulation. However, if water should condense on the filter, it will moisten the air and cause a temporarily high reading until it dries completely. Permanent drift can be caused by some contaminants such as SO_2. Finally, the electronic components used to convert capacitance to frequency and then to voltage may cause some errors. Combining all of these error sources, it is not surprising that these probes can have errors and may produce an output indicating a relative humidity in excess of 100%.

Another way of producing a voltage (or current) output is to use the affect of sorbed water on sensor resistance instead of capacitance. Two examples will be considered here. In the middle row of fig. 5-9 is a model of a bulk polymer resistive sensor. Bulk polymer simply means a relatively thick polymer layer (compared to the layer used in a capacitive sensor) and the resistance is measured through the volume of the polymer. As one might expect, sorbed water provides alternative conductive paths and thus the resistance decreases as relative humidity increases as shown in fig. 5-11. The resistance changes by five orders of magnitude between 0 and 100% RH. It is difficult to maintain and measure the very high resistance indicated for low values of RH. This type of sensor frequently is less accurate at values of RH below, say, 20%.

As indicated in the bottom row of fig. 5-9, the carbon hygristor experiences a dimensional change in response to a change in RH, as does the hair hygrometer described below. In this hygristor, the substance that experiences the dimensional change is impregnated with fine grains of carbon. The size, or linear dimension X, increases with RH and this increases the distance between the carbon particles, thus increasing the resistance as RH increases; see fig. 5-12. Carbon hygristors are subject to quite high drift rates and have been used only on radiosondes where the flight time is quite short and the sensor is not reused. Even then, the carbon hygristor must be kept in a sealed container until the flight and then the output must be adjusted with a baseline check just before the flight.

Electric hygrometers are usually small and may be relatively inexpensive and are suitable for remote measurements. They require calibration and some may have a significant temperature coefficient. Some have long lag times and many have a marked hysteresis. The lag of such sensors increases exponentially with decreasing

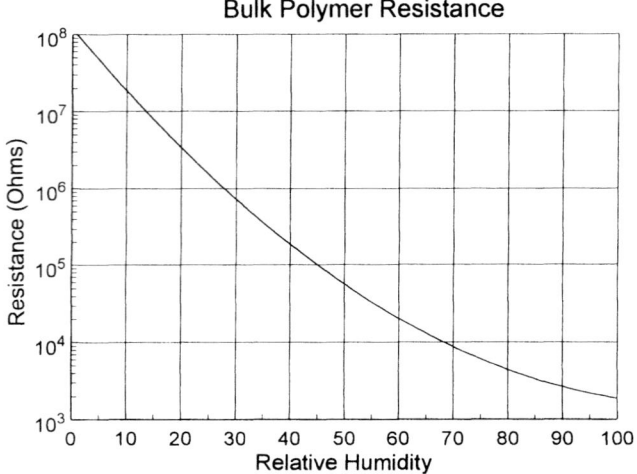

Fig. 5-11 Resistance of a bulk polymer sensor versus relative humidity.

temperature. The calibration can be affected by exposure to certain contaminants such as SO_2. Sensors that exhibit high electrical impedance at either end of the humidity scale are especially sensitive to calibration shift due to contamination.

5.3.3.2 Mechanical Hygrometers

Mechanical hygrometers are made from dimensionally variable materials mechanically coupled to an indicator or transducer. Materials such as human hair, goldbeater's skin, cotton, silk, nylon, paper, and wood have been used as the sensing element; see fig. 5-13. Their chief defects are drift, large hysteresis, and large lag times. In addition to the nonlinearity, shown in table 5-4, the calibration of a hair hygrometer may drift when exposed to very dry air, below 20% RH.

5.3.4 Measurement of Physical Properties of Moist Air

Physical properties of air such as the refractive index, radiative absorption, thermal conductivity, viscosity, density, and sonic velocity vary with the amount of water vapor present.

Table 5-4 Average elongation of human hair (as a percentage of the total elongation) as a function of relative humidity.

Relative humidity (%)	0	10	20	30	40	50	60	70	80	90	100
Elongation of hair (%)	0	21	39	53	64	73	79	85	90	95	100

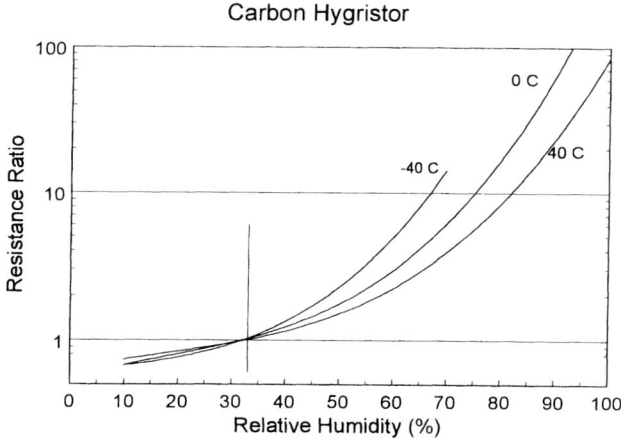

Fig. 5-12 Resistance of a carbon hygristor normalized at 33% RH.

5.3.4.1 Spectroscopic Hygrometer

A spectroscopic hygrometer measures the attenuation of certain bands in the spectrum due to water vapor absorption. These bands occur mostly in the ultraviolet and in the infrared. The Lyman-alpha line in the ultraviolet has been used by Buck (1976). Examples of infrared hygrometers are described by Hyson and Hicks (1975), Moore (1983), Ohtaki (1984), and Raupach (1978).

The fraction of incident radiation transmitted through an atmospheric path, τ, is given by Beer's law:

$$\tau = \frac{I}{I_0} = \exp(-k_\lambda \, d_v \, x) \tag{5.9}$$

where x = path length in m, d_v = absolute humidity in $\mathrm{kg\,m^{-3}}$ reduced to a standard atmosphere ($p = 1013.25$ hPa, $T = 273.15$ K) and k_λ = absorption coefficient in $\mathrm{m^2\,kg^{-1}}$. I_0 and I represent the source intensity and the intensity of the light after passing through the absorbing atmosphere. Absorption in the atmospheric path is given by $a = 1 - \tau$.

To apply this technique, we need a known I_0 or source strength in a known wavelength band, x, a fixed path length, and I, that is, a detector with known sensitivity. We also need to know the absorption coefficient k_λ in the wavelength interval determined by the bandwidths of the source and the detector. The ideal band of wavelengths would be where no other atmospheric gas is an absorber, where sources and detectors are readily available, and where transparent window materials (to enclose the source and detector) are available.

In the infrared portion of the spectrum the band from about 1000 nm to 3000 nm is attractive because there is less solar and earth background radiation. Glass is readily available that transmits up to about 2800 nm. Sources and detectors are available and there are strong water vapor absorption bands; see fig. 5-14.

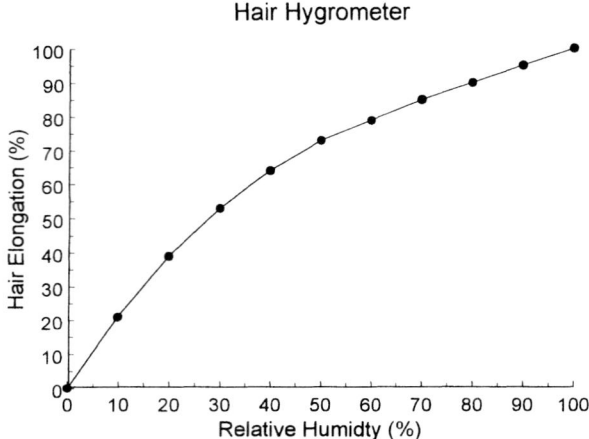

Fig. 5-13 Transfer plot for a hair hygrometer.

Water vapor absorption occurs primarily in distinct lines. There are strong vibration–rotation lines at wavelengths of 2663, 2734, 3163, and 6271 nm. Each of these lines is broadened by the total atmospheric pressure and by molecular motion (Doppler broadening) which is a function of temperature. Beer's law would hold if water vapor were the only absorbing gas and if the instrument wavelength resolution were small compared to the broadened absorption lines. The first condition can be almost completely satisfied at some wavelengths. The second condition could be satisfied by using a narrow band source, a laser.

Laser hygrometers are very expensive and not entirely suitable for field applications, so IR hygrometers use broadband sources and filters to define the wave bands. Real sources and detectors tend to drift with time, and windows change or get dirty, and all of these affect the apparent source strength, I_0. To compensate for this, IR hygrometers use two bands, one in the absorbing region, such as around 2600 nm, and a reference or non-absorbing band, say around 2300 nm. Figure 5-14 is a plot of the absorption from 1000 to 3000 nm for a 20 cm absorbing path and a mixing ratio of 20 g/kg.

A simple IR hygrometer schematic employing these concepts is shown in fig. 5-15. This is an example of a single beam device with one source and one detector, both working in the region 2300 to 2600 nm. Two filters, one in the reference band and one in the absorbing band, are rotated into the beam. The filter wheel is opaque except at the filters so the detector signal output, depicted in fig. 5-15, could be sampled at three different times during the filter wheel rotation. The three signals would be V_W, the signal sampled while the absorbing filter is in the beam, V_D, the signal sampled when neither filter is in the beam, and V_R, the signal obtained using the reference filter.

The transfer equation of the IR hygrometer is

$$V_\lambda = I_{0\lambda}\, \tau_{g\lambda}\, \tau_{a\lambda}\, R_\lambda + B \qquad (5.10)$$

where $I_{0\lambda}$ is the source intensity at wavelength λ (= reference or absorbing band), $\tau_{g\lambda}$ is the transmissivity of the optical components (windows and lens), $\tau_{a\lambda}$ is the transmissivity of the absorbing atmospheric path, and R_λ is the detector responsivity.

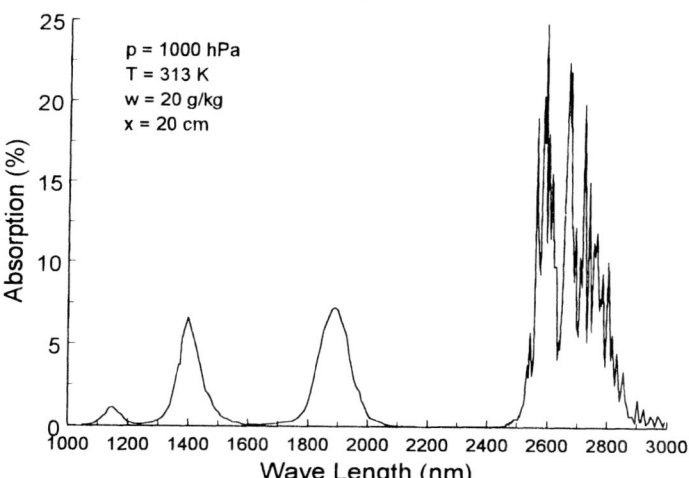

Fig. 5-14 Water vapor absorption of infrared radiation showing bands of little or no absorption and bands of strong absorption.

There may also be some bias B due to light leakage, detector dark current, and so on. The wavelength and spectral wavelength interval are determined by the filters.

The source intensity, the detector responsivity, and the sensor bias are functions of time, supply voltage, and sensor temperature. Even the transmissivity of the optical components cannot be regarded as constant as it will be affected by dirt accumulation on the windows. Let $\lambda = w$, d or r, depending on when the signal is sampled; then a normalized signal can be obtained:

$$V_n = \frac{V_w - V_d}{V_r - V_d} = \frac{I_{ow}\, \tau_{gw}\, \tau_{aw}\, R_w}{I_v\, \tau_{gr}\, \tau_{ar}\, R_r} = S\, \tau_{aw} \qquad (5.11)$$

where S is the sensor static sensitivity. The normalized signal eliminates drift and is insensitive to variations in source strength. Finally, it is necessary to measure air temperature and pressure to obtain the absolute humidity

$$d_v = f(T, p, \tau_{aw}) \qquad (5.12)$$

where the function can be obtained theoretically but must be confirmed by calibration.

The Lyman-alpha hygrometer (Buck, 1976) uses the Lyman-alpha emission line of atomic hydrogen at 121.56 nm (in the ultraviolet) as the source radiation. Glass and most other materials are opaque at this wavelength, so the windows are made of magnesium fluoride that transmits from 115 nm to 132 nm. Water vapor absorbs strongly at 121.56 nm, so a short path length of 0.2 to 5 cm is adequate. Both oxygen and ozone absorb 121.56 nm radiation. Oxygen is a weak absorber and the effect can be corrected using atmospheric temperature and pressure to estimate the oxygen density. The ozone contribution to absorption is small in the troposphere.

It is difficult to use a reference wavelength to cancel out drift, so the Lyman-alpha instrument design is simpler than the IR hygrometer shown in fig. 5-15; the motor

104   Meteorological Measurement Systems

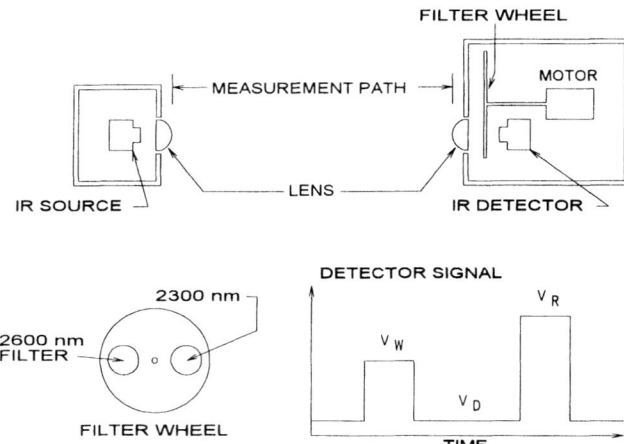

Fig. 5-15 Schematic of one possible implementation of a simple IR hygrometer.

and filter wheel are not used. While the source and detector are fairly stable, the transmission of the windows changes at a rate of about 0.5 to 5% per hour of operating time due to interaction of the magnesium fluoride with atmospheric constituents. This drift rate must be corrected by using a reference instrument, such as the chilled-mirror hygrometer, for periodic comparisons. The advantage of the Lyman-alpha hygrometer is that it is simpler, smaller, and much faster than the IR hygrometer (or chilled mirror sensor, see sect. 5.3.5). The speed comes from not having to rotate filters into and out of the absorbing path. The Lyman-alpha hygrometer is suitable for research aircraft and for tower measurements of turbulent fluctuation of humidity.

5.3.5 Attainment of Vapor–Liquid or Vapor–Solid Equilibrium

5.3.5.1 Dew- and Frost-Point Hygrometer

Atmospheric humidity can be determined by cooling a surface until vapor–liquid or vapor–solid equilibrium is achieved. The major components of a chilled-mirror hygrometer, shown in fig. 5-16, are an air intake system, a mirror surface, a method of heating and cooling the mirror, a method for detecting the formation of frost or dew, a sensor to measure the mirror surface temperature, and a control system to regulate the temperature and rate of heating or cooling of the mirror surface.

The air intake system provides a flow of ambient air at a uniform rate and may include filters to remove dust particles. Filters can be troublesome if they get wet as they will thereafter moisten the air. The air intake allows the measurement system to be removed from the point of air sampling by at least a few meters. All materials in the air intake system should be nonhygroscopic.

A small, front surface mirror is provided whose temperature can be controlled and the formation of dew or frost detected. The mirror is usually metallic.

Hygrometry 105

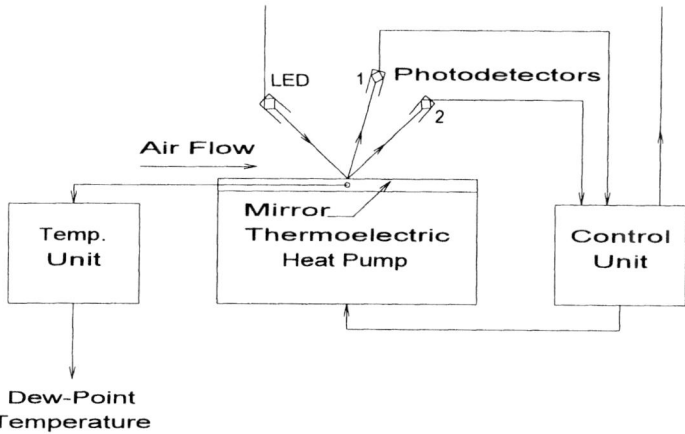

Fig. 5-16 Block diagram of a chilled-mirror hygrometer. The LED is the light source and the two photodetectors sense the scattered (#1) and the reflected light (#2).

The mirror is heated or cooled by a thermocouple heat pump utilizing the Peltier effect. This is a reasonably compact system and is completely reversible (e.g., the same system can either heat or cool the mirror). The speed with which the mirror can be heated or cooled determines the frequency response of the system.

Formation of frost or dew on the mirror is detected optically with an LED (light-emitting diode) and one or more photodetectors which sense the change in light scattering when frost of dew forms on the mirror. Two photodetectors are shown in fig. 5-16; detector #2 will receive the reflected light from a smooth surface but #1 will not be illuminated unless the light is scattered at the surface by dew or frost. In this scheme, the control unit uses the ratio of light received by the two detectors to determine if mirror heating or cooling is required. In a single detector scheme, the control unit uses the strength of the signal received from the detector (#1 in fig. 5-16).

A temperature sensor, usually a thermistor or a RTD, is embedded just under the mirror surface where it will not affect the optical or thermal properties of the mirror. It must be in the center of the mirror and very close to the mirror surface to minimize temperature gradients between the sensor and the active surface.

The control system accepts inputs from the optical detector, and sometimes from the temperature sensor, and controls current through the thermocouple heat pump. The control system must regulate the mirror temperature to the point where dew or frost just starts to form. This can be done by maintaining a uniform thickness of dew or frost. It must also regulate the rate or heating or cooling to avoid overshoot and still follow changes in ambient humidity quickly.

The dew-point method may seem to be a fundamental technique for determining humidity but this type of hygrometer is influenced by several factors that make calibration a necessity.

- It is difficult to measure the temperature of the front surface of the mirror without interfering with the detection of dew or frost. If the temperature sensor is buried in the mirror, there may be a temperature gradient between it and the layer where dew or frost forms. It should always be remembered that this

system is actually measuring the temperature of the mirror and that this is assumed to be the dew-point or frost-point. No fundamental measure of dew-point is actually taking place; the temperature of the mirror is the only measurement being made.
- The controlling mechanism must be sensitive to rate of change of ambient humidity and to air temperature.
- The presence of extremely small amounts of water-soluble matter on the condensing surface can lower the vapor pressure and hence the dew-point.
- If the condensate forms as extremely small droplets, the internal pressure, due to the surface tension of the curved surface of the droplets, increases. This increases the vapor pressure and the dew-point. One solution is to use a supply of water with a known concentration of condensation nuclei for cleaning the mirror. Some instruments incorporate a periodic, automatic self-cleaning mechanism to insure that the mirror does not get too dirty over time (however, it will not get any cleaner by itself).
- At temperatures below freezing, the initial condensate formed may be either dew or frost. Dew formation has been observed at temperatures down to $-27°C$. While supercooled dew usually changes quickly to frost, it has been observed to linger for hours in some conditions. The saturation vapor pressure of supercooled water at $-20°C$ corresponds to a frost point of $-18°C$, thus causing a $2°C$ error.
- The control system of some chilled-mirror hygrometers can fail when the ambient humidity is changing rapidly. When this happens the indicated dew-point can be grossly in error.

The chilled-mirror hygrometer is capable of continuous, unattended operation and, *with skilled maintenance*, is capable of high accuracy. Typical range of operation is $-70 \leq T_d, T_f \leq 60°C$. It is possible to achieve imprecision varying from $0.4°C$ for dew-point temperatures above freezing to $2°C$ for a frost-point temperature of $-70°C$ when the ambient atmospheric humidity is constant. Some chilled-mirror systems attempt to determine when the mirror surface is at the correct temperature and output a status byte that can be monitored for quality assurance purposes. However, it is often difficult to determine when or if a chilled mirror is operating correctly, particularly when ambient conditions are changing rapidly. Chilled-mirror devices typically have slow response and may not be suitable for aircraft measurements where the ambient vapor pressure can fluctuate very rapidly (e.g., when entering a cloud).

The objective is to measure the water vapor present in the free air. So the site should be upwind of any artificial vapor sources, such as a lawn sprinkler or a cooling tower. It should be away from paved surfaces that may be wet, and away from trees and buildings. Dust or salt particles adversely affect the instrument. It is essential to maintain a mirror cleaning schedule.

5.3.5.2 Saturated Salt Solution

The mixing ratio of moist air in equilibrium with a plane surface of a saturated aqueous salt solution is a function of temperature and pressure. The dewcel contains a saturated solution of lithium chloride (LiCl) applied to a glass cloth or other wick that surrounds a temperature measuring device. A heater with a simple control circuit raises the solution to, and maintains it at, the equilibrium temperature. Some factors that affect the performance of a dewcel are as follows.

- The dewcel can operate in the temperature range from -30 to $100°C$ provided the ambient relative humidity is between 100% and the equilibrium value of a saturated solution of LiCl (11% at $50°C$, 15% at $0°C$).
- The dewcel must be protected from high ventilation that might cause excessive heat loss, reducing its temperature and giving an erroneous reading.
- Loss of power may cause cooling and saturation with liquid water that may wash salt from the wick. The salt can be easily restored by applying a fresh LiCl solution.
- Speed of response depends on the rate at which the dewcel can be heated or can lose heat to the surroundings. This, in turn, affects the power consumption.

5.3.6 Chemical Reactions

Chemical methods are usually used only in the laboratory. One method is to remove the water vapor by use of a chemical reagent and then weigh the resulting water.

5.4 Choice of Humidity Sensor

Some factors to consider when choosing a hygrometer are cost, accuracy, maintenance required, applicability to application, speed of response, and power consumption. Response time is important when the instrument is used to make humidity measurements from an aircraft in flight or to measure turbulent fluctuations in humidity. Power consumption is critical when the power source is a battery or a combination of battery and solar panel.

Psychrometers are low-cost hand-held devices but are fairly high in cost when automated. Automation is relatively rare because of problems with wick contamination and the difficulty of providing an adequate supply of pure water to the wick. A psychrometer can provide high accuracy with proper attention to design and operation. But, for any design, inaccuracy will increase as the air temperature decreases. The great advantage of psychrometers is that most sources of error can be readily detected. If the temperature sensors are high-quality mercury-in-glass thermometers, there is little that can go wrong with them other than breakage. Given accurate and matched temperature sensors, other sources of error relate to keeping the wick clean and supplied with pure water and to providing adequate ventilation. Other types of humidity sensors are much more susceptible to drift or other relatively difficult-to-detect problems. Therefore, a high-quality hand-held psychrometer, such as the Assmann design, is ideal for checking other sensors in the field. In the typical Assmann design, ventilation is provided by a spring-driven fan so there is no electrical power required.

Sorption sensors are generally the lowest-cost automated sensors. The best of the sorption sensors combine small size, low cost, moderate accuracy, and very low power consumption. This makes them ideal for applications requiring multiple measuring sites, all operating on batteries with or without solar panels. Maintenance requirements are moderate, mainly to cope with drift. Unlike psychrometers, sorption sensor problems are rarely detectable by visual examination. Therefore, in-situ intercomparisons or, in the case of a fairly dense network, comparisons between stations are required to check for drift. Vendor-specified drift rates can be used to schedule periodic recalibrations unless the site is contaminated with a gas that causes sensor drift (such as SO_2 for many sensors).

Spectroscopic hygrometers are used for special applications as they have very high cost, fairly high power consumption and, depending upon the design, may have high maintenance requirements. The Lyman-alpha hygrometer has very fast response which is required for many aircraft applications. Unfortunately, it also is subject to severe drift which limits its applications. IR absorption hygrometers, to date, have very limited response times due to the need to mechanically switch filters into and out of the optical path. All spectroscopic hygrometers are high cost due to limited applications and therefore limited sales volume.

Chilled-mirror dew-point sensors combine high cost, high power consumption, slow response, and high maintenance with potentially high accuracy and a wide dew-point operating range. High accuracy can be achieved with skilled maintenance, so these devices are well suited to laboratory applications. They are also one of the few instruments that can perform well at very low dew-point temperatures.

It is difficult to compare hygrometer performance specifications since psychrometers, sorption sensors, spectroscopic sensors, and chilled-mirror dew-point sensors all produce very different outputs. The difficulty is illustrated in figs. 5.3 and 5.4 where, respectively, a constant dew-point temperature error is converted to an error in relative humidity and a constant RH error is converted to a T_d error. The conversion is neither simple nor linear but requires a family of curves.

5.5 Calibration of Humidity Sensors

A system for calibrating sorption sensors is illustrated in fig. 5-17. It shows an array of six sensors exposed to air with a known temperature (approximately room temperature) and relative humidity in the test chamber. Humidity in the test chamber is generated by controlling the mix of dry air and saturated air. Dry air can be obtained by pumping air from a tank of dry air through a tube of desiccant. A switch is used to divert some of this air through a water bubbler to saturate this part of the air stream. If the dry air is completely dry ($e = 0$) and the saturated air is completely saturated ($e = e_s$), and if the temperatures are uniform throughout the system, the relative humidity in the test chamber will be given by $U = 100 V_S/(V_S + V_D)$, where V_S is the volume of saturated air and V_D is the volume of dry air pumped into the chamber per unit time. The computer is programmed to generate a controlled time sequence of humidity in the test chamber using this relation. Relative humidity in the chamber can be programmed to change in steps from near 0% almost to 100% and back. Humidity is held constant at each step until the sensors settle at the new value.

As noted above, the controller uses certain assumptions that can never be fully satisfied. Therefore, the air exhausting from the test chamber is passed through a chilled-mirror dew-point temperature sensor. The computer uses the measured T_d

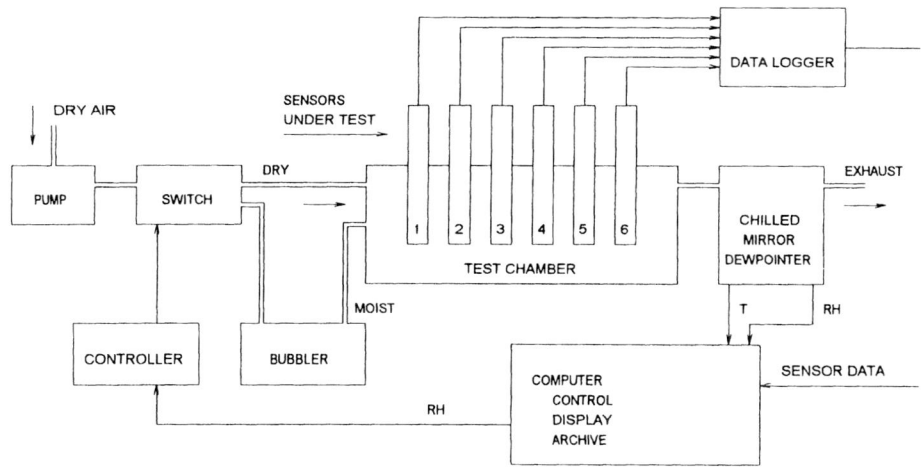

HUMIDITY CALIBRATION SYSTEM

Fig. 5-17 Humidity calibration system. The desired humidity is generated by mixing dry and moist air in a known ratio.

with the measured T in the test chamber to calculate the relative humidity in the chamber, and this value is used to calibrate the sensors.

One of the simplest methods of calibrating sorption-type humidity sensors is to expose the sensor in a confined space to an aqueous salt solution. The ASTM (1984) has developed a recommended practice for maintaining constant relative humidity using such solutions. Some useful salt solutions are listed in table 5-5. Measurement accuracy is strongly dependent on the ability to achieve and maintain the sensor and the salt solution at a uniform and known temperature. This temperature must be held constant to within ±0.1°C.

Table 5-5 Equilibrium relative humidity values for selected saturated aqueous salt solutions. Listed values are for temperatures of 20, 25, and 30°C.

Salt	20°C	25°C	30°C	Formula
Lithium chloride	11.1	11.3	11.3	$LiCl$
Potassium acetate	23.1	22.5	21.6	$KC_2H_3O_2$
Magnesium chloride	33.1	32.8	32.4	$MgCl_2 \cdot 6H_2O$
Potassium carbonate	43.2	43.2	43.2	K_2CO_3
Magnesium nitrate	54.4	52.9	51.4	$Mg(NO_3)_2 \cdot 6H_2O$
Sodium chloride	75.5	75.3	75.1	$NaCl$
Potassium chloride	85.1	84.3	83.6	KCl
Barium chloride	91.0	90.0	89.0	$BaCl_2 \cdot H_2O$
Potassium nitrate	94.6	93.6	92.3	KNO_3
Potassium sulfate	97.6	97.3	97.0	K_2SO_4

5.6 Exposure of Humidity Sensors

Measurement of humidity almost always entails measurement of air temperature in meteorological applications. In some sensors, air temperature measurement is routinely incorporated into the sensor, as with the psychrometer. In other sensors it may not be needed for the sensor itself but for subsequent conversion of the measured humidity quantity to another form. Therefore, the exposure of humidity sensors is closely related to the exposure of air temperature sensors, discussed in chap. 4.

Psychrometry always includes direct measurement of air temperature since computation of vapor pressure, or of any other humidity quantity, requires the wet-bulb depression, $T - T_w$. Forced aspiration of the wet- and dry-bulb sensors is normally a part of the technique which alleviates some of the problems associated with air temperature measurement. The sensors are still susceptible to radiation-induced error, so operation of a sling psychrometer in direct sunlight is not recommended. An Assmann psychrometer is an improved design that controls aspiration and shields the sensors from solar radiation.

Psychrometers are sensitive to wick contamination by salts which inhibit evaporation, therefore causing reduced wet-bulb depression. This is a major problem for continuously operating psychrometers, especially near the ocean. Even in a salt-contaminated environment, a sling or Assmann psychrometer can be operated successfully if the wick is flushed with distilled water and then operated for just a short period of time, as would normally be the case.

Since the static sensitivity of a psychrometer decreases as the air temperature decreases, psychrometer error will be larger in cold weather. A continuously operating psychrometer cannot work in freezing weather but a sling or Assmann type can be used, although with increased error. An ice-covered wick is acceptable as long as the ice completely encloses the wick.

Sorption sensors do not require air temperature per se but are useless for meteorological applications without it. If ambient vapor pressure is held constant while the air temperature is changed, the relative humidity and the sorption sensor output will change. A completely independent measure of air temperature is recommended, but it is not sufficient. Exposure problems for sorption sensors are similar to those of air temperature sensors and the usual solution is to enclose the sorption sensor in a radiation shield without forced aspiration. Since sorption sensors are sensitive to certain gaseous contaminants and to dust, it is usual to enclose the sorption sensor inside a dust filter. Optimum exposure of the air temperature sensor is outside the dust filter but inside the radiation shield where sensor ventilation will not be reduced by the dust filter. But then the temperature of the sorption sensor may differ markedly from the temperature indicated by the air temperature sensor. This will create an error in the subsequent conversion of the measured relative humidity to dew-point. The optimum solution would be to measure air temperature outside the dust shield and to use a second temperature sensor to measure the temperature of the sorption sensor. This sensor should be in direct contact with the sorption sensor, or in very close proximity. Then conversion from relative humidity to dew-point would use the sorption sensor temperature. Air temperature measurement would not be compromised by placing that sensor inside the dust shield where there is markedly reduced air flow and, therefore, increased probability of radiation-induced temperature error or increased lag error.

Sorption sensors are sensitive to contamination by water, dust, and certain gases (other than water vapor). Gaseous contamination is a function of the individual sensor construction. For example, some sensors are sensitive to SO_2. In general, gaseous contamination causes a permanent calibration shift. Such a sensor, located near a source of SO_2, may drift excessively. Dust, especially hygroscopic dust, may cause local elevation of the humidity, causing an exposure error. This is not a permanent shift as the sensor calibration can be restored by cleaning the sensor. In the same way, water on the sensor or on the dust filter, caused by condensation or blowing rain, causes a temporary error until the water evaporates.

Spectroscopic hygrometers can and should incorporate temperature compensation so they would need air temperature only for conversion of absolute humidity to other measures of humidity. In some designs, there is built-in compensation for optical attenuation caused by anything other than water vapor, such as dust accumulation on the lenses. A notable exception is the Lyman-alpha hygrometer which can experience such rapid drift that another kind of hygrometer must be used in conjunction with it to provide periodic drift compensation.

A chilled-mirror hygrometer does not require air temperature measurement, but an air temperature sensor is usually incorporated to provide air temperature for other needs or to convert dew-point temperature to relative humidity. Aside from air temperature measurement problems, a chilled-mirror device is sensitive to contaminants that affect drop formation on the mirror and to temporary liquid water contamination in the air ducts leading to the mirror. Mirror contamination problems seriously compromise the performance of this sensor in the field. To achieve optimum performance, a chilled-mirror sensor requires a lot of attention from skilled technicians, especially in field applications.

In addition to exposure problems, users of humidity data often perceive apparent problems due to lack of understanding of the limitations of the sensor used or of field conditions at the measurement site. For example, a well-calibrated sorption sensor with an imprecision of 3% RH could legitimately indicate a humidity of 102% even though relative humidity in the atmosphere only rarely can exceed 100%. Humidity at the measurement site may differ from the value at surrounding stations due to localized rainfall, proximity to a lake or river upwind of the site, and changes in nearby farming practice. Irrigation is the obvious example but plowing a field may produce a detectable effect. In addition, conversion of air temperature and relative humidity from a sorption sensor to dew-point can result in significant errors when the relative humidity is low (e.g., below 20%). Therefore, a user not familiar with the operation of both sensors may incorrectly conclude there are calibration errors when comparing a sorption sensor to a chilled-mirror sensor.

QUESTIONS

(More difficult questions are indicated with an asterisk.)

1. The addition of salt lowers the vapor pressure of water. Will the presence of salt on the wet bulb of a psychrometer cause the wet-bulb temperature to read too high or too low?

2. What is the effect of radiation error on the wet-bulb temperature (a) during a clear, hot day? (b) on a clear night? Discuss.

3. Why should the air not flow from the wet bulb to the dry bulb?

112 Meteorological Measurement Systems

4. Using the equations in the text, fill in the following table:

p (hPa)	800.0	1000	1000	1000	900.0
T (°C)	20.0	25.0	25.0	26.0	25.0
T_w (°C)	18.0	20.0	21.0	21.0	20.0
Vapor press. e	19.7				
Relative hum. U	84.0				
Mixing ratio w	0.0157				
Specific hum. q	0.0155				
Absolute hum. d_v	0.0146				
Dew-point T_d	17.2				

5. In the above table, if the column headed with $T = 25$ and $T_w = 20°C$, represents the true dry- and wet-bulb temperatures, then the next column represents a 1.0°C error in the wet-bulb depression. Contrast this with the column headed with $T = 26$ and $T_w = 21°C$ which would represent no error in the wet-bulb depression but a 1.0°C error in absolute temperature of both the wet and the dry bulb.

6. Consider a psychrometer, without forced ventilation, mounted on an aircraft. The dry bulb and the wet bulb are directly exposed to the air stream. There is a radiation shield. Discuss error sources. Would they be acceptable on (a) a slow aircraft (100 mph)? (b) a fast one (400 mph)? (c) an ultralight (40 mph)?

7. Would it be feasible to add antifreeze to the water supply of a psychrometer to operate in subfreezing weather? How would you interpret the results?

8. Plot the output versus the input of a resistive sensor from the data given below. Use semi-log paper. Is it more sensitive at high or low humidities? Would a fingerprint on the sensitive material affect the accuracy?

RH (%)	20	30	40	50	60	70	80	90	100
Resistance (kΩ)	1400	300	77	24	8.7	3.8	2.0	1.0	1.0

9. Make a plot of the human hair data in table 5-4. Fit a straight line to the data. What is the residual error? Would a polynomial of higher order be a better fit? How would you determine the coefficients of such a polynomial?

10. Would it be feasible to use saturated aqueous salt solutions to calibrate a psychrometer? Explain. How about a hair hygrometer?

11. Your assistant compares two sorption probes. Probe A reads 51% RH at 20°C while B reads 53% at 23°C. Assume these temperatures are correct. Do you think it likely that these probes agree? Explain.

12. A carbon hygristor is a plastic strip coated with a hygroscopic mixture containing a suspension of finely divided carbon particles. Its electrical resistance increases as the relative humidity increases. What physical mechanism causes this?

13.* Use your general knowledge of water vapor in the atmosphere and of humidity parameter definitions to explain the physical principles of a sorption sensor whose output is proportional to relative humidity. This question does *not* require specialized knowledge of any sensor. However, some examples of sorption-type RH sensors are the hair hygrometer, the carbon element used in radiosondes, and the capacitive hygrometer used in the Vaisala and Rotronic probes. State the definition of relative humidity and carefully define any other quantities referenced in your definition. The raw output of these sensors is elongation, resistance, and capacitance, respectively. What is the actual, physical input? That is, what causes the above changes? How is it possible for a simple sensor to respond to RH over a broad range of vapor pressures and temperatures?

14.* Given a combination temperature and relative humidity sensor exposed in a non-aspirated radiation shield where there is some radiation-induced temperature error, can the temperature error be corrected? Is there an error in the relative humidity? Can it be corrected?

15. If the thermometers in a psychrometer were perfect, would there be any source of error that would cause the wet bulb to read too low?

16.* The humidity generator in fig. 5-17 pumps a mixture of dry and saturated air into the humidity chamber which is at, or very near, ambient atmospheric pressure since it is open to the atmosphere. As a first approximation, assume that the temperature of the dry air, of the saturated air, and of the mixed moist air is the same, T, and that $T =$ room temperature which is constant in time. Also, as a first approximation, assume that all of the air streams are at ambient pressure p. For the dry air, the vapor pressure $e = 0$, and $e = e_s$ in the saturated air. $V_D =$ volume of dry air pumped into the chamber per second, $V_S =$ volume of saturated air pumped into the chamber per second, and $U =$ relative humidity (%) in the chamber.

Define $\delta \equiv V_S/(V_S + V_D)$. A controller regulates δ. Prove that $U = 100\delta$ for $0 \leq \delta \leq 1$. What if the dry air stream and the saturated air stream are not at the same temperature as the mixing chamber?

Hint: $U = 0\%$ when $= 0$ and $U = 100\%$ when $\delta = 1$. What is U for other values of δ?

17. A certain humidity sensor produces the following output:

Vapor press. (Pa)	1000	2000	3000	4000
Output Y_1 (V)	2.500	2.596	2.685	2.786

(a) What is the static sensitivity at the low end of the range?
(b) Is this sensor linear? Explain.

18. Why are linear thermistors inappropriate for use in psychrometers?
Hint: Psychrometers are sensitive to errors in wet-bulb depression. Examine errors that could occur due to the residual nonlinearity of linear thermistors.

19. How does a chilled-mirror dewpoint hygrometer detect condensate on the mirror?

20. Dry-bulb temperature $= 4°C$, wet-bulb temperature $= -3°C$, pressure $= 900\,\text{hPa}$. Show how to compute e.

21. Is it necessary to calibrate a psychrometer? Explain.

22. For all sorption sensors:
(a) What are measurands?
(b) What is raw output?
(c) Observed output changes are an indirect consequence of what?

23. Why does an IR hygrometer use two wavelengths?

LABORATORY EXERCISES

1. Use saturated salt solutions to calibrate a carbon hygristor, a hair hygrometer, a Humicap, or other humidity sensors whose active element can be placed in the calibration jar.

Prepare closed containers of saturated aqueous salt solutions to generate humidities between 10% and 90%. Do not expose a hair hygrometer to humidity less than 20%. The container and lid must be airtight, and must be corrosion resistant and nonhygroscopic. The container should be as small as possible. Use a small amount of distilled water to dissolve the salts to make the solutions. Make sure that not all of the salt is dissolved.

Obtain a thermometer to measure the air temperature inside the container. The humidity generated by the salt solutions is a function of temperature. If the temperature is outside the range given in table 5-5, consult the reference ASTM (1984).

The "lock-in" resistance of a carbon hygristor is its resistance at 33% RH. Humidity is determined from the ratio of the ambient hygrometer resistance to the lock-in resistance. The lock-in resistance may drift while the resistance ratio still provides a useful measurement. For the carbon hygristor, determine the lock-in resistance, then the resistance at other humidities and, finally, recheck the lock-in resistance.

Check the output of the other hygrometers as a function of temperature and humidity. Repeat all measurements.

Generate a table of salts used, observed temperature, nominal humidity, corrected humidity, and sensor outputs. Plot sensor output versus humidity. Comment on any discrepancies such as a failure of repeated measurements to reproduce exactly.

2. Compare psychrometers. Obtain one or more sling psychrometers, Assmann psychrometers, and any other psychrometers available. First, with a dry wick, operate the psychrometer and record "wet-bulb" and dry-bulb temperatures. Then wet the wick with distilled water and operate both indoors and outdoors. Compute vapor pressure, relative humidity, and dew-point. If the two thermometers in a psychrometer gave different readings when the wick was dry, what did you do about it? Did you attempt to adjust the data? Explain why the various psychrometer results differed. Which one is likely to be the most accurate? Why?

BIBLIOGRAPHY

Anderson, P.S., 1995: Mechanism for the behavior of hydroactive materials used in humidity sensors. *J. Atmos. Oceanic Technol.* 12, 662–667.

ASTM, 1984: Standard recommended practice maintaining constant relative humidity by means of aqueous solutions. *ASTM Designation E104.* American Society for Testing and Materials, Philadelphia, PA.

Auble, D.L., and T.P. Meyers, 1992: An open path, fast response infrared absorption gas analyzer for H_2O and CO_2. *Bound.-Layer Meteor.*, 59, 243–256.

Buck, A.L., 1976: The variable-path Lyman-alpha hygrometer and its operating characteristics. *Bull. Am. Meteor. Soc.*, 57, 1113–1118.

Buck, A.L., 1981: New equations for computing vapor pressure and enhancement factor. *J. Appl. Meteor.*, 20, 1527–1532.

Cerni, T.A., 1994: An infrared hygrometer for atmospheric research and routine monitoring. *J. Atmos. Oceanic Technol.*, 11, 445–462.

Chahuneau, F., R.L. Desjardins, E. Brach, and R. Verdon, 1989: A micrometeorological facility for eddy flux measurements of CO_2 and H_2O. *J. Atmos. Oceanic Technol.*, 6, 193–200.

Denton, D.D., J.B. Camou, and S.D. Senturia, 1985: Effects of moisture uptake on the dielectric permittivity of polyimide films. In *Moisture and Humidity, Int. Symp. on Moisture and Humidity*, Research Triangle Park, NC. Instrumentation Society of America, Washington, DC, pp. 505–513.

Eloranta, E.W., R.B. Stull, and E.E. Ebert, 1989: Test of a calibration device for airborne Lyman-α hygrometers. *J. Atmos. Oceanic Technol.*, 6, 129–139.

Fan, J., 1987: Determination of the psychrometer coefficient A of the WMO reference psychrometer by comparison with a standard gravimetric hygrometer. *J. Atmos. Oceanic Technol.*, 4, 239–244.

Friehe, C.A., R.L. Grossman, and Y. Pann, 1986: Calibration of an airborne Lyman-alpha hygrometer and measurement of water vapor flux using a thermoelectric hygrometer. *J. Atmos. Oceanic Technol.*, 3, 299–304.

Gates, R.S., 1994: Dew point temperature error from measuring dry bulb temperature and relative humidity. *Trans. Am. Soc. Agric. Engr.*, 37, 687–688.

Haman, K.E., and A.M. Makulski, 1985: Hygrometry with temperature stabilization. *J. Atmos. Oceanic Technol.*, 2, 448–467.

Heikinheimo, M.J., G.W. Thurtell, and G.E. Kidd, 1989: An open path, fast response IR spectrometer for simultaneous detection of CO_2 and water vapor fluctuations. *J. Atmos. Oceanic Technol.*, 6, 624–636.

Holbo, H.R., 1981. Dew-point hygrometer for field use. *Agric. Meteor.*, 24(2), 117–130.

Hyson, P., and B.B. Hicks, 1975. A single-beam infrared hygrometer for evaporation measurement. *J. Appl. Meteor.*, 14, 301–307.

Jensen, J.B., and G.B. Raga, 1993: Calibration of a Lyman-α sensor to measure in-cloud temperature and clear-air dewpoint temperature. *J. Atmos. Oceanic Technol.*, 10, 15–26.

Katsaros, K.B., J. DeCosmo, R.J. Lind, R.J. Anderson, S.D. Smith, R. Kraan, W. Oost, K. Uhlig, P.G. Mestayer, S.E. Larsen, M.H. Smith, and G. De Leeuw, 1994: Measurements of humidity and temperature in the marine environment during the HEXOS main experiment. *J. Atmos. Oceanic Technol.*, 11, 964–981.

Lind, R.J., and W.J. Shaw, 1991: The time-varying calibration of an airborne Lyman-α hygrometer. *J. Atmos. Oceanic Technol.*, 8, 186–190.

Mestayer, P., and C. Rebattet, 1985: Temperature sensitivity of Lyman-alpha hygrometers. *J. Atmos. Oceanic Technol.*, 2, 656–644.

Meteorological Office, 1981: *Measurement of Humidity*, Vol. 3 in *Handbook of Meteorological Instruments*, 2nd ed. Meteorology Office 919c. Her Majesty's Stationery Office, London, 28 pp.

Moore, C.J., 1983: On the calibration and temperature behavior of single-beam infrared hygrometers. *Bound.-Layer Meteor.*, 25, 245–269.

Mukode, S., and H. Futata, 1989: A semiconductive humidity sensor. *Sensors and Actuators*, 16, 1–11.

Muller, S.H., and P. J. Beekman, 1987: A test of commercial humidity sensors for use at automatic weather stations. *J. Atmos. Oceanic Technol.*, 4, 731–735.

Munro, D.S., 1980: A portable differential psychrometer system. *J. Appl. Meteor.*, 19, 206–214.

Ohtaki, E., 1984: Application of an infrared carbon dioxide and humidity instrument to studies of turbulent transport. *Bound.-Layer Meteor.*, 29, 85–107.

Ohtaki, E., and T. Matsui, 1982: Infrared device for simultaneous measurement of fluctuations of atmospheric carbon dioxide and water vapor. *Bound.-Layer Meteor.*, 24, 109–119.

Pike, J.M., F.V. Brokc, and S.R. Semmer, 1983: Integrated sensors of PAM II. *Preprints 5th Symp. on Meteorological Observations and Instrumentation*, Toronto, Canada. American Meteorological Society, Boston, MA, pp. 327–333.

Priestley, J.T., and R.J. Hill, 1985: Measuring high-frequency humidity, temperature and radio refractive index in the surface layer. *J. Atmos. Oceanic Technol.*, 2, 233–251.

Raupach, M.R., 1978: Infrared fluctuation hygrometry in the atmospheric surface layer. *Quart. J. Roy. Meteor. Soc.*, 104, 309–322.

Salasmaa, E., and P. Kostamo, 1975: New thin film humidity sensor. *Preprints 3rd Symp. on Meteorological Observations and Instrumention*, Washington, DC. American Meteorological Society, Boston, MA, pp. 33–38.

Smedman, A.-S., and K. Lundin, 1987: Influence of sensor configuration on measurements of dry and wet bulb temperature fluctuations. *J. Atmos. Oceanic Technol.*, 4, 668–673.

Spyers-Duran, P.A., 1991: An airborne cryogenic frost-point hygrometer. *Preprints 7th Symp. on Meteorological Observations and Instrumentation*, New Orleans, LA. American Meteorological Society, Boston, MA, pp. 303–306.

Stigter, C.J., and A.D. Welgraven, 1976: An improved radiation protected differential thermocouple psychrometer for crop environment. *Arch. Meteor. Geophys. Biokl.*, Ser. B, 24, 177–187.

Trevitt, A.C.F., 1986: An infrared hygrometer with on-line temperature compensation. *Bound.-Layer Meteor.*, 34, 157–169.

Wexler, A., 1970: Measurement of humidity in the free atmosphere near the surface of the earth. *Meteor. Monogr.*, 11(33): *Meteorological Observations and Instrumentation*, pp. 262–282.

Wexler, A., 1976: Vapor pressure formulation for water in the range 0° to 100° – A revision. *J. Res. Natl. Bur. Stand.*, 80A, 775 ff.

Wexler, A., 1977: Vapor pressure formulation for ice. *J. Res. Natl. Bur. Stand.*, 81A, 5–20.

Wylie, R.G., and T. Lalas, 1985: Accurate psychrometer coefficients for wet and ice cylinders in laminar transverse airstreams. In *Moisture and Humidity, Int. Symp. on Moisture and Humidity*, Research Triangle Park, NC. Instrumentation Society of America, Washington, DC, pp. 37–43.

NOTES

1. John Dalton (1766–1844) observed that the total pressure exerted by a mixture of gases which do not interact chemically is equal to the sum of the partial pressures of the gases.

2. The theoretical value of the psychrometer coefficient is $0.000\,65\,°C^{-1}$ at 0°C. The coefficient is slightly temperature dependent because the latent heat of water vapor is temperature dependent.

6

Dynamic Performance Characteristics, Part 1

When the input to a sensor is changing rapidly, we observe performance characteristics that are due to the change in input and are not related to static performance characteristics. In this chapter we will assume that a static calibration has been applied so that we can consider dynamic performance independently of static characteristics. The terms "linear" and "nonlinear" have been used in chap. 3 in the static sense. Now they are being used in the dynamic sense where "linear" connotes the applicability of the superposition property. A given sensor could be nonlinear in the static sense (e.g., a PRT is nonlinear in that is static sensitivity is not constant over the range) but could be linear in the dynamic sense (modeled by a linear differential equation).

We use differential equations to model this dynamic performance while realizing the models can never be exact. If the dynamic behavior of physical systems can be described by linear differential equations with constant coefficients, the analysis is relatively easy because the solutions are well known. Such equations are always approximations to the actual performance of physical systems that are often nonlinear, vary with time, and have distributed parameters. The justification for the use of simple, readily solved models must be the quality of the fit of the solution to the actual system output and the usefulness of the resulting analysis.

Dynamic performance characteristics define the way instruments react to measurand fluctuations. When a temperature sensor is mounted on an airplane these characterisics will indicate what the sensor "sees." If the airplane flies through a cloud with a slow sensor (where time constant is large) it may not register change of temperature or humidity. That would not be tolerable if we wanted to measure the cloud. Similarly, if the airplane flies through turbulence we would like to measure changes in air speed. Variations in temperature and humidity would be vital in the flight of a radiosonde, so again the time constant of the sensors would be considered. Fluxes of heat, water vapor, and momentum near the ground require fast sensors (with small time constants).

6.1 First-Order Systems

Differential equations describe the behavior of physical systems in which a redistribution of energy is taking place. In a mechanical system, a mass in motion stores kinetic energy and may store potential energy by virtue of its position in a force field. When a mechanical system does not store potential energy but does dissipate energy, the differential equation is first order in velocity, for example,

$$m\frac{dv}{dt} + Dv = F \tag{6.1}$$

where v = velocity, dv/dt = acceleration, m = mass, D = dissipation factor, and F = external force.

The above equation applies to a cup anemometer because the anemometer can store kinetic energy in the cup wheel as moment of inertia but, because the cup wheel has no preferred position with respect to the wind vector, it cannot store potential energy. It dissipates kinetic energy into the wind stream.

A thermal system, such as a thermometer, could be described by

$$mC dT = UA(T_i - T)dt$$

where m = mass in kg, C = specific heat in $J kg^{-1} K^{-1}$, T_i = input temperature, T = sensor output temperature, U = heat transfer coefficient in $J K^{-1} s^{-1} m^{-2}$, A = effective area for heat transfer in m^2, and t = time in s. For a temperature sensor immersed in a fluid such as air, the heat transfer coefficient U is a function of the type of fluid and its velocity. We can rearrange this equation as:

$$\frac{mC}{UA}\frac{dT}{dt} + T = T_i$$

where the coefficient mC/UA has the units of time and so is called a time constant, τ.

For a linear system, the response to a sum of inputs is simply the sum of the responses to these inputs applied separately. This is the superposition principle and can be taken as the defining property of linear systems. This is an extremely useful property because it allows analysis of the response to complex signals in the frequency domain by superposition of responses to individual frequencies. This is the justification for using linear models even when the fit is far from ideal.

A physical system is said to be in a static state when the distribution of energy within the system is constant with time. When there is an exchange of energy within the system, the system is in a dynamic state and its performance is described by a differential equation containing derivatives with respect to time. To determine static characteristics such as threshold, measurements of the output must be made for many different values of the input. Each measurement is made while the system is static (i.e., not changing). During the transition from one static state to another, the system is dynamic. We wait until the dynamic energy exchange has ceased before making the static measurement.

When forces are applied at discrete points and are transmitted by discrete components within the system, the system can be defined by lumped parameters. But when it is necessary to describe the variation across space coordinates of a physical component, the system must be described with distributed parameters and is modeled by a partial differential equation.

Dynamic performance analysis is concerned with modeling the performance of dynamic, lumped parameter systems with ordinary differential equations where time is the independent variable. The number of dynamic performance parameters is equal to the order of the system so, for a first-order system, the performance equation can be written in the canonical form

$$\tau \frac{dx}{dt} + x = x_i \tag{6.2}$$

where τ is the time constant, with units of time.

The solution to eqn. 6.2 is $x(t) = x_T(t) = x_S(t)$, where $x_T(t)$ is the transient solution and $x_S(t)$ is the steady-state solution. The transient response, or complementary function in mathematical terms, is obtained when the forcing function is set to zero and the system is released from some set of initial conditions at time $t = 0$. The distribution of energy in the system storage elements at the time of release must tend towards zero because of the always-present energy dissipation. In system terms, the output for a given initial energy distribution and driving input is the transient solution plus the steady-state solution. In mathematical terms, the equation solution for a given set of initial conditions and a forcing function is the complementary function plus the particular function.

The steady-state solution can be found by the method of undetermined coefficients. Given that the input is some function $x_i(t)$, repeatedly differentiate $x_i(t)$ with respect to t until the derivatives either go to zero or repeat the functional form of some lower-order derivative. This is also the test for the applicability of the method: if neither of the above conditions prevail, the method of undetermined coefficients cannot be used. Write the steady-state solution as

$$x_s(t) = k_1 \, x_i(t) + k_2 \, D x_i(t) + k_3 \, D^2 \, x_i(t) + \cdots \tag{6.3}$$

where D is the differential operator d/dt. The right-hand side of eqn. 6.3 must include one term for each functionally different form found by examining $x_i(t)$ and its derivatives. The constants k_i do not depend upon the initial conditions. They are found by substituting eqn. 6.3 into 6.2.

6.1.1 Step-Function Input

One of the simplest inputs to consider is the step function, a function which is zero for $t < 0$ and equal to some non-zero constant for $t > 0$. We can model this by setting $x_i = x_c$, a constant, and by setting the initial condition $x(0) = 0$. A first-order equation has only one initial condition, and so this specifies that the sensor is at rest with output equal to zero at time $t = 0$ and then responds to the input, which is constant for $t > 0$. One way of solving for the transient response of eqn. 6.2 (with $x_i(t) = 0$) is to integrate it directly; we can write it in the form

$$\frac{dx}{x} = -\frac{dt}{\tau} \tag{6.4}$$

The transient solution is

$$x = C e^{-t/\tau} \tag{6.5}$$

where C is an arbitrary constant. The steady-state solution of eqn. 6.2, with $x_i = x_c$, a constant, must be $x_S = x_c$, so the complete solution is given by

$$x(t) = x_T(t) + x_S(t) = x_c + C e^{-t/\tau} \tag{6.6}$$

The constant C can be determined by applying the initial condition, which is $x(0) = x_c + C = 0$, so $C = -x_c$; therefore

$$x(t) = x_c(1 - e^{-t/\tau}). \tag{6.7}$$

The first-order step-function response is shown in fig. 6-1 for two different time constants. Both systems exhibit 63.2% response to the step input when $t = \tau$, 86.5% response when $t = 2\tau$, and 95% response when $t = 3\tau$.

A first-order response to a decreasing step input is shown in fig. 6-2. This condition may be obtained by setting the input $x_i(t) = 0$ and the initial condition $x(0) = x_c$. The solution, similar to eqn. 6.6, is

$$x(t) = x_c e^{-t/\tau} \tag{6.8}$$

As before, 63.2% response is obtained when $t = \tau$.

A general step-function solution for any step from some initial state x_{IS} to a final state x_{FS} is given by

$$x(t) = x_{FS} - (x_{FS} - x_{IS})e^{-t/\tau} \tag{6.9}$$

Solutions (6.7) and (6.8) are special cases of (6.9).

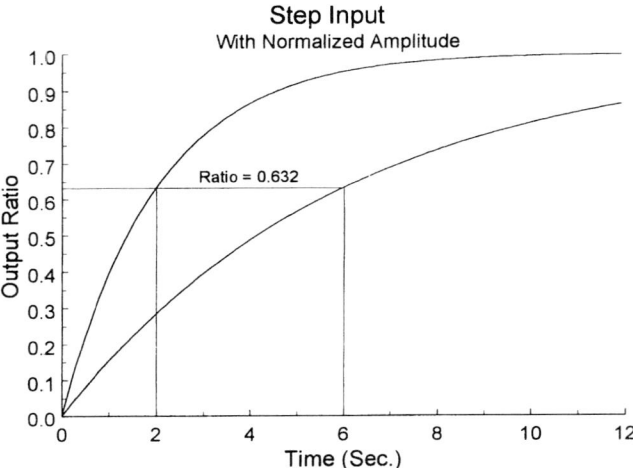

Fig. 6-1 Positive step response. In the top curve, $\tau = 2$ s, and $\tau = 6$ s in the bottom curve. The output ratio is $x(t)/x_c$.

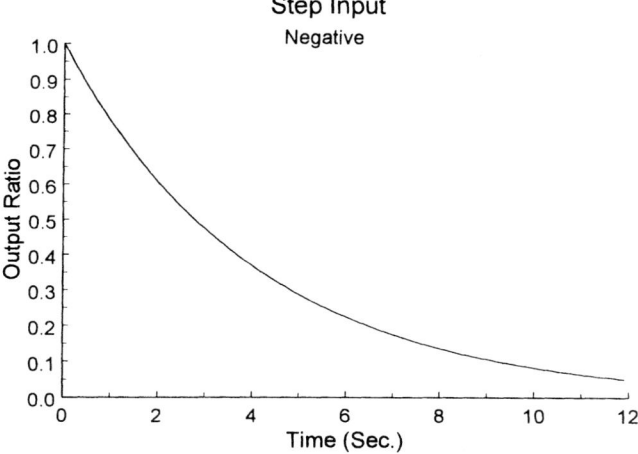

Fig. 6-2 Response to a negative step with $\tau = 4$ s.

EXAMPLE

A thermometer is moved rapidly from an ice bath (0°C) to room temperature (25°C). The following data are obtained:

Time (s)	0	1	2	3	4	5
Temperature (°C)	0	7.09	12.16	15.80	18.41	20.28

Determine the time constant.

SOLUTION

We know the initial and final temperatures and that the solution is of the form $T(t) = 25(1 - e^{-t/\tau})$. When $t = \tau$, $T(\tau) = 25(1 - e^{-1}) = 25 \times 0.632 = 15.80°C$. The solution was exactly this value when $t = 3$ s, so $\tau = 3$ s. This is too easy since, in a real experiment, we might not have collected data at the time that turned out to be the time constant or, if we had, that datum might have been contaminated with noise.

EXAMPLE

A more realistic step-function test of a thermometer produces the following results:

Time (s)	0	3	6	9	12	15
Temperature (°C)	20.00	35.54	39.00	39.78	39.95	39.99

SOLUTION

In this case, we know the initial value, $T_{IS} = 20.00°C$, and the final value is probably $T_{FS} = 39.99°C$ plus or minus a small uncertainty. Accepting this, the solution form is $T(t) = T_{FS} - (T_{FS} - T_{IS})e^{-t/\tau}$. When $t = \tau$, $T(\tau) = 40 - 20 \times 0.632 = 27.36$. There is no datum with this value but we can see that $0 < \tau \le 3$ s. We can solve for the time constant, at any point where data is available, by writing

$$\tau = \frac{-t}{\ln\left[\dfrac{T_{FS} - T(t)}{T_{FS} - T_{IS}}\right]} = \frac{-3}{\ln\left[\dfrac{40 - 35.54}{40 - 20}\right]} = 2.00 \text{ s}$$

We can repeat this calculation for $t = 6, 9, 12,$ and 15 s and obtain $\tau = 2.00, 2.00, 2.00,$ and 1.97 s respectively. One would expect to get the same value for the time constant each time if the system really is first order. The last value obtained differs slightly from the expected value of 2.00 because the output was recorded to only four significant digits.

6.1.2 Ramp Input

A ramp input is achieved when $x_i(t) = 0$ for $t < 0$ and $x_i(t) = at$ for $t \geq 0$, where a is constant and the initial condition is $x(0) = 0$. Equation 6.2 becomes

$$\tau \frac{dx}{dt} + x = at \tag{6.10}$$

The transient solution is given by eqn. 6.5 and the steady-state solution will be $x_s(t) = k_0 + k_1 t$. Solve for k_0 and k_1 by substituting the steady-state solution into eqn. 6.10, which yields $k_0 = -a\tau$ and $k_1 = a$, so that

$$x(t) = -a\tau + at + ce^{-t/\tau} \tag{6.11}$$

Apply the initial condition $x(0) = 0$ to obtain $c = a\tau$, and the solution of eqn. 6.10 is given by

$$x(t) = at - a\tau(1 - e^{-t/\tau}). \tag{6.12}$$

We can define the dynamic error (fig. 6-3) as

$$\varepsilon_d = x(t) - x_i(t) = -a\tau(1 - e^{-t/\tau}) \tag{6.13}$$

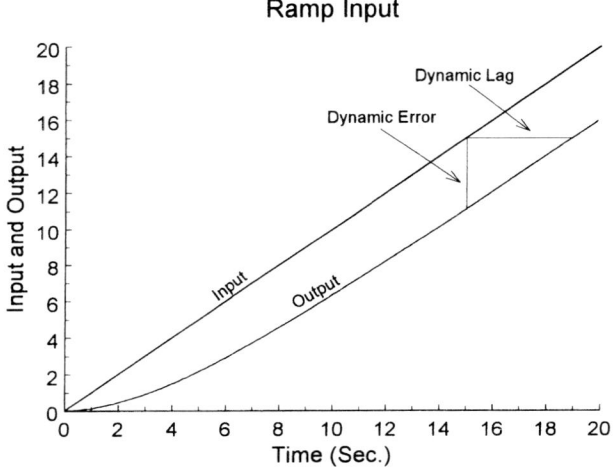

Fig. 6-3 Response to a ramp input.

which has a transient component due to the decaying exponential and a steady-state component equal to $-a\tau$. The dynamic lag could be defined as a value of Δt such that $x_i(t) = x(t + \Delta t)$. This gives a dynamic lag of $\Delta t = \tau$ in the steady state.

6.1.3 Sinusoidal Input

To examine the frequency response of a first-order sensor, let the input to eqn. 6.2 be $x_i(t) = A_i \sin \omega t$ and consider only the steady-state response; see fig. 6-4. The assumed solution is $x_s(t) = b_0 \sin \omega t + b_1 \cos \omega t$. The steady-state solution is

$$x_S(t) = A_i \left[\frac{1}{1 + (\tau\omega)^2} \sin \omega t - \frac{\tau\omega}{1 + (\tau\omega)^2} \cos \omega t \right] \quad (6.14)$$

With the aid of the trigonometric identities

$$K \sin(\alpha + \phi) = K \sin(\alpha)\cos(\phi) + K \cos(\alpha)\sin(\phi)$$

and

$$K^2 \sin^2(\phi) + K^2 \cos^2(\phi) = K^2$$

we can transform eqn. 6.14. Let $\alpha = \omega t$, $K \cos \phi = 1/[1 + (\tau\omega)^2]$, and $K \sin \phi = -\tau\omega/[1 + (\tau\omega)^2]$; then the solution can be written as

$$x_S(t) = \frac{A_i}{\sqrt{1 + (\tau\omega)^2}} \sin(\omega t + \phi) \quad (6.15)$$

where phase shift $\phi = \tan^{-1}(-\tau\omega)$. If we write $x_s(t) = A \sin(\omega t + \phi)$, then the amplitude ratio (see fig. 6-5) becomes

$$\frac{A}{A_i} = \frac{1}{\sqrt{1 + (\tau\omega)^2}} \quad (6.16)$$

Note that the amplitude ratio

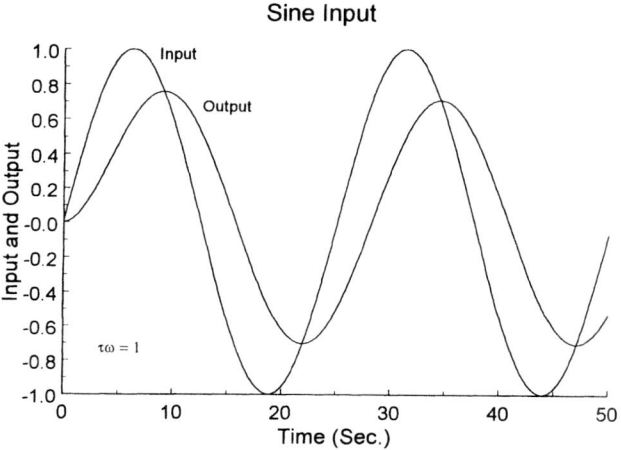

Fig. 6-4 Normalized sinusoidal input and output for $\tau\omega = 1$.

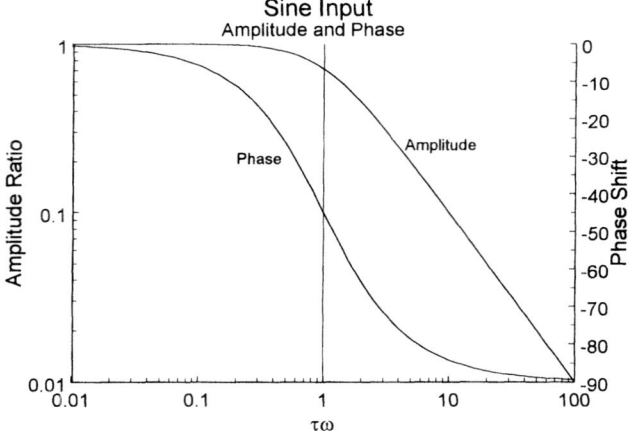

Fig. 6-5 Amplitude ratio (A/A_i) and phase versus normalized frequency ($\tau\omega$) for a first-order system.

$$\left.\begin{array}{l} \dfrac{A}{A_i} = \dfrac{1}{\sqrt{1+(\tau\omega)^2}} \leq 1 \text{ for all } \tau\omega \\[2mm] \lim_{\tau\omega \to 0} \dfrac{A}{A_i} = 1 \\[2mm] \lim_{\tau\omega \to \infty} \dfrac{A}{A_i} = 0 \end{array}\right\} \quad (6.17)$$

and when $\tau\omega = 1$, $A/A_i = 0.707$.

EXAMPLE
Given an input $x_i(t) = 2.3 \sin(0.50\,t)$ to a system with a time constant $\tau = 4.0\,\text{s}$, find the steady-state output.

SOLUTION
Input amplitude $A_i = 2.3$, $\omega = 0.50\,\text{s}^{-1}$, $\tau = 4.0\,\text{s}$ so $\tau\omega = 2.0$ (dimensionless).

Steady-state output:

$$x_S(t) = \dfrac{A_i}{\sqrt{1+(\tau\omega)^2}} \sin(\omega t + \phi)$$

Output amplitude ratio:

$$\dfrac{A}{A_i} = \dfrac{1}{\sqrt{1+(2.0)^2}} = 0.45$$

Output amplitude:

$$A = 2.3 \times 0.45 = 1.0$$

Phase shift:

$$\phi = \tan^{-1}(-\tau\omega) = \tan^{-1}(-2.0) = -1.1\,\text{rad} = -63°$$

Therefore:
$$x_S(t) = 1.0\sin(0.50t - 63°)$$

Alternative method. From fig. 6-5, find $\tau\omega = 2.0$ on the x-axis and read from the curve the y-axis value of about 0.45. From fig. 6-6, again with $\tau\omega = 2.0$ on the x-axis, read the y-axis value of about $-65°$.

EXAMPLE
Find the steady-state output of a linear, first-order system with a time constant of 4.0 s when the input is $x_i(t) = 21 + 0.6\cos(0.2t) + 2.3\sin(0.5t)$.

SOLUTION
Since the system is linear, we can invoke superposition and observe that the output will be the sum of the outputs that would have been obtained if each of the input terms had been applied separately. So we first find the outputs corresponding to each input term: $x_{i1}(t) = 21$, $x_{i2}(t) = 0.6\cos(0.2t)$, and $x_{i3}(t) = 2.3\sin(0.5t)$.

When the steady-state input is $x_{i1}(t) = 21$, a constant, the steady-state output must be the same. Remember that we are assuming a static calibration has been done, so if the input is a constant, the output must be the same constant, so $x_{S1}(t) = 21$.

What about when the input is $x_{i2}(t) = 0.6\cos(0.2t)$? No explicit solution was given for a cosine input but, if the input is a cosine, the output will be a cosine. In this case, $A_i = 0.6$, $\omega = 0.2$, and $\tau\omega = 0.8$. Then $A/A_i = [1 + (0.8)^2]^{-1/2} = 0.78$ and $\phi = \tan^{-1}(-0.8) = -39°$. Therefore, $x_{S2}(t) = 0.47\cos(0.2t - 39°)$.

In the previous example, we found the solution for $x_{i3}(t) = 2.3\sin(0.5t)$, so $x_{S3}(t) = 1.0\sin(0.50t - 63°)$. Now, invoking superposition.

$$x_S(t) = x_{S1}(t) + x_{S2}(t) + x_{S3}(t) = 21 + 0.47\cos(0.2t - 39°) + 1.0\sin(0.50t - 63°)$$

6.2 Experimental Determination of Dynamic Performance Parameters

The system parameter τ could be determined by applying a step-function input to a sensor and determining the time delay from applying the step to when the response has reached 63.2% of its steady-state value. This is practical only under ideal conditions. If the recorded signal is noisy or if the data are missing in this critical time period, this simple method fails. Figure 6-6 shows a first-order step response when the recorded signal is noisy and it would be difficult to determine the time constant. Another possible problem is that the system may not be first-order. It is not enough to simply assume that it is.

We could rearrange eqn. 6.7 and take logarithms of both sides to obtain

$$\ln\left(1 - \frac{x(t)}{x_c}\right) = -\frac{t}{\tau} \tag{6.18}$$

and we could plot the data on a semi-log axis. The data will plot as a straight line if the system really is linear and first-order, and then the slope of the line will be equal

126 Meteorological Measurement Systems

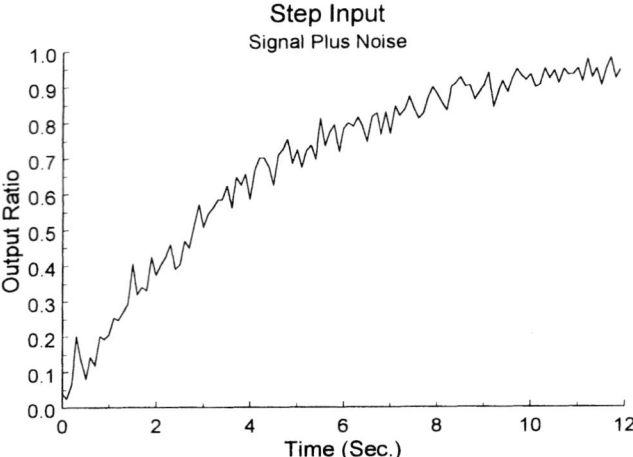

Fig. 6-6 Positive step with added noise ($\tau = 4$ s).

to $-1/\tau$. Since τ is determined from the slope of the line fitted to all of the data, one is not dependent upon data at any critical time and a better overall determination of τ is obtained. The noisy data shown in fig. 6-6 have been plotted in fig. 6-7 as described above. It is reasonable to conclude that a straight line could be fitted to the data and therefore the results are from a first-order system.

6.3 Application to Temperature Sensors

Each temperature sensor has one or more thermal energy storage reservoirs. If the sensor has only one, then it can be modeled with a first-order differential equation such as eqn. 6.2. Physically, every temperature sensor has more than one thermal energy storage reservoir but, if there is only one primary reservoir and the others are small in comparison, then it is reasonable to use the first-order model.

According to Tsukamoto (1986), the time constant of a small-diameter temperature sensor in light winds is given by

$$\tau = \frac{C_s \rho_s A_s}{K + \sqrt{2\pi K c_v \mu R_e}} \tag{6.19}$$

where C_s = sensor specific heat capacity, J kg^{-1} K^{-1}, ρ_s = sensor density, kg m^{-3}, A_s = sensor cross-sectional area, m^2, K = air thermal conductivity, 25.7×10^{-3} J K^{-1} m^{-1} s^{-1}, c_v = air specific heat at constant volume, 717 J K^{-1} kg^{-1}, μ = air viscosity, 18.18×10^{-6} kg m^{-1} s^{-1}, R_e = Reynolds number = $V d_s / \nu$, V = wind speed, m s^{-1}, d_s = sensor diameter, m, and ν = kinematic viscosity, 15.29×10^{-6} m^2 s^{-1}.

Some of the parameters referring to air are functions of temperature and, sometimes pressure, but we will treat them as constants here. Accordingly, we can write the above equation as

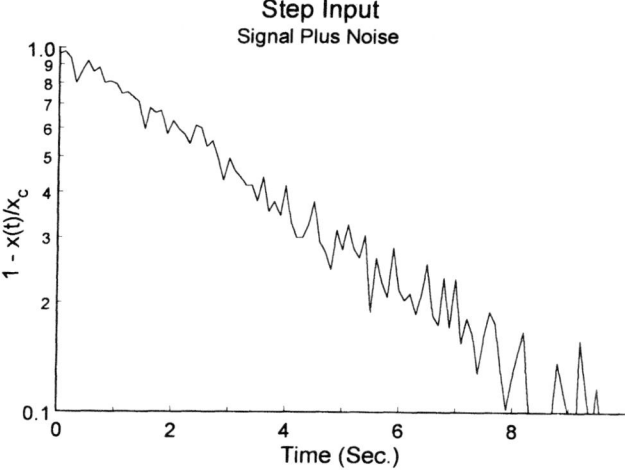

Fig. 6-7 Semilog plot of the data used in Fig. 6-6.

$$\tau = \frac{C_s \rho_s A_s}{0.0257 + \sqrt{137.7 V d_s}} \tag{6.20}$$

For copper wire, $C_s = 390 \, \text{J kg}^{-1} \text{K}^{-1}$ and $\rho_s = 8900 \, \text{kg m}^{-3}$. Let $V = 2 \, \text{m s}^{-1}$ and $d_s = 120 \, \mu\text{m}$; then the time constant $\tau = 189 \, \text{ms}$. This model does not include radiation or conduction heat transfer effects.

QUESTIONS

1. A temperature sensor with a 1-minute time constant reports a steady temperature of 13°C. Then a very well defined cold front moves rapidly over the station dropping the temperature to −4°C in a few seconds. Sketch the sensor output versus time from before the event until the sensor has settled to a new temperature. Label the axes and indicate the scale.
2. A temperature sensor is subjected to a step input from T_0 to T_∞. After t_1 seconds, the output is observed to be T_1. Show how to obtain the time constant.
3. The input frequency to a linear system is $2 \, \text{rad s}^{-1}$. What is the output frequency?
4. The input to a temperature sensor with time constant τ is $3\sin(2t) + 5\sin(7t)$. What is the steady-state output?
5. When will the dynamic error of a first-order system with a sinusoidal input approach zero?
6. Will a sensor with a small time constant respond faster or slower than one with a larger time constant?
7. Why is the dynamic error associated with a ramp input negative?
8. Show how to obtain eqn. 6.15 from 6.14.
9. $T_i(t) = 9\sin(7t) - 3\sin(2t) + 5$. $\tau = 30 \, \text{s}$. What is the steady state, $T_s(t)$?
10. Write the solution, and identify all symbols, for a dynamic test of a thermometer when it is moved rapidly (a) from an ice bath, T_C, to room temperature, T_R; (b) from room temperature, T_R, to an ice bath, T_C.
11. The half-life of a first-order sensor with a negative step input is H. Find the time constant in terms of H.

12. The input to a first-order sensor is a spike (rises to a large positive value and then falls rapidly). The maximum indicated output is 52 units. The actual maximum input peak was either < 52 or ≥ 52 units (choose one). Why?

13. The output of a first-order sensor is a sine wave with amplitude A. The actual input amplitude is either $< A$ or $\geq A$. Which and why?

14. If the input to a first-order system is at (a is constant), the steady-state output is $x_S(t) = at - a\tau$. Dynamic error is the output minus the input at any given time. Dynamic lag is the time interval that passes before the output is equal to the input. Define (a) dynamic error in terms of τ; (b) dynamic lag in terms of τ.

15. Your task is to measure the dynamic performance of a first-order, linear humidity sensor. Your assistant performs a step-function test by rapidly moving the sensor from a moist chamber (100% RH) to a dry chamber (0% RH). Unfortunately, he/she loses some data and damages the sensor so the test cannot be repeated. The available data (M = missing) are:

t (s)	0	3	6	9	12	15	18	21
U (%)	100	78	M	47	M	29	M	17

Show your assistant how to extract the dynamic performance parameter from the available data.

Hint: Do not attempt to use any kind of interpolation scheme to replace the missing data. That could introduce more error.

Extension: Suppose that your assistant also lost the $t = 0$ datum so that you don't know the humidity in the moist chamber. Solve this problem using just the data for $t = 3, 9, 15$ and 21 s.

BIBLIOGRAPHY

Badgley, F.I., 1957: Response of radiosonde thermistors. *Rev. Sci. Instr.*, 28, 1079–1084.

Fuehrer, P.L., C.A. Friehe, and D.K. Edwards, 1994: Frequency response of a thermistor temperature probe in air. *J. Atmos. Oceanic Technol.*, 11, 476–488.

MacCready, P.B., Jr, 1970: Theoretical considerations in instrument design. *Meteor. Monogr.*, 11, 202–210.

Snow, J.T., D.E. Lund, M.D. Conner, S.B. Harley, and C.B. Pedigo, 1989: The dynamic response of a wind measuring system. *J. Atmos. Oceanic Technol.*, 6, 140–146.

Tsukamoto, O., 1986: Dynamic response of the fine wire psychrometer for direct measurement of water vapor flux. *J. Atmos. Oceanic Technol.*, 3, 453–461.

7

Anemometry

The function of an anemometer (sometimes with a wind vane) is to measure some or all components of the wind velocity vector. It is common to express the wind as a two-dimensional horizontal vector since the vertical component of the wind speed is usually small near the earth's surface. In some cases, the vertical component is important and then we think of the wind vector as being three-dimensional. The vector can be written as orthogonal components (u, v, and sometimes w) where each component is the wind speed component blowing in the North, East, or vertically up direction. Alternatively, the vector can be written as a speed and a direction. In the horizontal case, the wind direction is the direction from which the wind is blowing measured in degrees clockwise from North. The wind vector can be expressed in three dimensions as the speed, direction in the horizontal plane as above, and the elevation angle.

Standard units for wind speed (a scalar component of the velocity) are $m\,s^{-1}$ and knots (nautical miles per hour). Some conversion factors are shown in table 7-1.

Wind velocity is turbulent; that is, it is subject to variations in speed, direction, and period. The wind vector can be described in terms of mean flow and gustiness or variation about the mean. The WMO standard defines the mean as the average over 10 minutes.

7.1 Methods of Measurement

The ideal wind-measuring instrument would respond to the slightest breeze yet be rugged enough to withstand hurricane-force winds, respond to rapidly changing turbulent fluctuations, have a linear output, and exhibit simple dynamic performance

Table 7-1 Wind speed units conversion.

Source	Multiply source by the factor below to obtain:				
	$m\,s^{-1}$	kt	$km\,h^{-1}$	$ft\,s^{-1}$	$mi\,h^{-1}$
$m\,s^{-1}$	1.0000	1.9438	3.6000	3.2808	2.2369
kt	0.5144	1.0000	1.8520	1.6878	1.1508
$km\,h^{-1}$	0.2778	0.5400	1.0000	0.9113	0.6214
$ft\,s^{-1}$	0.3048	0.5925	1.0973	1.0000	0.6818
$mi\,h^{-1}$	0.4470	0.8690	1.6093	1.4667	1.0000

characteristics. It is difficult to build sensors that will continue to respond to wind speeds as they approach zero or will survive as wind speeds become very large. Thus a variety of wind sensor designs and, even within a design type, a spectrum of implementations have evolved to meet our needs.

7.1.1 Wind Force

The drag force of the wind on an object, which we have all experienced, can be written as

$$F_d = \tfrac{1}{2} C_d \rho A V^2 \tag{7.1}$$

where C_d is the drag coefficient, a function of the shape of the device and of the wind speed. It is dimensionless and, in this context, $0 < C_d \leq 1$. The dependence of the drag coefficient on wind speed is weak over a wide range of wind speeds and is therefore often assigned a value that is a function of shape alone. The air density ρ has units of $kg\,m^{-3}$. The cross-section area of the sensor A is in m^2 while V is the wind speed in $m\,s^{-1}$. For some sensors, the wind speed must be taken as a vector quantity and then V^2 is replaced by $V|V|$. Wind sensors that respond to the drag force, or the closely related lift force, are treated in this section.

7.1.1.1 Cup and Propeller Anemometers

A cup anemometer turns in the wind because the drag coefficient of the open cup face is greater than the drag coefficient of the smooth, curved surface of the back of the cup, as shown in fig. 7-1. Before a cup anemometer starts to turn, the effective wind speed is just V_i. Then, as the cup wheel rotates, the effective speed is the relative speed $V_i - s$ for the cup on the left of fig. 7-1 and $V_i + s$ for the cup on the right. But the difference in the drag coefficients dominates, so the cup continues to turn.

The raw output of a cup or propeller anemometer is the mechanical rotation rate of the cup wheel (and of the supporting shaft); see fig. 7-2. The static sensitivity, nearly constant above the threshold speed, is a function of the cup wheel or propeller design. Some typical values are 30 to 60 rpm/$(m\,s^{-1})$ for a cup wheel and 180 to 210 rpm/$(m\,s^{-1})$ for a propeller design. A propeller always rotates faster than a cup wheel in the same wind. While the cup wheel responds to the differential drag force, both the drag and lift forces act to turn a propeller.

Anemometry 131

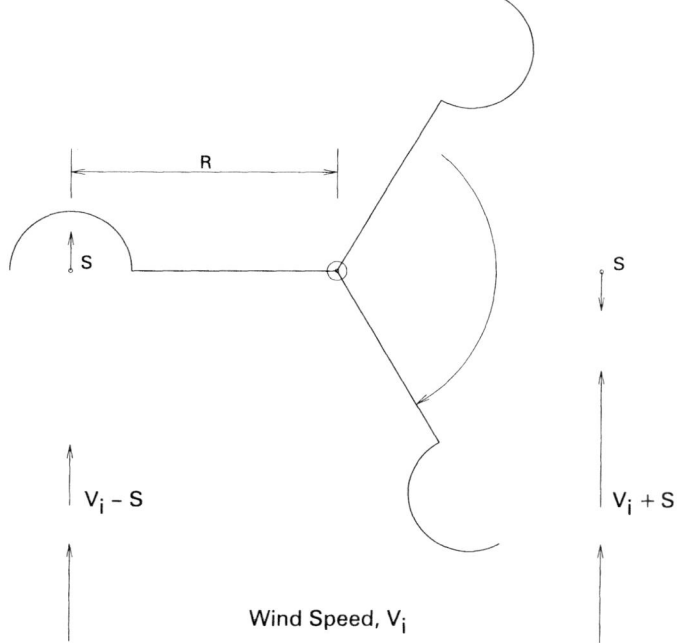

Fig. 7-1 Line drawing of a cup anemometer showing the wind force acting on the cups.

The shaft is coupled to an electrical transducer which produces an electrical output signal, typically a dc voltage proportional to shaft rotation rate and therefore to wind speed. An ac transducer may be used which produces an ac voltage with amplitude and frequency proportional to rotation rate. Still another option is an optical transducer that generates a series of pulses as an optical beam is interrupted. The pulse rate is proportional to rotation rate.

Cup and propeller anemometers are linear over most of their range, with a notable exception at the lower end of the range. Since these anemometers are driven by wind force which is proportional to the square of the wind speed, there is very little wind force available to overcome internal friction when the wind speed approaches zero. Consequently, there is a wind speed, called the threshold wind speed, below which the anemometer will not turn. Figure 7-3 shows the threshold effect for increasing and decreasing wind speeds. The starting threshold for wind speed slowly increasing from zero in much higher than the stopping threshold. This is because running friction is much less than static friction. Despite this, the lower range limit is often defined to be zero. The upper limit is the maximum wind speed the anemometer can sustain without damage.

Anemometer static performance specifications include the range, usually from zero to some maximum wind speed, and the threshold, usually taken as the wind speed at which the cup or propeller starts turning and continues to turn. The nonlinear threshold effects, which extend well above the stated threshold speed, are usually ignored and no correction is attempted. The threshold speed is a function of vibration and of

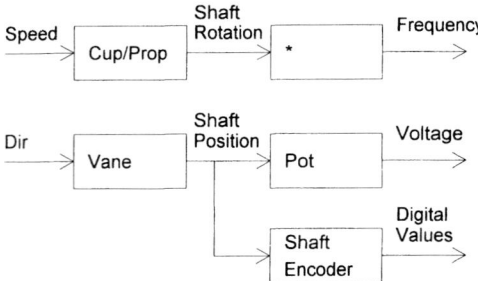

Fig. 7-2 Anemometer functional model.

bearing friction, which increases as the anemometer ages. Sometimes it is possible to detect the threshold effect in wind speed data.

Ideally, a cup anemometer would respond to just the horizontal component of the wind vector and the indicated speed would be proportional to the cosine of the angle of the wind vector with respect to the horizontal. As shown in fig. 7-4, generated from data given by MacCready (1966), the actual response function is sometimes greater than an ideal cosine response and this causes overestimation of the horizontal component in turbulent flow or when the mean vertical component is not zero, as may occur in complex terrain or around buildings. The ideal cosine response is marked in the figure; the lines above the ideal curve represent the non-ideal response of two different cup anemometer designs. Not all cup anemometer designs overestimate the cosine response.

This type of error is a function of anemometer design and can be a significant source of error in some anemometers. The magnitude of this error is seldom, if ever, given in static performance specifications although the wind tunnel testing procedure is fairly easy to perform.

Propeller anemometers have a different form of cosine response, as shown in fig. 7-5. They underestimate the magnitude of off-axis wind components. If a propeller is used in conjunction with a vane, the propeller is oriented into the wind (on the average) and so the underestimation of the horizontal component of the wind vector is not much of a problem. Propellers are sometimes used in fixed 2-D or 3-D configurations and then the off-axis underestimation is a problem and must be corrected. When a propeller is oriented vertically, to measure the vertical component of the wind speed, the average wind vector is horizontal to the propeller axis and the instantaneous wind vector seldom deviates much from the normal, so the deviation from the ideal cosine law can be compensated by slightly altering the calibration coefficient.

The response of propeller anemometers is a function of the design of the propeller and, for a fixed design, will also depend upon the materials used. Lightweight propellers have faster response but are more fragile and can be easily damaged in high winds. However, propellers can be quite rugged and suitable for harsh marine environments including use on buoys (Michelena and Holmes, 1983).

In a steady wind, as in a wind tunnel, the dynamic performance of the cup anemometer can be reasonably approximated with a first-order linear differential equation

$$\tau \frac{dV}{dt} + V = V_i \qquad (7.2)$$

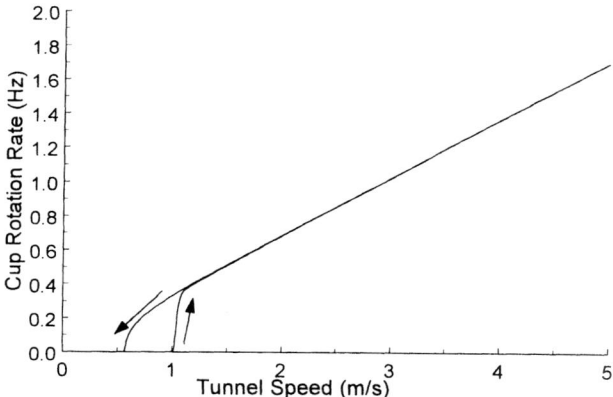

Fig. 7-3 Raw output of a cup anemometer as the wind tunnel speed is slowly increased from 0 to 5 m/s and then slowly decreased back to zero.

where V_i is the wind speed in m s^{-1}, V is the indicated wind speed after applying the calibration equation, and τ is the time constant given by

$$\tau = \frac{I}{\rho R^2 C A V_i} \qquad (7.3)$$

where I is the cup wheel moment of inertia in kg m^2, R is the cup wheel radius as shown in fig. 7-1, ρ is the air density, A is the cross-sectional cup area, and C is a constant related to C_d.

When the time constant is measured in wind tunnel tests performed at several wind speeds, it is found that the time constant decreases as the wind speed increases: $\tau = \lambda/V_i$, where λ is called the distance constant in m. This is a source of some difficulty with eqn. 7.3, as the time constant is not a constant and the simple solutions for the step and sinusoidal input found in chap. 6 are, strictly speaking, not applicable. Since

$$I = \sum_i m_i R_i^2 \approx m_c R^2 \qquad (7.4)$$

where m_c is the mass of the cup wheel, most of the inertia is in the cup wheel. Then the distance constant can be written as

$$\lambda = \frac{m_c}{\rho C A} \qquad (7.5)$$

so, to minimize the distance constant, we need to make m_c small and A large. This analysis shows that the length of the cup radius arm is irrelevant. However, other practical considerations, such as the need for a rugged cup wheel, dictate a small radius arm.

Although the time constant is a function of the wind speed, the usual approximation is to assume that the wind speed fluctuations are small compared to the mean and, therefore, to substitute the mean wind speed for V_i in the time constant equation.

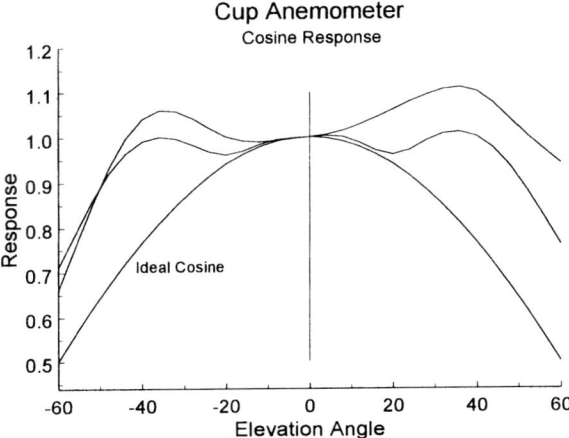

Fig. 7-4 Cosine response of a cup anemometer.

Then the simple solutions can be used. The dynamic performance specification for cup and propeller anemometers is the distance constant, λ, not the time constant.

The sinusoidal response given in the previous chapter is applicable to cup and propeller anemometers but the amplitude and phase response functions were given in terms of $\tau\omega$. Now $\tau = \lambda/V$. And ω, the input frequency, can be expressed in terms of an input wavelength, yielding $\omega = 2\pi f = 2\pi V/\lambda_i$. Then

$$\tau\omega = 2\pi \frac{V}{\lambda_i}\frac{\lambda}{V} = 2\pi \frac{\lambda}{\lambda_i} \tag{7.6}$$

where the term λ_i can be thought of as a gust wavelength. Note that when $2\pi\lambda/\lambda_i = 1$, $A_0/A_i = 0.707$; that is, the amplitude of the response has been reduced to approximately 70% of the input amplitude.

The anemometer amplitude response is less than 0.707 when $\lambda_i < 2\pi\lambda$. Typical anemometer distance constants are $1\,\text{m} \leq \lambda < 10\,\text{m}$. So, if a typical anemometer has a distance constant of 3 m it will attenuate the amplitude of all gusts whose wavelength is less than $2\pi\lambda = 19\,\text{m}$ to less than 70% of the input amplitude. This is surprising in that the cup wheel radius R is always less than 15 cm. One might expect that a cup anemometer would respond well to all gusts whose wavelength is greater than the physical size of the cup wheel but, as shown here, the typical anemometer sharply attenuates all gusts whose wavelengths are less than 19 m. Even a relatively fast response anemometer with a distance constant of about 1 m will attenuate all gusts whose wavelength is less than 6 m.

To obtain the conventional solutions for the step, ramp, and sinusoidal response, it was necessary to assume that the time constant was a function of the mean wind speed rather than the instantaneous wind speed. If we allow the time constant to vary inversely as the wind speed, the differential equation becomes nonlinear and it is more difficult to obtain the solutions. Consider a simple case, shown in fig. 7-6, where the wind speed is a rectangular wave.

Since the time constant is inversely proportional to the wind speed, when the wind speed is low, the time constant is high and, when the wind speed is high, the time

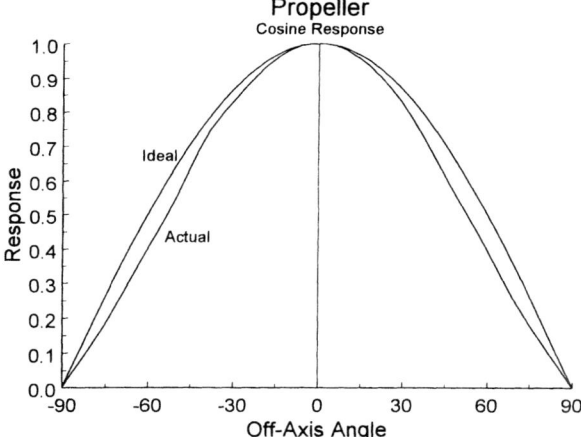

Fig. 7-5 Cosine response of a propeller anemometer.

constant is lower. Thus the anemometer responds more rapidly to an increasing step than to a decreasing step. The result is that the average wind speed reported by the anemometer is higher than the actual average wind speed. This overspeeding error must occur in all cup and propeller anemometers but the magnitude of the error may be small. The error is a function of the time or distance constant of the anemometer, the average wind speed, and the turbulence intensity, the ratio of the wind speed standard deviation to the average wind speed.

All cup anemometers are susceptible to overestimation of the mean wind speed and there are two causes. The first, and most significant, is the static overestimation error due to the lack of cosine response. The second source of this error is a dynamic effect that will be greater for anemometers with larger distance constants.

Propeller anemometers are also susceptible to overestimation of the mean, but this is due only to the dynamic effect; therefore, the overestimation of the mean is far less for propeller anemometers.

Air density has an effect upon the threshold speed and distance constant of cup and propeller anemometers. The threshold speed is inversely proportional to the density because it represents a balance between the aerodynamic force and the friction forces in the anemometer bearings. It has also been shown that the time constant, and therefore the distance constant, are inversely related to the air density. At speeds well above the threshold speed, the static sensitivity should be independent of air density because the static sensitivity is a function of the balance of aerodynamic forces.

7.1.1.2 Wind Vanes

A wind vane, as shown in fig. 7-7, is a flat plate or airfoil that can rotate about a vertical shaft and, in static equilibrium, is oriented along the wind vector. There is usually a counterweight, not shown in the figure, to balance the vane about the vertical shaft.

136   Meteorological Measurement Systems

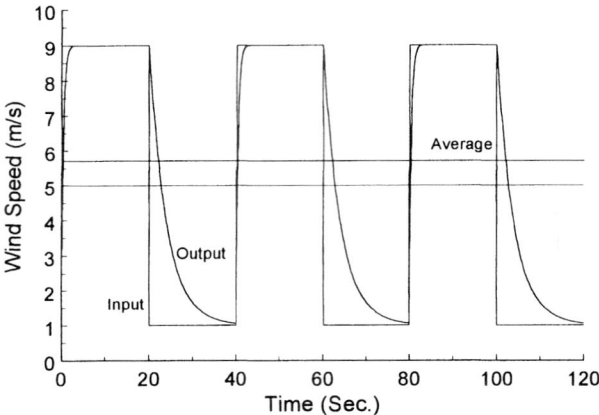

Fig. 7-6 Response of a cup or propeller to a simple wind speed input.

The most common electrical transducer is a simple pot mounted concentrically with the vertical shaft to convert the azimuth angle (0° to 360°) to a voltage proportional to that angle. The pot, of necessity, has a dead zone of 3° to 5° usually oriented to North. The effect causes vane angles of North ±3° to be reported as North. Busch et al. (1980) used a shaft digitizer to convert shaft angle to a digital signal, thus eliminating the dead zone error.

The only source of static error is misalignment of the vane. While it is fairly straightforward to align a vane to North, human error frequently causes misalignment. The vane input is the wind direction and the raw output is the vane shaft alignment relative to North. With a simple pot transducer, the transducer output is a voltage proportional to the shaft angle.

A vane uses a combination of the lift and drag forces on the vane to align itself with the wind vector. Since the vane has a moment of inertia and aerodynamic damping, there is a dynamic misalignment error, $\varepsilon_d = \theta - \theta_i$, due to changing wind direction, θ_i. The equation of motion is

$$I\frac{d^2\theta}{dt^2} + \frac{NR}{V}\frac{d\theta}{dt} = -N\varepsilon = N(\theta_i - \theta) \qquad (7.7)$$

where I is the vane moment of inertia, kg m^2, N is the aerodynamic torque per unit angle, kg m^2 s^{-2}, and R is the distance from the axis of rotation to the effective center of the aerodynamic force on the vane, m. Aerodynamic torque is given by

$$N = \tfrac{1}{2} C_L \rho A V^2 R \qquad (7.8)$$

where C_L is the lift coefficient. As before, ρ is the air density, A the vane area and V the wind speed. The second-order dynamic performance equation will be discussed in the next chapter.

The ideal wind vane will have the following characteristics:

- low friction bearings;
- statically balanced (using the counterweight);

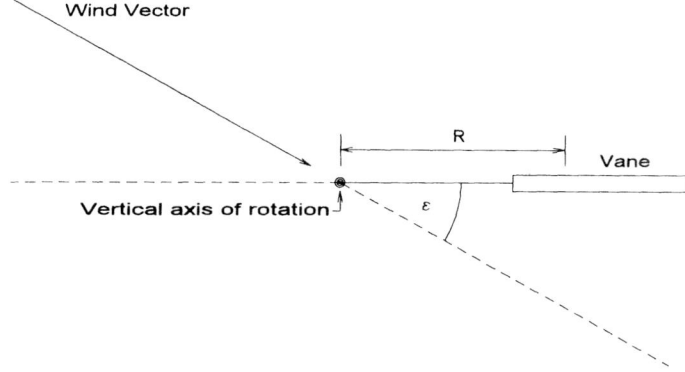

Fig. 7-7 Schematic of a wind vane.

- maximum wind torque and minimum moment of inertia;
- damping ratio (to be defined later) ≥ 0.3;
- low threshold wind speed (about 0.5 m s^{-1}); and
- rugged design capable of withstanding wind speeds up to 90 m s^{-1} (for hurricane survival).

Maintenance requirements for a wind vane are fairly simple:

- verify low bearing friction;
- verify mechanical integrity (check for bent vane arm);
- verify alignment to North; and
- verify proper operation of the transducer.

7.1.1.3 Drag Cylinder or Sphere

The drag cylinder or sphere anemometer is a sensor which measures wind velocity by measuring the drag force on an object in the flow. The drag cylinder is used to measure two-dimensional flow whereas the sphere can measure the three-dimensional wind vector.

While cup and propeller anemometers rotate in the wind and are subject to mechanical friction, the drag sphere anemometer does not rotate and its motion is extremely limited as it responds to the drag force of the wind. This anemometer has been described by Doebelin (1983) and Smith (1980a). The drag force acting on the sphere is given by

$$\vec{F} = \tfrac{1}{2} \rho A \, C_d \, \vec{V} \left| \vec{V} \right| \tag{7.9}$$

where vector notation is used to represent the three-dimensional wind velocity and force vectors. As usual, ρ is the air density and A is the cross-sectional area. Each velocity component depends upon all three force components. If we define the velocity vector as

138 Meteorological Measurement Systems

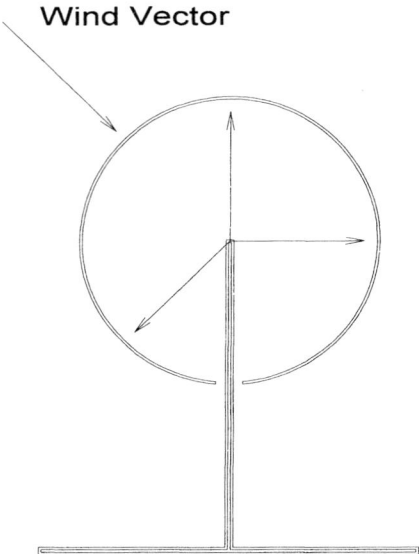

Fig. 7-8 Schematic of a drag sphere anemometer. The arrows represent the three-dimensional support and transducer system.

$$\vec{V} = u\vec{i} + v\vec{j} + w\vec{k} \tag{7.10}$$

then the absolute magnitude of the vector is

$$|V| = \sqrt{u^2 + v^2 + w^2} \tag{7.11}$$

To simplify the following, let $K = C_D \rho A/2$. Then

$$\vec{F} = K\sqrt{u^2 + v^2 + w^2}(u\vec{i} + v\vec{j} + w\vec{k}) \tag{7.12}$$

Each velocity component will be a function of all of the force components, for example,

$$u = F_u K^{-1/2} \left(F_u^2 + F_v^2 + F_w^2\right)^{-1/4} \tag{7.13}$$

These computations can be readily performed using a microprocessor integrated into the instrument.

Unlike a cup anemometer or a wind vane, the dynamic response is not a function of the wind speed. It is determined by the spring torque of the supporting members used to hold the cylinder or sphere in position. These supports can be quite stiff to make the anemometer rugged and to increase the frequency response. But, with stiff supports, it can be quite difficult to measure light winds since there is very little deflection. On the other hand, if the supports are not so stiff, the anemometer may not be able to withstand strong winds. And the resonant frequency (see chap. 8) will be lower, so the anemometer will not be able to resolve the higher wind frequencies. Strain gauges, used to detect displacement, may be temperature sensitive and require high-gain amplifiers to generate a reasonable voltage signal. They may also be susceptible to drift. The cylinder or sphere can be affected by an accumulation of snow or ice which would change the aerodynamics and could create an offset.

7.1.1.4 Pitot-Static Tube

The pitot-static tube is actually a pair of concentric tubes, as shown in fig. 7-9. The stagnation port, at the end of the tube, is a blunt obstacle to airflow and therefore the drag coefficient is unity. The static port is located at a point far enough back along the tube to have no dynamic flow effects at all, so the pressure observed there is just the ambient atmospheric pressure.

The pitot-static tube must be oriented into the airflow. A typical tube will tolerate misalignment errors up to ±20° but the alignment problem makes them virtually unsuitable for atmospheric work. They are ideal for wind tunnels and are frequently used to calibrate other anemometers. As noted above, p-static $= p$, the ambient atmospheric pressure, whereas p-stagnation $= 0.5\rho V^2 + p$, thus the differential pressure $\Delta p = (p\text{-stagnation}) - (p\text{-static}) = 0.5\rho V^2$. The transfer equation is plotted in fig. 7-10 and the calibration equation is

$$V = \sqrt{\frac{2\Delta p}{\rho}} = \sqrt{2RT\frac{\Delta p}{p}} \qquad (7.14)$$

the calibration being a function of atmospheric pressure and temperature since $\rho = p/RT$, where R is the gas constant $= 287\,\text{J}\,\text{kg}^{-1}\,\text{K}^{-1}$ for dry air, T is the air temperature in K, while p and Δp can be in any consistent pressure units. Since R is the gas constant for dry air, humidity will have an affect, but less than 1%. Compressibility of the air is also a factor at speeds over $100\,\text{m}\,\text{s}^{-1}$, where it affects the results by about 1%.

The pitot-static probe is inexpensive but requires a high-quality differential pressure sensor to convert the Δp to a usable signal. It is insensitive to light winds as the

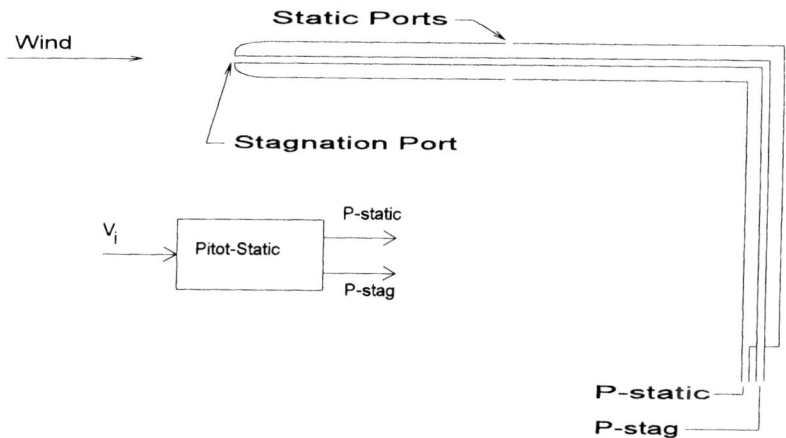

Fig. 7-9 Pitot-static tube.

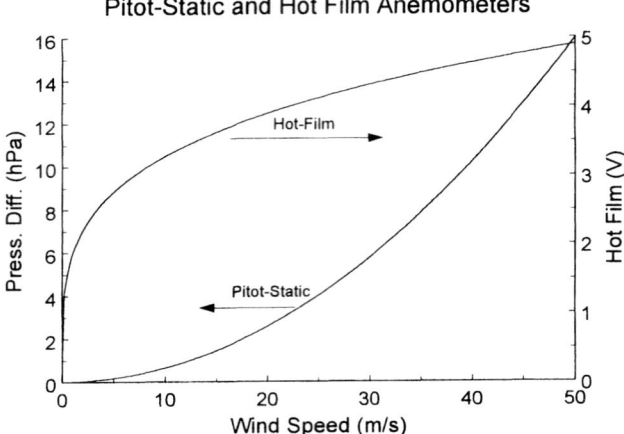

Fig. 7-10 Static transfer of a Pitot-static tube and of a hot-film anemometer.

static sensitivity goes to zero as the wind speed goes to zero. The probe is stable and needs no calibration; however, the associated pressure sensor does need calibration and may be subject to drift and temperature effects.

7.1.1.5 Definitions

Distance constant is the distance air flow past a rotating anemometer during the time it takes the cup wheel or propeller to reach 63.2% of the equilibrium speed after a step change in the input wind speed. Also, the distance constant $\lambda = \tau V_i$.

Starting threshold of a cup or propeller anemometer is the lowest wind speed at which the anemometer, initially stopped, starts and continues to turn and produces a measurable signal. This applies to a propeller anemometer oriented into the wind and to a cup anemometer when the wind vector is in the plane of the cup wheel.

Wind run is the average of the scalar wind speed. Direction is ignored.

7.1.2 Heat Dissipation

Hot-wire and hot-film anemometers are used to infer the wind speed from the cooling of a heated wire or film, which is dependent on the mass flow rate (speed and density of flow) past the sensing element. The response speed of wires and films is a function of the thermal mass of the element. Hot wires are the fastest conventional wind sensors available since they can use very fine platinum wires, down to 5 µm in diameter. These sensors are well suited to measurement of atmospheric turbulence or for use on an aircraft. Film sensors are made by depositing a thin film of platinum on a cylindrical quartz or glass core and then insulating it with a very thin quartz or ceramic coating. The rod diameter may be 50 µm or more in diameter, thus inhibiting the frequency response to some extent. For a general treatment, see Doebelin (1983), Hasse and Dunckel (1980), and Perry (1982).

In a wire operated in the constant temperature mode, the current I through the sensor is related to the wind speed by King's law

$$I^2 = A + B\sqrt{V} \tag{7.15}$$

where A and B are constants. This equation is applicable above some threshold flow rate that can be as great as 5 m s^{-1}. The calibration is a function of the air density and of the wire (or film) characteristics, including possible atmospheric contamination. It is not uncommon to monitor the mean air speed simultaneously with a cup anemometer to provide an ongoing calibration check.

The dynamic response characteristics can be quite complex but it is relatively easy to increase the frequency response by decreasing the probe size so that the details of the frequency response are seldom needed. However, a hot-film anemometer with a large probe may have rather poor frequency response, in the 10 to 100 Hz range.

Probe configurations are available to sense the three-dimensional wind vector. They are susceptible to atmospheric contamination that affects the calibration. Larger hot-film probes are less susceptible and can be cleaned to restore the calibration. Also, larger probes are more rugged than the small hot-wire probes. Rain produces spikes in the data that, if averaged, produce an apparent increase in the wind speed. Attempts to shield a 2-D probe from the rain are not completely successful since strong winds will carry small drops into the probe housing and affect the data. It is possible to devise a scheme to filter out the raindrop effect.

Hot-wire and hot-film probes are expensive and power hungry. They must be calibrated and may be susceptible to drift. It is difficult to resolve low wind speeds with a thermal anemometer. The problem is that they are too sensitive at low wind speeds; the static sensitivity becomes very large as the speed goes to zero. While we would usually like to have high static sensitivity, it should be more or less uniformly distributed over the range.

7.1.3 Speed of Sound

The sonic anemometer measures the time required to transmit an acoustic signal across a fixed path to determine the wind velocity component along that path. Its frequency response is limited by the spatial averaging along the path. It responds linearly to the wind speed. See Coppin and Taylor (1983), Kaimal (1980), and Kaimal et al. (1991).

Figure 7-11 is a schematic of a single-axis sonic anemometer showing the wind vector V. The wind component parallel to the sound path is V_d and that normal to the path is V_n. Transmitters T_1 and T_2 emit periodic pulses of sound energy directed towards receivers R_1 and R_2, respectively. If there were no wind, $V = 0$, a sound pulse would travel from T_1 to R_1 (and from T_2 to R_2) in time $t = d/C$, where d is the separation distance between a transmitter and a receiver and C is the speed of sound given by $C^2 = \gamma R T$, where $\gamma = 1.4$, the ratio of specific heats, $R = 287 \text{ J K}^{-1} \text{ kg}^{-1}$, the gas constant for dry air, and T is the air temperature in K.

EXAMPLE

In a certain sonic anemometer the transmitters and receivers are separated by 20 cm. If the air temperature is 20°C, the speed of sound will be $C =$

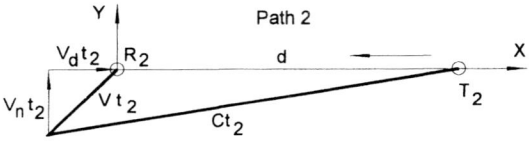
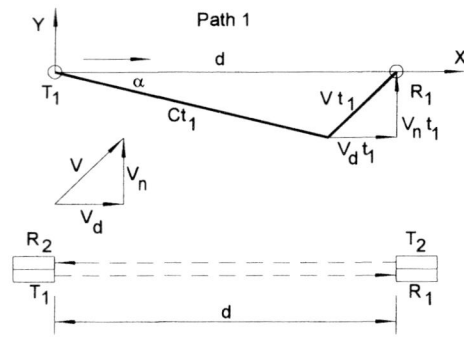

Fig. 7-11 Sonic anemometer vector relation.

$(1.4 \times 287 \times 293.15)^{1/2} = 343.2 \text{ m s}^{-1}$ and the time required for the sound to travel from transmitter to receiver is $t = 0.2/343.2 = 582.7 \text{ μs}$.

Humidity affects sound speed since the gas constant for moist air is different from that of dry air. This is often expressed by using the gas constant for dry air but substituting virtual temperature, T_v, for air temperature, T. At $p = 1000$ hPa and $T = 20°C$, a change in humidity from dry air to 100% relative humidity would change the speed of sound by 0.4%.

When the wind component parallel to the path $V_d \neq 0$, the travel time will be affected because acoustic pulses are carried along by the air flow. If this wind component is blowing from T_1 to R_1 then the T_1–R_1 travel time will be reduced since the apparent speed of the acoustic pulse will be $C + V_d$ whereas the T_2–R_2 travel time will increase as the apparent speed becomes $C - V_d$. Wind components normal to the path have a much smaller effect on the travel time. They increase the apparent travel distance or decrease the speed of sound from C to $C \cos \alpha$, where $\alpha = \sin^{-1}(V_n/C)$. When $V_n = 20 \text{ m s}^{-1}$, $\alpha = 0.058 \text{ rad} = 3.3°$ and $\cos \alpha = 0.9983$, so the effect is to reduce the apparent speed of sound by 0.17%.

The travel times are

$$t_1 = \frac{d}{C \cos \alpha + V_d}; \quad t_2 = \frac{d}{C \cos \alpha - V_d} \qquad (7.16)$$

and the sum of the inverse travel times is

$$\frac{1}{t_1} + \frac{1}{t_2} = \frac{2C \cos \alpha}{d} = \frac{2 \cos \alpha}{d} \sqrt{\gamma RT} \cong \frac{2}{d} \sqrt{\gamma RT} \quad \text{if } \cos \alpha \cong 1 \qquad (7.17)$$

If the wind component normal to the sonic path is 20 m s^{-1} and the speed of sound is 343 m s^{-1} (for an air temperature of $20°C$), then $\alpha = \sin^{-1}(V_n/C) = 0.0583 \text{ rad}$ and

$\cos\alpha = 0.998$; thus the assumption that $\cos\alpha = 1$ is reasonable except for extremely high wind speeds. The sonic virtual temperature is given by

$$T_v = \frac{1}{\gamma R}\left[\frac{d}{2}\left(\frac{1}{t_1}+\frac{1}{t_2}\right)\right]^2 \tag{7.18}$$

if the gas constant for dry air is used for R. Taking the difference between the transit times yields

$$\frac{1}{t_1} - \frac{1}{t_2} = \frac{C\cos\alpha + V_d}{d} - \frac{C\cos\alpha - V_d}{d} = \frac{2V_d}{d} \tag{7.19}$$

so the velocity component parallel to the path is

$$V_d = \frac{d}{2}\left(\frac{1}{t_1}-\frac{1}{t_2}\right) \tag{7.20}$$

EXAMPLE
Determine the velocity resolution of a sonic anemometer given that a particular instrument can measure transit time to $0.1\,\mu s$ and that the distance $d = 20\,cm$.

SOLUTION
Let $\Delta t = t_2 - t_1 = 2dV_d/C^2$. If the speed of sound is $343\,m\,s^{-1}$ at $20°C$, the velocity resolution is

$$\Delta V_d = \Delta t C^2/(2d) = 10^{-7} \times (343)^2/(2 \times 0.2) = 3\,cm\,s^{-1}$$

EXAMPLE
A sonic anemometer, with $d = 20\,cm$, measures $t_1 = 563.4\,\mu s$ and $t_2 = 593.5\mu s$. Find the velocity parallel to the path and the sonic virtual temperature.

$$V_d = \frac{0.2}{2}\left(\frac{1}{563.4 \times 10^{-6}} - \frac{1}{593.5 \times 10^{-6}}\right) = 9\,m s^{-1}$$

$$T_v = \frac{1}{1.4 \times 287}\left[\frac{0.2}{2}\left(\frac{1}{563.4 \times 10^{-6}} + \frac{1}{593.5 \times 10^{-6}}\right)\right]^2 = 297.9\,K$$

In one implementation, a sonic anemometer makes 200 measurements per second and averages over ten samples to yield less noisy results at 20 per second. It achieves a resolution of $1\,cm\,s^{-1}$ and has a range of $\pm 30\,m\,s^{-1}$.

The sonic anemometer has a space resolution limitation imposed by the path length d. It measures a wind speed averaged over the path length but this is much shorter than the equivalent space resolution limit of $2\pi\lambda$ of cup and propeller anemometers.

The sonic anemometer is fairly expensive, compared to a simple mechanical sensors, and requires considerably more power. There can be signal loss due to heavy rain or wet snow. It has somewhat greater bandwidth than the mechanical anemometers but considerably less than hot-wire or hot-film anemometers.

144 Meteorological Measurement Systems

7.2 Calibration

The preferred calibration method for wind sensors is to place them in a wind tunnel along with a suitable reference instrument. The wind tunnel must be large enough to accommodate the sensors and provide a steady flow that is uniform across the tunnel.
When the calibration of cup and propeller anemometers has been established, the performance of the rest of the instrument can be readily checked by spinning the anemometer shaft with a motor turning at a constant speed. This is an adequate check for in-situ calibrations as long as the cup wheel or propeller has not been damaged or replaced with one of different design or dimensions.

7.3 Exposure

The World Meteorological Organization (WMO) standard exposure height for surface winds is 10 m. In addition, the anemometer must have good exposure in all directions within about 3 km. No obstruction to wind flow should be more than 3° above the horizon, which means that the distance from the anemometer to an obstruction should be at least 20 times the height of the obstruction. Since the wind flow is perturbed by obstructions, including buildings, the anemometer should not be located on top of a building. It is very difficult to find measuring sites that conform to these standards; most sites are compromised to some extent, in some directions. It is a good practice to take photographs at each measuring site of the surrounding terrain in all directions to document the fetch. Ideally, this should be done in winter and in summer to show the seasonal effect, and the site characterization should be repeated whenever there are major changes in the vicinity of the station and at intervals of five years or less.
Other exposure problems include snow accumulation, icing, extreme low and high temperatures, high wind gusts, deterioration of plastic parts due to ultraviolet radiation, breakage due to fatigue failure in turbulent wind, corrosion due to rain and high humidity, wind loading (which affects both the anemometer itself and the mounting platform), and lightning-induced surges.

7.4 Wind Data Processing

The WMO standard averaging time is 10 minutes. Typical processing generates an average speed and average direction every 10 minutes. Within each 10-minute averaging period, some systems calculate the maximum 3-second wind speed and direction, the maximum one-minute average speed and direction, and the standard deviation of the speed and of the direction over the averaging period. The 3-second wind speed is the average of the wind speed over a 3-second interval.
For wind engineering applications, such as calculating the force of wind gusts on structures, it is sometimes useful to average the square of the wind speed, while the average of the cube of the wind speed is needed for wind power applications.
In general, one must be cautious about averaging wind direction because of the discontinuity at $0° = 360°$. The simple average wind direction, when the wind fluc-

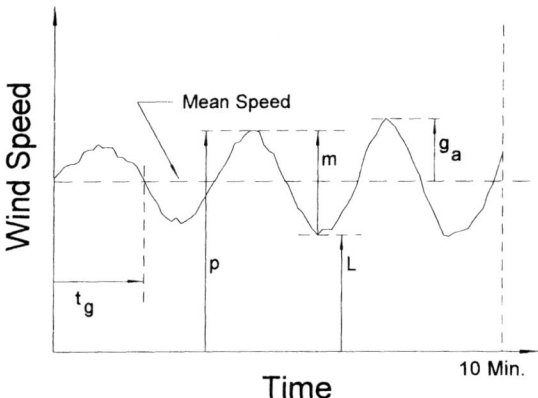

Fig. 7-12 Illustration of some WMO gust definitions.

tuates about North, will be South; for example, $(5° + 335°)/2 = 180°$. This will be a problem whenever a vane is used. The usual solution is to convert the instantaneous speed and direction (polar coordinates) to u and v components (rectangular coordinates) before averaging. Then, if desired, the averaged vector in rectangular coordinates can be converted back to polar coordinates. This process yields the vector average wind speed and direction.

It is possible to calculate an average wind direction by compensating for the discontinuity at North. The simple average of the wind speed is sometimes called the wind run.

The WMO has a set of gust definitions, illustrated in fig. 7-12.

- *Gust peak speed*, p = wind speed associated with a positive gust amplitude.
- *Gust duration*, t_g = time interval from the beginning of a gust to the end of the gust.
- *Gust magnitude*, m = the scalar difference between a gust peak speed and the adjacent lull speed.
- *Gust frequency* = number of positive gusts which occur per unit time.
- *Gust amplitude*, g_a = maximum scalar of the gust from the mean wind speed.
- *Gust lull speed*, L = wind speed associated with a negative gust amplitude.

QUESTIONS

(More difficult questions are indicated with an asterisk.)

1. Is a cup anemometer linear near the threshold speed? Explain.
2. Can a cup anemometer overestimate a gust magnitude? How about a propeller anemometer?
3. Can a cup anemometer overestimate the mean wind speed in a gusty wind? How about a propeller anemometer? Explain.
4. Describe the optimum siting for an anemometer. Would you put one on top of a building?

5. Define the threshold speed of a cup anemometer. Is the threshold speed a function of air density? Determine the threshold of the cup anemometer from the data plotted in fig. 7-3.

6. How does the design of a sonic anemometer compensate for the horizontal component of the wind vector in the measurement of the vertical component?

7. If a cup anemometer in a wind tunnel takes 2.5 s to increase speed from $0.28\,V_T$ to $0.74\,V_T$ (where V_T is the tunnel speed), what is the anemometer time constant? If $V_T = 6\,\mathrm{m\,s^{-1}}$, what is the distance constant?

8. What is the output of a pitot-static tube, when the wind speed is $20\,\mathrm{m\,s^{-1}}$ and the air density is $1.2\,\mathrm{kg\,m^{-3}}$?

9. What is the static sensitivity equation for a sonic anemometer? What is the calibration equation?

10. Why does a propeller anemometer rotate faster than a cup anemometer?

11.* In the geophysical coordinate system (see Appendix D) used in meteorology, the positive x-axis points East and positive y-axis points North. Angular rotation is clockwise from North (0°). In the mathematical coordinate system used in computer languages (FORTRAN, Pascal, C, BASIC), angular rotation is counter-clockwise from the positive x-axis (0°). Write and test a computer program to convert between winds in the ddff format (dd = direction from which the wind is blowing in 10s of degrees and ff = wind speed in knots) and u, v wind components (u and v are the vector components of the wind to the East and North directions respectively and the units are in $\mathrm{m\,s^{-1}}$). Your program should be able to read ddff or u,v values, convert them to the other format and then back to the original format.

Test your program with the following ddff values: 0713, 1614, 3140, 0915 and 3604.

Test your program with the following u, v values: (10, 10); (−19, 23); (−13, −1); (20, −10); (0, 13); and (17, 0).

Hint: The u,v values corresponding to ddff = 0713 are −6.26, −2.29 m/s.

12. Write an expression for the gust wavelength where the amplitude ratio of a cup anemometer falls to 0.7.

13. How could one modify a cup wheel to improve dynamic response? Is the calibration of a cup anemometer a function of air density?

14. Write the steady-state solution for a propeller anemometer with a distance constant of 2 m when $V_i(t) = 5 + \sin(2\pi t) + \cos(4\pi t)$, in units of $\mathrm{m\,s^{-1}}$.

15. A mission to Mars plans to set a meteorological measuring station on the surface to measure wind speed and direction. Would a cup anemometer and wind vane be acceptable? Why? How would you expect the static and dynamic performance characteristics to change from Earth environment? Assume Mars surface atmospheric density $= 0.005\,\mathrm{kg\,m^{-3}}$. What kind of wind sensors would you recommend?

16. Derive the calibration equation of the pitot-static tube from the drag-force equation.

17. Discuss the dynamic performance of a sonic anemometer. What is the dynamic performance parameter? Are their similarities to that of a cup anemometer?

18. We use a first-order, linear model for a cup anemometer. What evidence do we have for the failure of the model? Why do we continue to use it? What assumption is required to make it work?

19. A cup anemometer is mounted in a wind tunnel running at constant speed V_C. A stick is poked through the side of the tunnel to prevent rotation. The stick is removed at time $t = 0$. Write the expected dynamic test solution.

20. Show how to convert $\tau\omega$ for a propeller anemometer to an expression independent of the wind speed even though τ is inversely proportional to the wind speed.

21. What are the dynamic performance parameters (a) for a propeller anemometer? (b) for a wind vane?

22. List some common sources of wind sensor exposure errors that can be avoided.

23. For a cup anemometer:
  (a) What is the dynamic performance parameter?

(b) What is its energy storage reservoir?
(c) How would you change the design to improve this parameter?
(d) What dynamic performance characteristic causes overestimation of the mean?
(e) What static performance characteristic causes overestimation of the mean?
(f) Write an expression for the gust wavelength where the amplitude ratio falls to 0.7.
(g) What causes the threshold effect?
(h) Is the calibration a function of air density?
(i) Is the threshold a function of air density?
(j) Is the distance constant a function of air density?
(h) Can a cup anemometer, in perfect static calibration, overestimate the magnitude of a gust?
24. What causes the threshold effect in propeller anemometers?
25. Identify the following terms for a sonic anemometer: V_d, V_n, C, d, t_1, t_2.

BIBLIOGRAPHY

Acheson, D.T., 1988: Comments on Anemometer Performance determined by ASTM methods. *J. Atmos. Oceanic Technol.*, 5, 381.
Andreas, E.L.. and B. Murphy, 1986: Calibrating cylindrical hot-film anemometer sensors. *J. Atmos. Oceanic Technol.*, 3, 283–298.
Atakturk, S.S., and K.B. Katsaros, 1989: The K-Gill: a twin propeller-vane anemometer for measurements of atmospheric turbulence. *J. Atmos. Oceanic Technol.*, 6, 509–515.
Baker, J.M., C.F. Reece, and E.J. Sadler, 1993: Shear stess estimation with a bivane anemometer. Preprints 8th Symp. on Meterological Observations and Instrumentation, Anaheim, CA. American Meterological Society, Boston, MA, pp. 11–14.
Baynton, W.W., 1976: Errors in wind run estimates from rotational anemometers. *Bull. Am. Meteor. Soc.*, 57, 1127–1130.
Bernier, P.Y., 1988: Low-cost wind speed measurements using naphthalene evaporation. *J. Atmos. Oceanic Technol.*, 5, 662–665.
Bowen, A.J., and H.W. Teunissen, 1986: Correction factors for the directional response of Gill propeller anemometers. *Bound.-Layer Meteor.*, 37, 407–413.
Busch, N.E., and L. Kristensen, 1976: Cup anemometer overspeeding. *J. Appl. Meteor.*, 15, 1328–1332.
Busch, N.E., O. Christensen, L. Kristensen, L. Lading, and S.E. Larsen, 1980: Cups, vanes, propellers, and laser anemometers. In *Air–Sea Interaction: Instruments and Methods*, ed. F. Dobson, L. Hasse and R. Davis. Plenum Press, New York, 801 pp.
Camp, D.W., and R.E. Turner, 1970: Response tests of cup, vane, and propeller wind sensors. *J. Geophys. Res.*, 75, 5265–5270.
Chang, S., and P. Frenzen, 1990: Further consideration of Hayashi's "Dynamic response of a cup anemometer." *J. Atmos. Oceanic Technol.*, 7, 184–194.
Coppin, P.A., and K.J. Taylor, 1983: A three-component sonic anemometer/thermometer system for general micrometeorological research. *Bound.-Layer Meteor.*, 27, 27–42.
Dilger, H., and P. Thomas, 1975: Cup anemometer testing device for low wind speeds. *J. Appl. Meteor.*, 14, 414–415.
Doebelin, E.O., 1983: *Measurement Systems: Application and Design*, 3rd ed. McGraw-Hill, New York, 876 pp.
Dyer, A.J., 1981: Flow distortion by supporting structures. *Bound.-Layer Meteor.*, 20, 243–251.
Finkelstein, P.L. J.C. Kaimal, J.E. Gaynor, M.E. Graves, and T.J. Lockhart, 1986: Comparison of wind monitoring systems. Part I: in situ sensors. *J. Atmos. Oceanic Technol.*, 3, 583–593.

Freitag, H.P., M.J. McPhaden, and A.J. Shepard, 1989: Comparison of equatorial winds as measured by cup and propeller anemometers. *J. Atmos. Oceanic Technol.*, 6, 327–332.

Gaynor, J.E., and C.A. Biltoft, 1989: A comparison of two sonic anemometers and fast-response thermometers. *J. Atmos. Oceanic Technol.*, 6, 208–214.

Gill, G.C., 1973: The helicoid anemometer. *Atmosphere*, 4, 145–155.

Gill, G.C., L.E. Olsson, J. Sela, and M. Suda, 1967: Accuracy of wind measurements on towers or stacks. *Bull. Am. Meteor. Soc.*, 48, 665–674.

Grant, A.L.M., and R.D. Watkins, 1989: Errors in turbulence measurements with a sonic anemometer. *Bound.-Layer Meteor.*, 46, 181–194.

Green, A.E., M.J. Judd, J.K. McAneney, M.S. Astill, and P.T. Prendergast, 1991: A rapid-response 2-D drag anemometer for atmospheric turbulence measurements. *Bound.-Layer Meteor.*, 57, 1–15.

Hasse, L., and M. Dunckel, 1980: Hot wire and hot file anemometers. In *Air–Sea Interaction: Instruments and Methods*, ed. F. Dobson, L. Hasse and R. Davis. Plenum Press, New York, 801 pp.

Hayashi, T., 1987: Dynamic response of a cup anemometer. *J. Atmos. Oceanic Technol.*, 4, 281–287.

Hignett, P., 1992: Corrections to temperature measurements with a sonic anemometer. *Bound.-Layer Meteor.*, 61, 175–187.

Holmes, R.M., G.C. Gill, and H.W. Carson, 1964: A propellor-type vertical anemometer. *J. Appl. Meteor.*, 3, 802–804.

Horst, T.W., 1973: Corrections for response errors in a three-component propeller anemometer. *J. Appl. Meteor.*, 12, 716–725.

Hyson, P., 1972: Cup anemometer response to fluctuating wind speeds. *J. Appl. Meteor.*, 11, 843–848.

Izumi, Y., and M.L. Barad, 1970: Wind speeds as measured by cup and sonic anemometers and influenced by tower structure. *J. Appl. Meteor.*, 9, 851–856.

Jonsson, H.H., and B. Vonnegut, 1986: Oscillatory anemometer. *J. Atmos. Oceanic Technol.*, 3, 737–739.

Kaganov, E.I., and A.M. Yaglom, 1976: Errors in wind-speed measurements by rotation anemometers. *Bound.-Layer Meteor.*, 10, 15–34.

Kaimal, J.C., 1980: Sonic anemometers. In *Air–Sea Interaction: Instruments and Methods*, ed. F. Dobson, L. Hasse and R. Davis. Plenum Press, New York, 801 pp.

Kaimal, J.C., and J.E. Gaynor, 1991: Another look at sonic thermometry. *Bound.-Layer Meteor.*, 56, 401–410.

Kaimal. J.C., J.E. Gaynor, H.A. Zimmerman, and G.A. Zimmerman, 1990: Minimizing flow distortion errors in a sonic anemometer. *Bound.-Layer Meteor.*, 53, 103–115.

Kirwan, A.D., Jr, G. McNally, and E. Mehr, 1974: Response characteristics of a three-dimensional thrust anemometer. *Bound.-Layer Meteor.*, 8, 365–381.

Kraan, C., and W.A. Oost, 1989: A new way of anemometer calibration and its application to a sonic anemometer. *J. Atmos. Oceanic Technol.*, 6, 516–524.

Kunkel, K.E., and C.W. Bruce, 1983: A sensitive fast-response manometric wind sensor. *J. Clim. and Appl. Meteor.*, 22, 1942–1947.

Lapworth, A.J., and P.J. Mason, 1988: The new Cardington balloon-borne turbulence probe system. *J. Atmos. Oceanic Technol.*, 5, 699–714.

Larsen, S.E., J. Hojstrup, and C.W. Fairall, 1986: Mixed and dynamic response of hot wires and cold wires and measurements of turbulence statistics. *J. Atmos. Oceanic Technol.*, 3, 236–247.

Lockhart, T.J., 1977: Evaluation of rotational anemometer errors. *Bull. Am. Meteor. Soc.*, 58, 962–964.

Lockhart, T.J., 1985: Some cup anemometer testing methods. *J. Atmos. Oceanic Technol.*, 2, 680–683.

Lockhart, T.J., 1987: Anemometer performance determined by ASTM methods. *J. Atmos. Oceanic Technol.*, 4, 160–169.
MacCready, P.B. Jr, 1966: Mean wind speed measurements in turbulence. *J. Appl. Meteor.*, 5, 219–225.
MacCready, P.B. Jr, and H.R. Jex, 1964: Response characteristics and meteorological utilization of propeller and vane wind sensors. *J. Appl. Meteor.*, 3, 182–193.
Massman, W.J., and K.F. Zeller, 1988: Rapid method for correcting the non-cosine response errors of the Gill propeller anemometer. *J. Atmos. Oceanic Technol.*, 5, 862–869.
Meteorological Office, 1981: *Measurement of Surface Wind*, Vol. 4 of *Handbook of Meteorological Instruments*, 2nd ed. Meteorology Office 919d. Her Majesty's Stationery Office, London, 38 pp.
Michelena, E.D., and J.F. Holms, 1983: A rugged, sensitive, and light-weight anemometer used by NDBC for marine meteorology. *Preprints 5th Symp. on Meterological Observations and Instrumentation*, Toronto, Canada. American Meterological Society, Boston, MA, pp. 573–577.
Moriarty, W.W., 1992: Tether corrections for tethered balloon wind measurements. *Bound.-Layer Meteor.*, 61, 407–417.
Ower, E., and R.C. Pankhurst, 1977: *The Measurement of Air Flow*, 5th ed. Pergamon Press, Oxford, 363 pp.
Perry, A.E., 1982: *Hot-Wire Anemometry*. Clarendon Press, Oxford, 184 pp.
Powell, M.D., 1993: Wind measurement and archival under the Automated Surface Observing System (ASOS): user concerns and opportunity for improvement. *Bull. Am. Meteor. Soc.*, 74, 615–623.
Quaranta, A.A., G.C. Aprilesi, G. De Cicco, and A. Taroni, 1985: A microprocessor based, three axes, ultrasonic anemometer. *J. Phys. E: Sci. Instr.*, 18, 384–387.
Ramachandran, S., 1969: A theoretical study of cup and vane anemometers. *Quart. J. Roy. Meteor. Soc.*, 95, 163–180.
Ramachandran, S., 1970: A theoretical study of cup and vane anemometers, Part II. *Quart. J. Roy. Meteor. Soc.*, 96, 115–123.
Redford, T.G., S.B. Verma, and N.J. Rosenburg, 1981: Drag anemometer measurements of turbulence over a vegetated surface. *J. Appl. Meteor.*, 20, 1222–1230.
Skibin, D., 1984: A simple method for determining the standard deviation of wind direction. *J. Atmos. Oceanic Technol.*, 1, 101–102.
Skibin, D. J.C. Kaimal, and J.E. Gaynor, 1985: Measurement errors in vertical wind velocity at the Boulder Atmospheric Observatory. *J. Atmos. Oceanic Technol.*, 2, 598–604.
Smith, S.D., 1980a: Dynamic anemometers. In *Air–Sea Interaction: Instruments and Methods*, ed. F. Dobson, L. Hasse and R. Davis. Plenum Press, New York, 801 pp.
Smith, S.D., 1980b: Evaluation of the Mark 8 thrust anemometer-thermometer for measurement of boundary-layer turbulence. *Bound.-Layer Meteor.*, 19, 273–292.
Snow, J.T., M.E. Akridge, and S.B. Harley, 1989: Basic meteorological observations for schools: surface winds. *Bull. Am. Meteor. Soc.*, 5, 493–508.
Van Oudheusden, B.W., and J.H. Huijsing, 1991: Microelectronic thermal anemometer for the measurement of surface wind. *J. Atmos. Oceanic Technol.*, 8, 374–384.
Wolfson, M.M. and T.T. Fujita, 1989: Correcting wind speed measurements for site obstructions. *J. Atmos. Oceanic Technol.*, 6, 343–352.
Wucknitz, J., 1980: Flow distortion by supporting structures. In *Air–Sea Interaction: Instruments and Methods*. ed. Dobson, F., L. Hasse, and R. Davis. Plenum Press, New York, 801 pp.
Wyngaard, J.C., 1981a: Cup, propeller, vane, and sonic anemometers in turbulence research. *Ann. Rev. Fluid Mech.*, 13, 399–423.
Wyngaard, J.C., 1981b: The effects of probe-induced flow distortion on atmospheric turbulence measurements. *J. Appl. Meteor.*, 20, 784–794.

Wyngaard, J.C., and S.-F. Zhang, 1985: Transducer-shadow effects on turbulence spectra measured by sonic anemometers. *J. Atmos. Oceanic Technol.*, 2, 548–558.

Wyngaard, J.C., L. Rockwell, and C.A. Friehe, 1985: Errors in the measurement of turbulence upstream of an axisymmetric body. *J. Atmos. Oceanic Technol.*, 2, 605–614.

Zhang, S.F., J.C. Wyngaard, J.A. Businger, and S.P. Oncley, 1986: Response characteristics of the U.W. sonic anemometer. *J. Atmos. Oceanic Technol.*, 3, 315–323.

8

Dynamic Performance Characteristics, Part 2

The first-order model discussed in chap. 6 is inadequate when there is more than one energy storage reservoir in the system to be modeled. If the sensor is linear it can be modeled with a higher-order dynamic performance model.

Here the term 'system' refers to a physical device such as a sensor while the equation refers to the corresponding mathematical model. There exists a dual set of terms corresponding to consideration of the physical system or of the mathematical model. For example, a_n are coefficients of the mathematical model (see eqn. 8.1) but they also represent some physical aspect of the sensor being modeled; thus they can also be called system parameters.

8.1 Generalized Dynamic Performance Models

The general dynamic performance model is the linear ordinary differential equation

$$a_n \frac{d^n x}{dt^n} + a_{n-1} \frac{d^{n-1} x}{dt^{n-1}} + \cdots + a_1 \frac{dx}{dt} + x = x_i(t) \tag{8.1}$$

where t = time, the independent variable, x = the dependent variable, a_n = equation coefficients or system parameters, and $x_i(t)$ = input or forcing function.

This equation is ordinary because there is only one independent variable. It is linear because the dependent variable and its derivatives occur to the first degree only. This excludes powers, products, and functions such as sin(x). If the system parameters a_n are constant, the system is time invariant.

We can define the differential operator $D = d/dt$; then eqn. (8.1) can be written as

$$(a_n D^n + a_{n-1} D^{n-1} + \cdots + a_1 D + 1)x = x_i(t) \tag{8.2}$$

As before, the solution is $x(t) = x_T(t) + x_s(t)$, where $x_T(t)$ is the transient solution and $x_s(t)$ is the steady-state solution. The transient solution has n arbitrary constants which may be numerically evaluated by imposing n initial conditions on eqn. (8.2). The first step in obtaining the transient solution is to calculate the roots of the characteristic equation

$$a_n r^n + a_{n-1} r^{n-1} + \cdots + a_1 r + 1 = 0 \tag{8.3}$$

where the operator D has been replaced with a simple algebraic variable, r. The roots of the characteristic equation, $r_1, r_2, \ldots r_n$ are used to obtain the solution with the following rules:

(1) *Real roots, unrepeated.* For each real unrepeated root r one term of the solution is written as Ce^{rt}, where C is an arbitrary constant.
(2) *Real roots, repeated.* For each real root r which appears p times, the solution is written as $(C_0 + C_1 t + \cdots + C_{p-1} t^{p-1})e^{rt}$.
(3) *Complex roots, unrepeated.* A complex root has the form $a \pm ib$. If the coefficients of eqn. (8.3) are real, which we usually expect, then for each root pair the corresponding solution is $e^{at}(c_1 \cos(bt) + c_2 \sin(bt))$.
(4) *Complex roots, repeated.* For each pair of complex roots which appears m times, the solution is

$$c_0 e^{at} \cos(bt + \phi_0) + c_1 t e^{at} \cos(bt + \phi_1) + \cdots + c_{m-1} t^{m-1} e^{at} \cos(bt + \phi_{m-1}) \tag{8.4}$$

The transient solution is simply the sum of the individual terms. To evaluate the constants c_i, there must be n initial conditions specified.

As with the first-order equation, the steady-state solution can be found by the method of undetermined coefficients. It does not work in all conditions but is adequate for the present purpose. Given that the input is some function $x_i(t)$, repeatedly differentiate $x_i(t)$ until the derivatives go to zero or repeat the functional form of some lower-order derivative. This is also the test for the applicability of the method; if neither of the above conditions prevails, the method of undetermined coefficients cannot be used. Write the steady-state solution as

$$x_S(t) = k_1 x_i(t) + k_2 D x_i(t) + k_3 D^2 x_i(t) + \cdots \tag{8.5}$$

where the right-hand side includes one term for each functionally different form found by examining $x_i(t)$ and its derivatives. The constants k_i do not depend upon the initial conditions. They are found by substituting eqn. (8.5) into eqn. (8.2).

8.2 Energy Storage Reservoirs

Differential equations describe the behavior of physical systems in which a redistribution of energy is taking place. In a mechanical system, a mass in motion stores kinetic energy and may store potential energy by virtue of its position in a force field.

When a mechanical system stores potential energy and dissipates energy, the differential equation is second order in position, for example

$$m\frac{d^2 x}{d t^2} + f_d \frac{dx}{dt} + x = F(t) \tag{8.6}$$

where x = position, dx/dt = velocity, m = mass, f_d = dissipation factor, and F = external force.

The above equation applies to a wind vane because the vane can store kinetic energy in the vane arm as moment of inertia and stores potential energy by virtue of its orientation relative to the wind vector. It dissipates kinetic energy into the wind stream.

The order of a differential equation is always equal to the number of energy storage reservoirs. In a mechanical system, these reservoirs comprise the kinetic energy storage elements plus the potential energy storage elements. Capacitors and inductors are energy storage elements in electrical systems. In thermal systems, energy storage is in thermal masses.

The transient response, or complementary function in mathematical terms, is obtained when the forcing function is set to zero and the system is released from some set of initial conditions at time $t = 0$. The distribution of energy in the system storage elements at the time of release must tend towards zero due to the always-present energy dissipation. In system terms, the output for a given initial energy distribution and driving input is the transient solution plus the steady-state solution. In mathematical terms, the equation solution for a given set of initial conditions and a forcing function is the complementary function plus the particular function.

For a linear system, the response to a sum of inputs is simply the sum of the responses to these inputs applied separately. This is the superposition principle and can be taken as the defining property of linear systems. This is an extremely useful property because it allows analysis of the response to complex signals in the frequency domain by superposition of responses to individual frequencies. This is the justification for using linear models even when the fit is far from ideal.

A physical system is said to be in a static state when the distribution of energy within the system is constant. When there is an exchange of energy within the system, the system is in a dynamic state and its performance is described by a differential equation containing derivatives with respect to time. To determine static characteristics such as threshold, measurements of the output must be made for many different values of the input. Each measurement is made while the system is static. During the transition from one static state to another, the system is dynamic. We wait until the dynamic energy exchange has ceased before making the static measurement.

When forces are applied at discrete points and are transmitted by discrete components within the system, the system can be defined by lumped parameters. But when it is necessary to describe the variation across space coordinates of a physical component, the system must be described with distributed parameters and is modeled by a partial differential equation.

Dynamic performance analysis is concerned with modeling the performance of dynamic, lumped parameter systems with ordinary differential equations where time is the independent variable.

8.3 Second-Order Systems

The number of dynamic performance parameters is equal to the order of the system; so, for a second-order system, the performance equation, 8.1, can be reduced to the canonical form

$$\left(\frac{1}{\omega_n^2}D^2 + \frac{2\zeta}{\omega_n}D + 1\right)x = x_i \qquad (8.7)$$

where ζ and ω_n are constants, called the damping ratio and the undamped natural frequency, and the characteristic equation is

$$\frac{1}{\omega_n^2}r^2 + \frac{2\zeta}{\omega_n}r + 1 = 0 \qquad (8.8)$$

The roots of the characteristic equation are

$$r = -\omega_n(\zeta \pm \sqrt{\zeta^2 - 1}) \qquad (8.9)$$

Physically, there are only two possible cases for the roots:

(1) Real and unrepeated when $\zeta > 1$; then

$$x_T(t) = c_1 e^{r_1 t} + c_2 e^{r_2 t} \qquad (8.10)$$

(2) Complex and unrepeated when $\zeta < 1$; then

$$x_T(t) = e^{-\zeta \omega_n t}\left[c_1 \cos(\omega_n \sqrt{1-\zeta^2}\, t) + c_2 \sin\left(\omega_n \sqrt{1-\zeta^2}\, t\right)\right] \qquad (8.11)$$

In the first case, the solution is non-oscillatory and similar to the first-order step function response. The second case is far more interesting since it exhibits oscillatory response. This solution is a decaying cosine and, if ζ is small, shows considerable overshoot and oscillation. The constant ω_n is called the undamped natural frequency as it is the frequency at which the system would oscillate if the damping ratio $\zeta = 0$.

If ζ could be zero, the response to a step function input would be a cosine with no damping; it would oscillate indefinitely. This implies that the system does not dissipate any energy and this is impossible to realize. Thus the case $\zeta = 0$ is not physically possible.

The case $\zeta = 1$ exists only as a mathematical abstraction because it would require that some system parameter(s) be set to exact values, which is never possible. Therefore, all second-order systems fall into the two categories mentioned above: $\zeta > 1$ or $0 < \zeta < 1$.

8.3.1 Step Function Input

Let the step function input be $x_i(t) = x_c$ for $t < 0$ and $x_i(t) = 0$ for $t \geq 0$ where x_c is a constant. The two initial conditions required for a second-order equation are $x(0) = x_c$ and $D_x(0) = 0$. To obtain the transient response, we will use the characteristic equation method. The characteristic equation is given by eqn. 8.8 and, if we assume that $\zeta < 1$ the transient solution is eqn. 8.11. The steady-state solution is $x_S(t) = 0$, so the complete solution is given by eqn. 8.11.

We can evaluate the constants c_1 and c_2 using the initial conditions:

$$X(0) = c_1 = x_c$$
$$Dx(0) = -c_1 \zeta \omega_n + c_2 \omega_n \sqrt{1-\zeta^2} = 0 \qquad (8.12)$$
$$c_2 = \frac{x_c \zeta}{\sqrt{1-\zeta^2}}$$

Then the solution is

$$x(t) = x_c e^{-\zeta \omega_n t}\left[\cos\left(\omega_n \sqrt{1-\zeta^2}\, t\right) + \frac{\zeta}{\sqrt{1-\zeta^2}} \sin\left(\omega_n \sqrt{1-\zeta^2}\, t\right)\right] \qquad (8.13)$$

$$x(t) = \frac{x_c}{\sqrt{1-\zeta^2}} e^{-\zeta \omega_n t} \cos\left(\omega_n \sqrt{1-\zeta^2}\, t + \phi\right) \qquad (8.14)$$

where the phase shift ϕ is given by eqn. 8.15.

$$\phi = \tan^{-1} \frac{-\zeta}{\sqrt{1-\zeta^2}} \qquad (8.15)$$

This solution is plotted in fig. 8-1 for the damping ratio $\zeta = 0$, 0.2, 0.4, 0.6, and 0.8, where the amplitude ratio is $x(t)/x_c$.

This solution can be generalized for any step function by using the same notation as in eqn. 6.9

$$x(t) = x_{FS} - (x_{FS} - x_{IS})\frac{e^{-\zeta \omega_n t}}{\sqrt{1-\zeta^2}} \cos\left(\omega_n \sqrt{1-\zeta^2}\, t + \phi\right) \qquad (8.16)$$

where x_{FS} and x_{IS} are the final and initial states of the step function and $x_c = x_{FS} - x_{IS}$.

Note that, for $\zeta = 0$, the solution is a cosine wave. This solution is included for reference; as noted above, it is not realized by any actual system. The various solu-

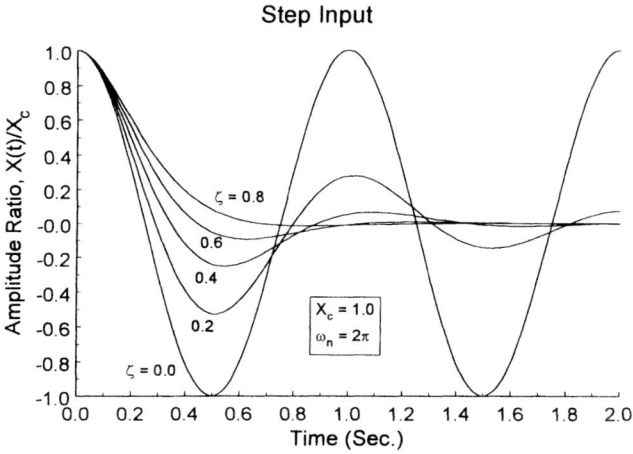

Fig. 8-1 Amplitude ratio, $x(t)/x_c$, for damping various ratios.

tions shown in fig. 8-1 are for a device released from some initial position x_c and coming to rest at a final position of $x(t) = 0$. For $\zeta < 1$, the system overshoots; that is, it goes from the position $x(t) = x_c$ to $x(t) = 0$ and then past that point to some position $x(t) < 0$. While some overshoot may be tolerated, excessive overshoot is unacceptable, therefore, sensors are usually designed to have a damping ratio > 0.7.

If the damping ratio $\zeta > 1$, the roots are real and there is no overshoot. The solution is not oscillatory so it would be inappropriate to express the solution in terms of a frequency such as ω_n. The transient solution will be of the form

$$x_T(t) = c_1 e^{r_1 t} + c_2 e^{r_2 t} \tag{8.17}$$

$$\begin{aligned} r_1 &= -\omega_n\left(\zeta + \sqrt{\zeta^2 - 1}\right) \\ r_2 &= -\omega_n\left(\zeta - \sqrt{\zeta^2 - 1}\right) \end{aligned} \tag{8.18}$$

But, as noted above, it is inappropriate to express the solution in terms of frequencies, so let us define

$$\tau_1 = \frac{1}{\omega_n(\zeta + \sqrt{\zeta^2 - 1})}; \quad \tau_2 = \frac{1}{\omega_n(\zeta - \sqrt{\zeta^2 - 1})} \tag{8.19}$$

where $\tau_2 > \tau_1$ since $\zeta > 1$. Then

$$\tau_1 \tau_2 = \frac{1}{\omega_n^2}; \quad \tau_1 + \tau_2 = \frac{2\zeta}{\omega_n} \tag{8.20}$$

and eqn. 8.7 can be written in the form

$$\left(\tau_1 \tau_2 D^2 + (\tau_1 + \tau_2)D + 1\right)x = x_i(t) \tag{8.21}$$

Using the step function previously employed, where $x_i(t) = 0$ with the initial conditions $x(0) = x_c$ and $Dx(0) = 0$, we can obtain the solution

$$x(t) = x_c \left[\frac{\tau_2}{\tau_2 - \tau_1} e^{-t/\tau_2} - \frac{\tau_1}{\tau_2 - \tau_1} e^{-t/\tau_1}\right] \tag{8.22}$$

Note that if $\tau_2 \gg \tau_1$, then

$$x(t) \approx x_c e^{-t/\tau_2} \tag{8.23}$$

that is, the first-order step response.

EXAMPLE

Consider a step function input with the step from $X_{IS} = 0.00$ to $X_{FS} = -5.00$, where $\omega_n = 0.50$ and $\zeta = 0.30$. Then the phase shift is

$$\phi = \tan^{-1} \frac{-0.3}{\sqrt{1 - (0.3)^2}} = -0.3047$$

and the undamped period is $T_n = 12.5664$ while the damped period is $T_d = 13.1731$. We can explore the solution, using eqn. 8.16, that will be similar to those plotted in fig. 8-1. Note that the cosine term in 8.16 will be zero when the argument of the cosine term,

$$\omega_n\sqrt{1 - \zeta^2}\,t + \phi = i\frac{\pi}{2}$$

where $i = 1, 2, \ldots$. This can be solved for

$$t = \frac{i\frac{\pi}{2} - \phi}{\omega_n\sqrt{1 - \zeta^2}}$$

where time and the corresponding solution are listed in the table below.

Time t	$X(t)$	Condition
0.000	0.000	Initial $X(0) = 0$
3.932	−5.000	$X(t) = X_{FS}$
7.225	−6.773	$X(t) =$ local max.
10.519	−5.000	$X(t) = X_{FS}$
13.812	−4.340	$X(t) =$ local min.
17.105	−5.000	$X(t) = X_{FS}$
20.398	−5.246	$X(t) =$ local max.
23.692	−5.000	$X(t) = X_{FS}$

8.3.2 Ramp Input

A ramp input occurs when $x_i(t) = at$. Following the method of undetermined coefficients, the trial steady-state solution will be $x_S(t) = k_1 + k_2 t$ and substitution into eqn. 8.7 yields $k_1 = -2a\zeta/\omega_n$ and $k_2 = a$. The steady-state solution is therefore

$$x_S(t) = a\left(t - \frac{2\zeta}{\omega_n}\right) \tag{8.24}$$

where a is the constant slope of the input.

A ramp input is shown as the upper line in fig. 8-2 and the complete solution, transient plus steady-state, is plotted as the lower line. In this case, $\omega_n = 0.5$, $\zeta = 0.7$, and $a = 1$.

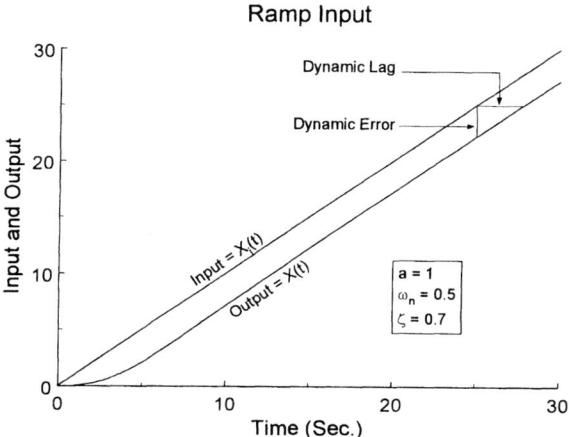

Fig. 8-2 Ramp input and output.

The dynamic error, after the transient has faded, is

$$\varepsilon_D = x_S(t) - x_i(t) = -\frac{2\zeta a}{\omega_n} \tag{8.25}$$

and the dynamic lag is the value of Δt that satisfies the equation $x_i(t) = x_S(t + \Delta t)$; this yields $\Delta t = 2\zeta/\omega_n$. The vertical line in fig. 8-2 at $t = 25$ s is the dynamic error and the horizontal line is the dynamic lag.

EXAMPLE

The system has $\omega_n = 0.50$ and $\zeta = 0.30$. The input is a ramp with a ramp rate of 2. Then the dynamic lag $= 2\zeta/\omega_n = 1.20$ and the dynamic error $= -2\zeta a/\omega_n = -2.40$.

8.3.3 Sinusoidal Input

For the sinusoidal input, $x_i(t) = A_i \sin \omega t$, we will develop only the steady-state solution. The trial form of the solution is $x_S(t) = k_1 \sin \omega t + k_2 \cos \omega t$. Substitution into eqn. 8.7 yields

$$x_S(t) = \frac{A_i}{\sqrt{\left(1 - \left(\frac{\omega}{\omega_n}\right)^2\right)^2 + 4\zeta^2 \left(\frac{\omega}{\omega_n}\right)^2}} \cos(\omega_n \sqrt{1 - \zeta^2} t + \phi) \tag{8.26}$$

where the phase shift ϕ is given by eqn. 8.27.

$$\phi = \tan^{-1}\left(-\frac{2\zeta \omega/\omega_n}{1 - \left(\frac{\omega}{\omega_n}\right)^2}\right) \tag{8.27}$$

A sinusoidal input, with $A_i = 1$, is shown in fig. 8-3. The output is shown for $\omega = \omega_n = 0.25$ and $\zeta = 0.4$. Note that the output lags behind the input and that the amplitude of the output is greater than that of the input. Phase lag is normal; one should expect the sensor output to lag behind the input.

In a first-order system, we saw that the output amplitude was always less than or equal to the input amplitude. It could never exceed the input. Evidently it is possible for the output amplitude to exceed the input amplitude in a second-order system. The amplitude ratio is the ratio of the output amplitude A_o to the input amplitude A_i, and this is given by

$$\frac{A_o}{A_i} = \frac{1}{\sqrt{\left(1 - \left(\frac{\omega}{\omega_n}\right)^2\right)^2 + 4\zeta^2 \left(\frac{\omega}{\omega_n}\right)^2}} \tag{8.28}$$

Equation 8.28 is plotted as a function of the normalized frequency (ω/ω_n) in fig. 8-4. The top curve is for $\zeta = 0.1$ and curves below it represent $\zeta = 0.2, 0.4, 0.6, 0.8, 1.0$, and 1.2.

Figure 8-4 shows that output amplification greater than 1 is possible for second-order sine response if the damping ratio is less than 0.707. Amplitude peaks, owing to a resonance phenomenon which occurs when the input frequency is very nearly equal to the natural system frequency and the damping ratio is sufficiently low.

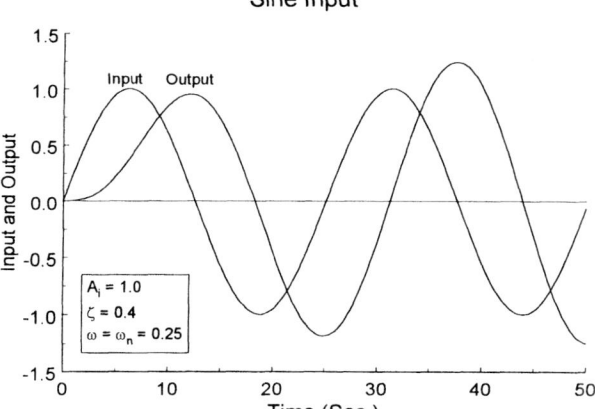

Fig. 8-3 Sinusoidal input and output.

$$\lim_{\frac{\omega}{\omega_n} \to 0} \frac{A_o}{A_i} = 1$$

$$\lim_{\frac{\omega}{\omega_n} \to \infty} \frac{A_o}{A_i} = 0$$
(8.29)

The phase shift given in eq. 8.27 is plotted in fig. 8-5 as a function of the normalized frequency ω/ω_n. When the normalized frequency is 0.5, the top curve is for a damping ratio of 0.1. The phase lag is always negative and ranges from 0° to −180°.

EXAMPLE

The system has $\omega_n = 0.50$ and $\zeta = 0.30$. The input is frequency $\omega = 0.40$ and amplitude $= 5.00$. Then the steady-state output has an amplitude of 8.33 with a phase shift of −53.13 degrees.

8.4 Application to Sensors

A wind vane is a second-order system because it has two energy storage reservoirs. It can store kinetic energy in the angular momentum of the vane and it can store potential energy in the position of the vane relative to the wind vector. The latter can be appreciated by a simple thought experiment. If the wind speed and direction were constant a wind vane would quickly align itself with the wind vector and remain in a stable position. It would take application of some force to move it away from this stable position and it would return to the normal position when released. This indicates that there is potential energy stored in its position relative to the wind vector.

The equation for a wind vane was mentioned in section 7.1.1.2. There eqn. 7.7, divided by N the aerodynamic torque, was

Sine Input

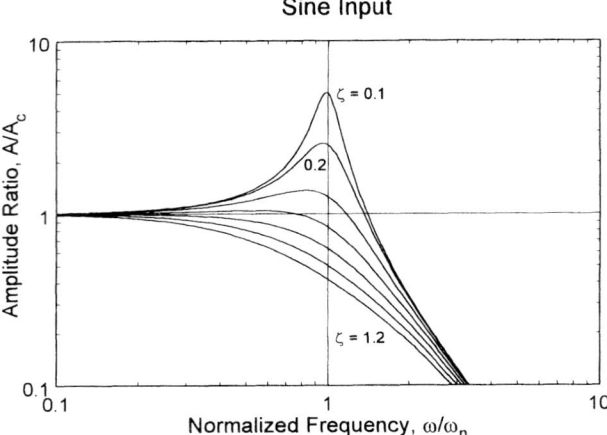

Fig. 8-4 Amplitude ratio versus normalized frequency for $\zeta = 0.1, 0.2, 0.4, 0.6, 0.8, 1.0$, and 1.2.

$$\frac{I}{N}\frac{d^2\theta}{dt^2} + \frac{R}{V}\frac{d\theta}{dt} + \theta = \theta_i \qquad (8.30)$$

and this equation can be written as

$$\frac{1}{\omega_n^2}\frac{d^2\theta}{dt^2} + \frac{2\zeta}{\omega_n}\frac{d\theta}{dt} + \theta = \theta_i \qquad (8.31)$$

where $\omega_n^2 = N/I$ and $\zeta = (R/2V)\sqrt{N/I}$. In these equations, N, (aerodynamic torque per unit angle) I (vane moment of inertia), and R (aerodynamic vane lever arm) are assumed to be constants for a given vane since they represent physical parts of the vane. The $\omega_n = f(V)$ but ζ is not a function of V. But the wind speed V is not a constant except in the wind tunnel.

In the atmosphere, the wind speed is a variable; therefore ω_n is not a constant, and that violates the assumption that the parameters are time-invariant. Then the solutions in sect. 8.3, especially the sine input, will not be exactly correct. Normally the wind speed is a fluctuation about the average, so we can use the constant-parameter solutions, above, as an approximation. Then $\omega_n = 2\pi V/\lambda_n$, where λ_n is called the undamped wavelength which is a constant for a particular wind vane. We will use V as the average wind speed. Then we can use wavelength ratio in place of frequency ratio

$$\frac{\omega}{\omega_n} = \frac{2\pi V/\lambda}{2\pi V/\lambda_n} = \frac{\lambda_n}{\lambda} \qquad (8.32)$$

in figs. 8-4 and 8-5. We have used λ as the wind input wavelength.

The damped natural frequency and the damping ratio of a wind vane can be determined from the step function response (see fig. 8-1) in a wind tunnel. The tunnel flow is held constant at speeds well above the threshold speed of the vane. To perform a step-function test, the initial vane deflection must be $\leq 10°$ from the tunnel centerline. Beyond $10°$ deflection the vane will likely be in aerodynamic stall where the lift no longer increases with increasing deflection angle from the wind flow and the vane

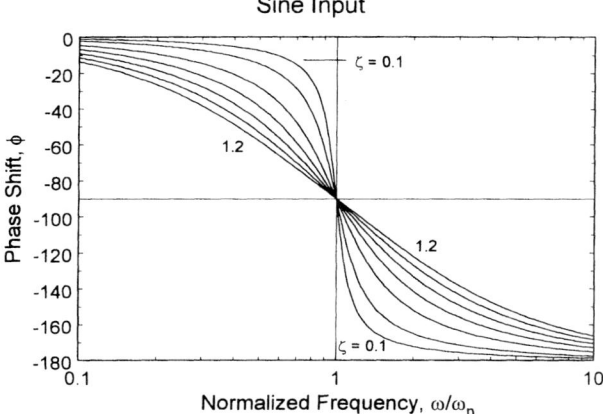

Fig. 8-5 Phase shift as a function of normalized frequency for $\zeta = $ 0.1, 0.2, 0.4, 0.6, 0.8, 1.0, and 1.2.

response will be nonlinear and therefore not predictable from eqn. 8.14. The test is usually performed at two speeds: 5 and $10\,\mathrm{m\,s}^{-1}$. Then the dependence on wind speed can be observed.

We used a number of terms that have developed in this chapter, such as the time constants τ_1, τ_2, and the damping ratio ζ.

Damped natural frequency, $\omega_d = \omega_n\sqrt{1-\zeta^2}$
Damped natural period, $T_d = 1/f_d = 2\pi/\omega_d$
Undamped natural frequency, $\omega_n = 2\pi f_n = 2\pi/T_n$
Undamped natural period, $T_n = 1/f_n = 2\pi/\omega_n$.

Now there are some special terms related to wind vanes because of their dependence on the wind speed:

Damped natural frequency, $\omega_d = 2\pi V/\lambda_d$
Damped natural wavelength, $\lambda_d = T_d V = V/f_d$
Undamped natural frequency, $\omega_n = 2\pi V/\lambda_n$
Undamped natural wavelength, $\lambda_n = T_n V = V/f_n$.

Temperature sensors are portrayed as simple first-order sensors with a single energy storage reservoir – the thermal mass of the sensor itself. While it is possible to build a temperature sensor with a single thermal mass, it is difficult. Typically, the raw sensor is enclosed in a protective shield and, frequently, there are shields used to protect the sensor from mechanical damage and from moisture. Unless these protective shields are fused together, which is not usually the case, there are two or more coupled thermal masses and then the sensor is a second-order or possibly even higher-order system. If there are just two thermal masses, the resulting second-order system will have a damping ratio greater than unity and can be represented by eqn. 8.21. The step function solution would be eqn. 8.22.

8.5 Experimental Determination of Dynamic Performance Parameters

The dynamic performance parameters, ζ and T_d (or ω_d or ω_n), can be determined from a step-function test. When $\zeta < 0.6$, one can measure the overshoot ratio and the damped period. The overshoot ratio is the ratio of the amplitudes of the two successive deflections of a sensor as it oscillates about the equilibrium position. This can be illustrated as in fig. 8-7. Then ζ and ω_n can be found from two equations.

$$\zeta = \frac{1}{\sqrt{\left(\frac{\pi}{\ln |X_{n+1}/X_n|}\right)^2 + 1}} \tag{8.33}$$

$$\omega_n = \frac{2\pi}{T_d\sqrt{1-\zeta^2}} \tag{8.34}$$

Equation 8.33 is plotted in fig. 8-6.

For example, a step-function test is shown in fig. 8-7 that illustrates some measurements which could be used to get the period and overshoots. Use sequential maximum (minimum) and then minimum (maximum) amplitude. Table 8-1 is an example of the use of experimental data, and corresponds to fig. 8-7.

EXAMPLE

Determine ζ and ω_n from the data plotted in fig. 8-7. We can use maximum and minimum to get ζ. Use points 1 and 3. $X_n = 0-5 = -5$ and $X_{n+1} = 6.55-5 = 1.55$, so $|X_{n+1}/X_n| = 0.310$. Then $\zeta = 0.35$.

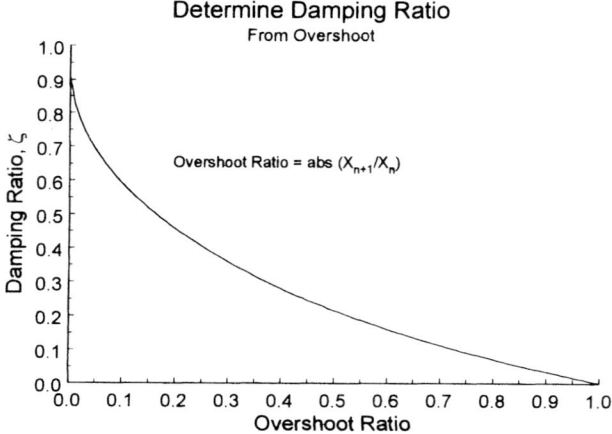

Fig. 8-6 Damping ratio as a function of overshoot for a second-order system with step input.

Dynamic Performance Characteristics, Part 2

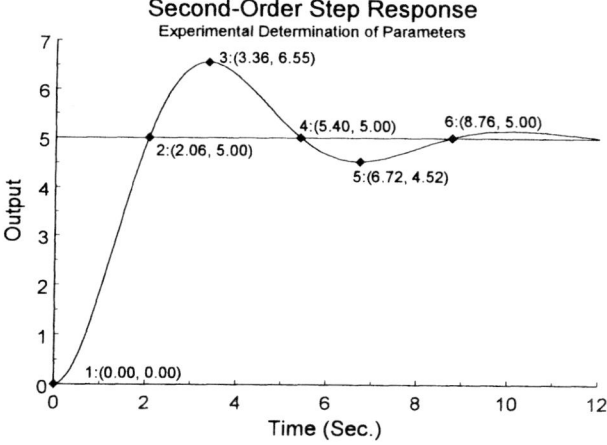

Fig. 8-7 Experimental determination of damping ratio and damped period.

$$\zeta = \frac{1}{\sqrt{\left(\frac{\pi}{\ln 0.310}\right)^2 + 1}} = 0.35$$

If use points 3 and 5, then $X_n = 6.55-5 = 1.55$ and $X_{n+1} = 4.52-5 = -0.480$ so $|X_{n+1}/X_n| = 0.310$. Again, $\zeta = 0.35$.

$$\omega_n = \frac{2\pi}{6.70\sqrt{1-(0.35)^2}} = 1.00$$

Since we have estimated ζ, we can obtain ω_n if know the period of the oscillation. Using points 2 and 6, the period $T_d = 8.76-2.06 = 6.70$ seconds.

These equations do not always work so well. Problems occur when the damping ratio > 0.6; then the amplitude of the oscillations drops so quickly that is difficult to measure the damping ratio and the period. There are many systems that are approximately linear but contain some nonlinearity or time-varying parameters, such as wind vanes. Then the period may be measured in one section but the trace may show a different value in another section. The methods outlined above are then not useful. In addition, when the damping ratio > 1.0, the system is usually defined with dual time

Table 8-1 Experimental data used for dynamic performance characteristics.

Item	Time (s)	Amplitude	Comments
1	0.00	0.00	Initial conditions
2	2.06	5.00	
3	3.36	6.55	Use amplitude at 1 & 3 to get ζ
4	5.40	5.00	
5	6.72	4.52	Use amplitude at 3 & 5 to get ζ
6	8.76	5.00	Use period from 2 & 6 to get ω_n

164 Meteorological Measurement Systems

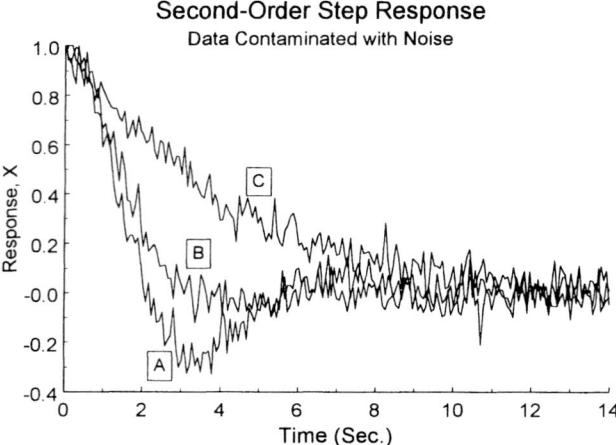

Fig. 8-8 Three step functions with noise added.

constants, τ_1 and τ_2, instead of ω_n and ζ. In any case, when noise contaminates the signal, which is the usual situation, measurement of the parameters is subject to error. There is a numerical algorithm that works reasonably well in many cases. It is the downhill simplex method (Press et al., 1992). Given that a data set of time and amplitude form a second-order system with step-function input, such as the data plotted in fig. 8-1, the object of the algorithm is to find the optimum pair of parameters, T_d and ζ, or τ_1 and τ_2. Since the algorithm uses all of the data in the set, it is less sensitive to noise, and small nonlinearities, and will work with a wide range of parameters (with ζ from almost 0 to some point much greater than 1.0).

In one implementation, the program reads a file of data that lists times and amplitudes $(t, X(t)$ for each time recorded). For example, in fig. 8-8 there are three traces. Suppose one, labeled "B", was read. The program requires three sets of starting values in (ζ, T_d) space to form a simplex, or triangle. In this case, the values were ζ, $T_d = (0.4, 8.7)$, $(0.6, 9.1)$ and $(0.5.8.1)$ respectively. Then the optimum was found at $\zeta = 0.72$ and $T_d = 8.82$ (and $\omega = 1.02$), whereas the correct values should have been $\zeta = 0.70$ and $T_d = 8.80$ (and $\omega_n = 1.00$). The program worked about as well with the "A" and "C" files. Note that this program worked with $\zeta = 2.0$ (curve "C") even though the curve was not oscillating.

QUESTIONS

1. A wind vane has an initial deflection of 10° in a wind tunnel and is released at time $t = 0$. It has a damping ratio of 0.5 and an undamped natural frequency of 1.0 rad s^{-1}. At what time will the output first equal zero? Assume the vane is aligned with the tunnel such that $x_I(t) = 0$.

Hint: In this case, $x(t) = x_T(t)$ and is given by eqns. 8.14 and 8.15. Then $x(t)$ will be equal to zero when $\omega_n \sqrt{1 - \zeta^2} t + \phi = \pi/2$.

2. Identify the energy storage reservoirs of a wind vane.

3. How many dynamic performance parameters are required for a second-order system?

4. Both the cup anemometer and the wind vane are rotating mechanical systems. Why is one a first-order system and the other a second-order system?

5. Let $\beta = \omega/\omega_n$, then find an expression for β in terms of ζ (other than $\beta = 0$) that maximizes the amplitude ratio when the input is a sinusoid.

6. Step input: $x(0) = 2$, $x_c = 4$, $\omega_n = 0.25$. Find $x(8)$ for $\zeta = 0.5$ and for $\zeta = 1.5$.

7. Why is it impossible to build a second-order sensor, such as a wind vane, with a damping ratio of zero?

8. Consider a silicon-diaphragm aneroid sensor. What physical properties determine its dynamic response? What model would be most appropriate, 1st or 2nd order? Identify the energy storage reservoir(s). Would you expect its frequency response to limit meteorological applications? Explain?

9. Test a wind vane in a wind tunnel to determine the dynamic performance characteristics of the vane. The test should be conducted in accordance with the American Society for Testing and Materials (ASTM) Standard Test Method for Determining the Dynamic Performance of a Wind Vane (1989). This test method specifies that the wind tunnel should be capable of producing constant wind speeds (± 0.2 m s^{-1}) in the range 0 to 10 m s^{-1}. Set the tunnel speed to 5 m s^{-1}, deflect the vane to an initial position of 10°, and release it at time $t = 0$. Repeat the test with an initial deflection of $-10°$. Repeat the above tests at a speed of 10 m s^{-1}. Figure 8-7 shows a typical output. If you do not have access to a wind tunnel, analyze Fig. 8-7.

Determine the vane overshoot ratio, the damping ratio, the delay distance, the undamped natural wavelength, and the damped natural wavelength. Why should the initial deflection be $\leq 10°$?

10. Show that $\zeta \neq f(V)$ in a wind vane.

BIBLIOGRAPHY

Finkelstein, P.L., 1981: Measuring the dynamic performance of wind vanes. *J. Applied Meteor.*, 20, 588–594.

Kristensen, L., and D.H. Lenschow, 1988: The effect of nonlinear dynamic sensor response on measured means. *J. Atmos. Oceanic Technol.*, 5, 34–43.

MacCready, P.B., Jr, 1970: Theoretical considerations in instrument design. *Meteor. Monogr.*, 11, 202–210.

MacCready, P.B., Jr, and H.R. Jex, 1964: Response characteristics and meteorological utilization of propeller and vane wind sensors. J. Applied Meteor., 3(2), 182–193.

Press, W.H., Teukolsky, S.A., Vetterling, W.T., and Flannery, B.P., 1992: *Numerical Recipes in FORTRAN*. Cambridge University Press, Cambridge, 963 pp.

9

Precipitation Rate

Accurate rainfall measurements are required, usually over broad areas because of the natural variability of rain. Coverage of a large area can be achieved using many distributed point measurement instruments or a remote sensor with large areal coverage, such as radar, or both.

This chapter describes several methods for measuring precipitation, both liquid and frozen types. Point measurements, e.g., rain gauges, are emphasized although a section on weather radar is included because this is a very important method of estimating precipitation.

9.1 Definitions

Precipitation rate could be specified as the mass flow rate of liquid or solid water across a horizontal plane per unit time: M_w in kg m^{-2} s^{-1}. Water density is a function of temperature but that can be ignored in this context; then the volume flow rate, or precipitation rate, becomes $R = M_w/\rho_w$ in m s^{-1} or, more conveniently, in units of mm hr^{-1} or mm day^{-1}. Precipitation rate is the depth to which a flat horizontal surface would have been covered per unit time if no water were lost by run-off, evaporation, or percolation. Precipitation rate is the quantity used in all applications but, in many cases, the unit of time is not specified, being understood for the application, commonly per day or per storm period. Some gauges measure precipitation, rain, snow and other frozen particles, while others measure only rain.

9.2 Methods of Measurement

Rainfall can be measured using point measurement techniques which involve measuring a collected sample of rain or measuring some property of the falling rain such as its optical effects. The other general technique is to use remote sensing, usually radar, to estimate rainfall over a large area. Both ground-based and space-based radars are used for rain measurement.

9.2.1 Point Precipitation Measurement

A precipitation gage (US) or gauge (elsewhere) could be a simple open container on the ground to collect rain, snow, and hail. However, this is not a practical method for estimating the amount of precipitation because of the need to avoid wind effects, enhance accuracy and resolution, and make a measurement representative of a large area. These issues will be discussed in sect. 9.2.1.6.

9.2.1.1 Accumulation Gauges

Accumulation gauges collect precipitation and hold it, usually as water, until it is emptied either manually or automatically. These gauges may be recording or non-recording. In the latter case, the depth of accumulation is recorded manually when the gauge is periodically emptied.

An open container, the simplest possible accumulation gauge, is shown in fig. 9-1(a). The measurand is R, the precipitation rate, and the output is the depth of water (hatched area) in the gauge, h_1 in m.

$$h_1(t) = \int_0^t R(g)\mathrm{d}g \qquad (9.1)$$

The transfer function is an integral since the open bucket is a near-perfect realization of the mathematical operation of integration. It falls short of perfection only in that evaporation causes $h_1(t)$ to decrease slightly with time when $R = 0$. A dummy variable of integration, g, is used to avoid confusion since the integration is from the

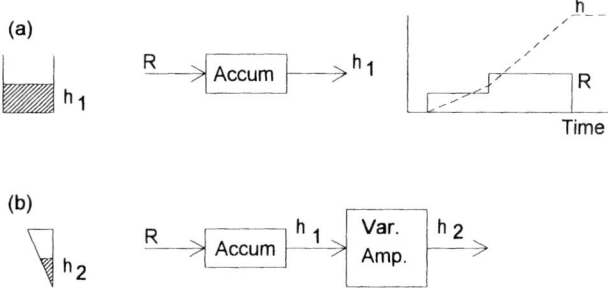

Fig. 9-1 Simple accumulation gauges.

last time the gauge was emptied, $t = 0$, to the present time, t. Integration is continuous from the time the bucket is emptied and the output $h_1(t)$ is a continuous variable, as shown on the right-hand side of fig. 9-1(a). Rain rate is shown as zero, a constant low rate, a constant high rate, and then returning to zero. Water depth in the gauge remains zero until R goes from zero to low, then increases as a ramp function. When R goes high, $h_1(t)$ is still a ramp but with increased slope. When R returns to zero, $h_1(t)$ stays constant, apart from evaporation. Depth should be measured at periodic intervals since the rain rate is estimated by differentiation of $h_1(t)$:

$$R_1 = \frac{dh_1(t)}{dt} \cong \frac{h_1(t + \Delta t) - h_1(t)}{\Delta t} \qquad (9.2)$$

High-resolution measurement of h_1 can be quite difficult with a simple bucket gauge because it is difficult to determine h_1 with sufficient accuracy, so various methods are used to enhance resolution. One common solution, the fence post gauge, is shown in fig. 9-1(b). The cross-section is rectangular. One face is vertical and the opposite face leans out at some angle. A scale is printed on the side. A small amount of rain falling into this gauge makes h_2 larger than the equivalent h_1, thus easier to read. This is variable amplification since the amplification ratio decreases as the gauge fills. This type of gauge has high sensitivity for small rain amounts and yet retains a fairly large capacity.

$$h_2 = \sqrt{2h_1 H} \qquad (9.3)$$

where h_1 is the depth that would have been obtained if the gauge area were uniform with depth, and H is the maximum gauge depth.

Another way of improving resolution is shown in fig. 9-2(a). Precipitation is caught in the large gauge catch area, A_c, then drains into the small-diameter inner tube that has a measurement area A_m. Since A_m is constant with depth, the amplification factor is constant, so $h_3 = A_c h_1 / A_m$. Typically, $A_c = 10 A_m$. An expanded scale is provided and it is possible to estimate h_1 to 0.25 mm (0.01 inch). When the inner cylinder fills, the overflow is contained in the outer cylinder so the capacity is very large. An 8-inch diameter gauge is commonly used in the US and is read to 0.01 inches. As above, this is a non-recording gauge so a person must read it periodically, usually daily.

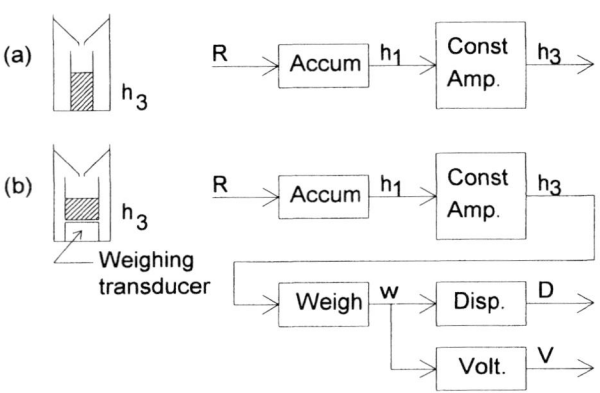

Fig. 9-2 More accumulation-style gauges.

Gauge output can be recorded automatically if a transducer is provided to convert the gauge output h_3 to a displacement or voltage. Figure 9.2(b) shows an accumulation gauge with such a transducer, a device to weigh the inner cylinder and its water content. This is commonly called a weighing gauge; in its original form, springs and levers converted the weight to displacement of a pen used to record gauge output on a strip of paper wound over a rotating drum. In more modern forms, the displacement is converted to a voltage that can be sampled, digitized, and recorded by a data logger. The weight of the inner cylinder is $w = (m_c + m_w)g = m_c g + A_c \rho_w g h_1$, where m_c is the cylinder mass, m_w is the water mass, ρ_w is the density of water, and g is the acceleration due to gravity. The offset due to the weight of the inner bucket is of little concern since one must take the derivative of the output to get an estimate of the rain rate and that makes the offset vanish. Following eqn. 9.2, a rain rate estimate can be obtained from $R_1 = dw/dt = A_c \rho_w g \, dh_1/dt$.

All of the accumulation gauges described so far have to be manually emptied periodically but there are some accumulation gauges that empty themselves automatically when they get full. These are the pressure gauge, the siphon gauge, and the tipping bucket gauge.

Pressure gauge. Water depth in a gauge such as the one shown in fig. 9-1(a) could be measured with a gauge-type (differential with respect to ambient atmospheric pressure) pressure sensor connected via tubing to the bottom of the gauge. Aneroid sensors are available with appropriate pressure ranges. Transducers are used to produce voltage outputs. To make this gauge fully automatic, some provision must be made to empty the gauge when it fills up. In one version, a microprocessor monitors the pressure transducer to detect the "gauge full" condition. When full, the microprocessor turns on a pump to empty the gauge. In cold weather, the gauge must have some antifreeze added to prevent freezing. After the microprocessor pumps out the gauge, it pumps some antifreeze into it.

EXAMPLE

Let the gauge-type pressure sensor used with a rain gauge have a range of 0 to 2000 Pa (20 hPa), very much less than the range needed for absolute atmospheric pressure applications. Signal output is a voltage proportional to pressure. Assume it has an inaccuracy of 0.1% of full scale, or 2 Pa. The maximum depth of water that can be allowed in the gauge is given by $\Delta h_{max} = p/(\rho_w g) = $ 2000 Pa/(1000 kg m^{-3} 9.8 m s^{-2}) = 0.2 m = 200 mm. The inaccuracy in the pressure sensor is 2 Pa which translates to an inaccuracy = Δh = 0.2 mm height. If the aneroid sensor has any temperature sensitivity (it usually does), the signal output will exhibit a small diurnal cycle when there is no rain.

Siphon gauge. Water depth in the inner cylinder of a gauge (see fig. 9-2(a)) could be measured with a capacitive sensor and the cylinder could be emptied using the siphon effect. A combination capacitive sensor and siphon is shown in fig. 9-3. In the siphon gauge, rain that collects in the funnel drains into a measuring chamber. When the chamber is full, the addition of more water starts a siphon that drains the water out of the chamber. The measuring chamber is equipped with a capacitive transducer that senses the depth of water in the gauge. A capacitive depth sensor is a pair of vertically oriented coaxial cylinders coated with an insulator such as Teflon. There are holes in the outer cylinder to allow water to flow in and out. The inner and

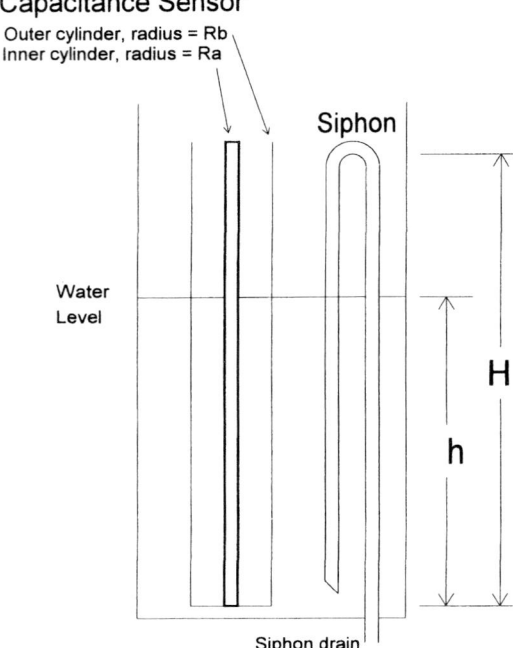

Fig. 9-3 Capacitance depth sensor and automatic siphon.

outer cylinders comprise a capacitor whose dielectric is air when empty and is water when full. Capacitance C of a cylinder filled with air is given by

$$C = \frac{2\pi\varepsilon_0 H}{\ln r_b/r_a} \qquad (9.4)$$

where ε_0 = free space permittivity = $8.8542 \times 10^{-12}\,\mathrm{C^2\,N^{-1}\,m^{-2}}$, H = cylinder length in m, r_a and r_b = radius of the inner and outer cylinders, respectively, and C is the capacitance in F (farads). When a dielectric is present, such as water, this equation is modified by the dielectric constant κ (dimensionless) of water. If the water height is some fraction $f (0 \leq f \leq 1)$ of H ($h = fH$), then eqn. 9.4 becomes

$$C = \frac{2\pi\varepsilon_0 H}{\ln r_b/r_a}\left[1 + f(\kappa - 1)\right] \qquad (9.5)$$

where $\kappa = 78.5$ at 25°C and is a function of temperature, as shown in fig. 9-4. The gauge will include an electronic circuit to convert capacitance to frequency and, usually, frequency to voltage in the form $V = a_0 + a_1 h$, where h = depth of water in the gauge.

The gauge output is much like that of a weighing gauge, in that the signal output increases linearly with accumulated rain. But when the gauge is full the siphon starts and automatically empties the gauge.

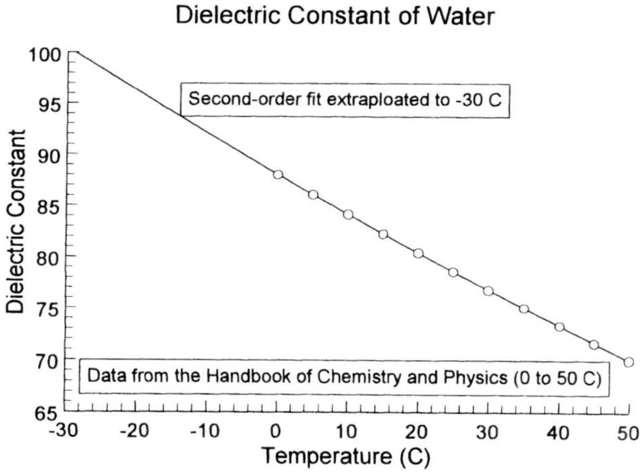

Dielectric Constant of Water

Fig. 9-4 Dielectric constant of water as a function of temperature.

Siphon gauges have no moving parts and could be used on buoys or ships. The capacitance depth sensor can be located in the center of the gauge to minimize the effect of tilt. It loses any rain that falls during the period when the siphon is emptying the gauge, on the order of 30 s. A siphon gauge must be kept very clean as dirt or oil could affect the siphon process, and it must be heated to protect the siphon chamber and the capacitive transducer from freezing.

Tipping-bucket gauge. In this type of gauge, a twin metallic or plastic bucket resting on a knife-edge support is mounted under the collecting funnel; see fig. 9-5. The incoming water falls into one of the buckets. When the bucket is full, its center of gravity is outside the point of support and it tips, dumping the water collected and bringing the other bucket into position to collect water. The bucket volume, $V_B = A_C \Delta h$ where, as before, A_C is the gauge collection area. Both 8-inch and 12-inch tipping-bucket gauges are used in the US. Δh is the depth increment of one bucket tip. Each tip can correspond to $\Delta h = 0.01$ in, 0.1 mm, 0.2 mm, or 0.25 mm of rain, depending upon the gauge design. When the bucket tips it momentarily closes a reed or mercury switch. The raw output is a brief contact closure whenever the bucket tips, as shown in fig. 9-6. The time required to fill a bucket, T_B, is given by

$$A_C \int_0^{T_B} R(t) \mathrm{d}t = V_B \tag{9.6}$$

When $R = $ constant, then $A_C R T_B = V_B$ and $T_B = \Delta h / R$.

EXAMPLE

Let the rain rate increase with time as shown in the top panel of fig. 9-6. The left and right buckets fill at an ever-increasing rate as the rain rate increases. When a bucket fills, it tips and brings the other bucket into position. The tipping action triggers a relay that is momentarily closed. A microprocessor in a data logger counts the number of times the relay is closed during some time period clocked

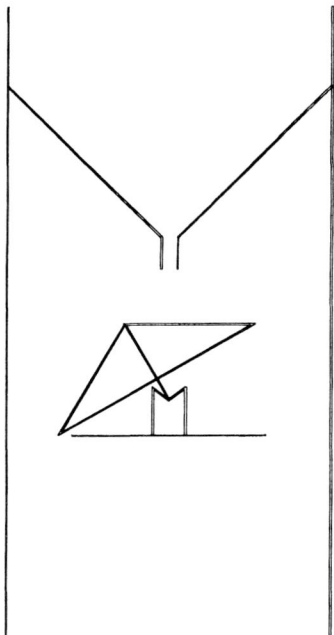

Fig. 9-5 Tipping-bucket gauge.

by the microprocessor. Let the time period be 5 minutes = 300 s and let $\Delta h = 0.2$ mm. Then, if there are two contact closures in a 5-minute period, the estimated rain rate must be $2\Delta h = 0.4$ mm/5 min = 4.8 mm hr^{-1} and that is the value of the points plotted in the top panel of fig. 9-6 for the 5th, 6th, and 7th 5-minute periods (elapsed time of 20 through 35 minutes). Thus the rain rate estimates exhibit both temporal and amplitude granularity. Temporal granularity is due to the time period over which pulses are counted, five minutes in this case. Amplitude granularity is due to the bucket size, expressed as Δh, and the time period used.

The frequency of contact closure, $f_c = 1/T_B = R/\Delta h = A_c R/V_B$, is proportional to the rain rate. Since it varies continuously with rain rate, it follows that the rain gauge output signal is an analog signal.

This gauge is easily automated since it is only necessary to count the pulses over some period of time and multiply by Δh to obtain the rainfall estimate R_1. The process of counting contact closures over a precise time base is a form of analog-to-digital conversion.

EXAMPLE

Let 50 1-mm diameter drops fall every second into a tipping-bucket rain gauge with a 12-inch diameter orifice. Rain rate is $R = NV_D/A_c$, where N is the number of drops and V_D is the drop volume. Then

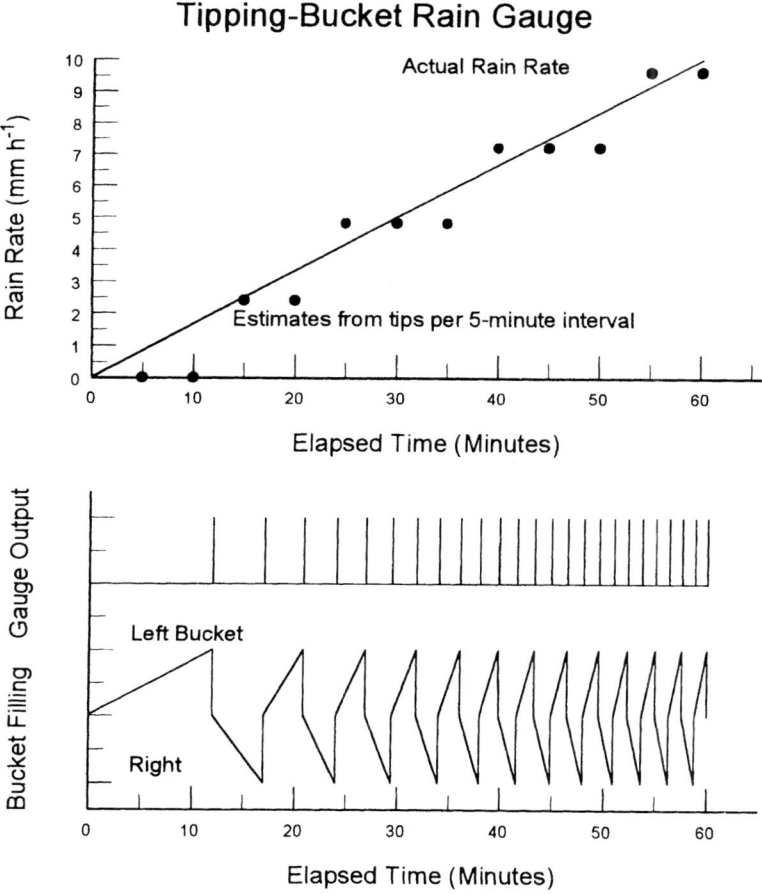

Fig. 9-6 Tipping-bucket gauge with increasing rain rate. Each tip represents 0.2 mm rain.

$$R = \frac{N\frac{4}{3}\pi\left(\frac{d}{2}\right)^3}{A_c} = \frac{50\frac{4}{3}\pi\left(\frac{10^{-3}}{2}\right)^3}{\pi\left(\frac{12 \times 0.0254}{2}\right)^2} = 3.59 \times 10^{-7}\,\mathrm{m\,s^{-1}} = 1.29\,\mathrm{mm\,hr^{-1}}$$

$$f_c = \frac{1}{T_B} = \frac{R}{\Delta h} = \frac{1.29\,\mathrm{mm\,hr^{-1}}}{0.2\,\mathrm{mm}} = 6.5\,\mathrm{tips\,hr^{-1}}$$

and the bucket will tip at the rate of 6.5 tips hr^{-1} on the average if the bucket tips for each 0.2 mm.

This gauge is subject to errors due to under-reporting of rainfall when the rain rate is too light (water evaporates before filling the bucket) or too heavy. In the latter case, water splashes out of the bucket and is not collected during the time required to tip. In a well-designed gauge this source of error is small, about 1% at a rainfall rate of 50 mm hr^{-1}. An 8-inch tipping bucket gauge may underestimate rainfall when the rain rate < 10 mm hr^{-1} and when the rain rate > 200 mm hr^{-1}.

When sufficient power is available, a heater is sometimes installed in a tipping-bucket gauge to melt frozen precipitation that would otherwise simply clog the gauge. This is not an adequate solution for snow measurement since wind will cause greater snow loss than rain loss and the heater will increase evaporative loss.

While tipping-bucket gauges are often heated, it is not necessary to do so. Without a heater, the tipping bucket gauge simply fails to measure frozen precipitation but the gauge itself is not damaged by cold weather. When the air temperature gets above 0°C, the frozen precipitation will melt and cause rain to be reported (delayed).

9.2.1.2 Snow Gauges

Even when heated, tipping-bucket and weighing gauges do not work well for snow; the snow can plug the funnel and snow is more susceptible to be blown into or out of the gauge by strong winds. A snow ruler is used to measure the snow depths in a variety of representative locations. Snow gauges attempt to measure the liquid-water content using several techniques which include collecting snow in an open gauge and melting it, measuring the weight of snow over a snow pillow, or detecting the water attenuation of gamma radiation.

An acoustic transmitter–receiver can be used to measure snow depth. The device, mounted up to 10 m above the ground, transmits an ultrasonic pulse towards the ground and measures the time required to receive the echo. Since the speed of sound in air is a function of the air density, it is necessary also to measure air temperature. The measurement range is 0.6 to 10 m with a typical inaccuracy of 2.5 cm. Another device used to measure snow depth is the snow pillow; an air-filled, nearly flat "balloon" on the ground that converts the weight of snow above to pressure that is measured with a pressure sensor.

9.2.1.3 Optical Rain Gauge

A rain gauge can be designed using optical techniques to detect the passage of rain and snow through a beam of light; one implementation is shown in fig. 9-7. The light source is an infrared light-emitting diode (LED). It is modulated, that is, powered by a voltage source that oscillates at about 50 kHz. After being focused by the lens, the light passes through 1 m of atmosphere and is then focused on a photodetector. The signal from the photodetector is amplified and filtered in the box labeled "ASC" (analog signal conditioning) before being digitized and sent to the microprocessor. While the photodetector is shielded to keep direct sunlight out, it is impossible to exclude all ambient light. Since the source is modulated at a high frequency, it is possible to use a high-pass filter to eliminate ambient light.

Precipitation particles (rain, snow, etc.) passing through the beam cause scintillation (rapid fluctuation) of the light received at the detector. When a drop falls through the beam the intensity of light detected fluctuates slightly. The amplitude and frequency of scintillation is a function of the drop size, the fall speed of the drop, and of

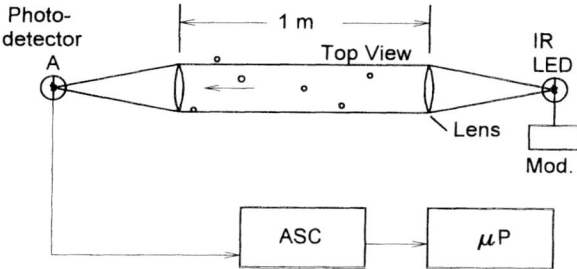

Fig. 9-7 Top view of an optical precipitation gauge using an infrared LED as the source.

the number of drops in the beam at any instant. If there were only a single drop in the beam at a time, a large drop would pass quickly through the beam and cause a large, rapid fluctuation in detected light amplitude. A small drop would have a lower fall speed and so would pass through the beam more slowly, producing smaller, lower frequency scintillation. The terminal fall speed is a monotonic function of the drop diameter or drop volume (see eqn. 9.11). When there are multiple drops in the beam, each produces a light fluctuation as it enters and leaves the beam. Thus the scintillation frequency is a function of the drop volume and of the number of drops in the beam, and the product of drop volume and number of drops in the beam or drops falling through per unit time is the rain rate.

The instrument calculates the spectral variance of the scintillation in two wave bands: 1 kHz to 4 kHz and 25 Hz to 250 Hz. The magnitude of the fluctuations in the first band is proportional to the rain rate, and the ratio of the signal strength in the two bands is used to discriminate between rain and snow. Snow particles fall more slowly, thus producing lower frequency fluctuations. The ratio can also be used to compensate for signal strength degradation due to source changes, dirt on the lens, etc. Failure of the source or blockage of the beam can also be detected and reported.

The device illustrated in fig. 9-7 is capable of detecting the presence of precipitation, determining whether it is rain or snow, and measuring the precipitation rate. Not shown in fig. 9-7 is a horizontal slot between the lens and the photodetector that makes the gauge sensitive only to the vertical component of the precipitation particle velocity. This makes the optical gauge insensitive to horizontal winds that contribute error to tipping-bucket and weighing gauges. One implementation is capable of measuring rates from 0.01 mm hr^{-1} to 3000 mm hr^{-1} for rain and from 0.005 mm hr^{-1} to 300 mm hr^{-1} for snow water equivalent.

Optical rain gauge simulation (for more advanced study). It is possible to simulate an optical rain gauge (ORG), in a vastly simplified way, to illustrate the operating principles. This simulation is based on the following assumptions:

- The ORG beam is perfectly parallel.
- Raindrops are spherical. Real drops, especially large ones, are flattened by air pressure.
- There is never more than one drop in the beam at a time. In reality, multiple drops can be in the beam simultaneously and may interfere (cast shadows) with each other.

- Beam edge effects are ignored. When a drop falls through the beam side edge, it causes only partial obscuration of the beam.
- The beam is 1 m long, 50 mm wide, and 1 mm high. Because of the above assumptions, the length and width of the beam do not matter.

In this simulation, we will allow drops to fall through the beam and partially obscure it. Each drop will fall through the beam at its terminal fall speed. As the drop falls through the beam, we can calculate the obscuration of the beam (the amount of shadow cast onto the photodetector) as a function of time. Each drop starts its fall from a position one drop radius above the beam, and the next drop is started as soon as the previous drop falls to a point one drop radius below the beam. Note that some of the drop diameters are larger than the beam height. Beam transmission, or simulated photodetector output, is plotted in fig. 9-8 for the test case where drops of diameter 1, 1.3, 1.6, ..., 3.7 mm are dropped in succession. The amount of obscuration increases with drop size and the amplitude of the higher frequencies also increases. The latter is not so obvious.

A more realistic test is to select drop diameters randomly from an exponential distribution similar to actual observed drop size distributions. A Marshall–Palmer drop size distribution will be used, with drop diameters constrained to range from 1 mm to 8 mm:

$$\left. \begin{array}{l} N(D) = N_0 \exp(-\Lambda D) \\ \Lambda = 4.1 R^{-0.21} \\ N_0 = 8000 \end{array} \right\} \qquad (9.7)$$

In this distribution, $N(D)$ is the number density in $\text{m}^{-3}\,\text{mm}^{-1}$, N_0 is a constant, $\text{m}^{-3}\,\text{mm}^{-1}$, Λ has units of mm^{-1}, R is the rain rate in $\text{mm}\,\text{hr}^{-1}$, and D is the drop diameter in mm.

Since this is an exponential distribution, there are many more small drops than large ones. For example, if the rain rate were $10\,\text{mm}\,\text{hr}^{-1}$, there should be 2260 drops

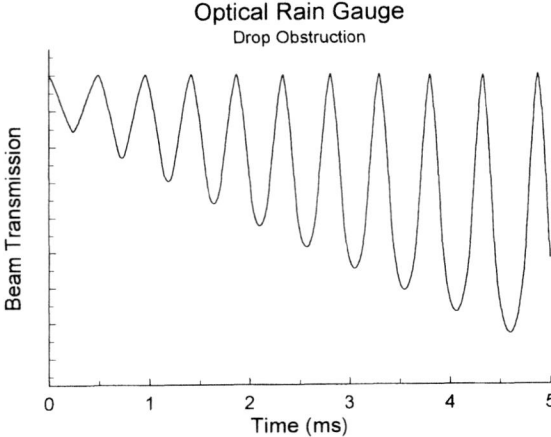

Fig. 9-8 Simulated beam obscuration for drop diameters of 1, 1.3, 1.6, ... 3.7 mm.

of 0.5 mm diameter and only 1.32×10^{-5} drops of 8 mm diameter per cubic meter. In the simulation, a rain rate is selected and drops is generated randomly according to this distribution. Since only a finite number of drops are generated, the actual rain rate will be determined by counting the drops and accumulating their volume as they fall. The initial drop position can be chosen randomly also, subject only to the constraint that the drop must be initially above the beam.

As these drops fall through the beam, the amount of light reaching the photodetector will be attenuated and the output signal will be decreased. This attenuated signal is sampled at 50 μs intervals and stored. The next step is to perform a spectrum analysis of this signal as shown in fig. 9-9 for seven runs with seven different rain rates. The top curve represents an actual rain rate of 289 mm hr^{-1} and the curves below it are the result of rain rates of 192, 49.7, 14.0, 6.33, 3.46, and 2.27 mm hr^{-1}. An ORG does not perform spectral analysis. Rather, it filters the signal in two bands as noted above. The high-frequency band, 1 to 4 kHz, is dominated by rain-generated signals. Then the ORG measures the signal power (variance) in each band. The rain signal is the time-varying power in the 1 to 4 kHz band. In the simulation we can take the average spectral power between these frequency limits to obtain an output similar to that of the ORG. The simulated power in the rain band is plotted in fig. 9-10 versus the observed rain rate. The output is very nearly a linear function of rain rate. It is not clear whether the slight nonlinearity is real or an artifact of the simulation. The quality of the simulation could be improved in a number of ways, perhaps most notably by using a larger number of drops and taking the spectrum analysis over a longer length of record.

9.2.1.4 Calibration

The simplest method of calibrating a tipping-bucket gauge is to pour a measured amount of water slowly into the gauge and count the number of bucket tips produced. A more comprehensive test involves the use of a digital balance as shown in fig. 9-11.

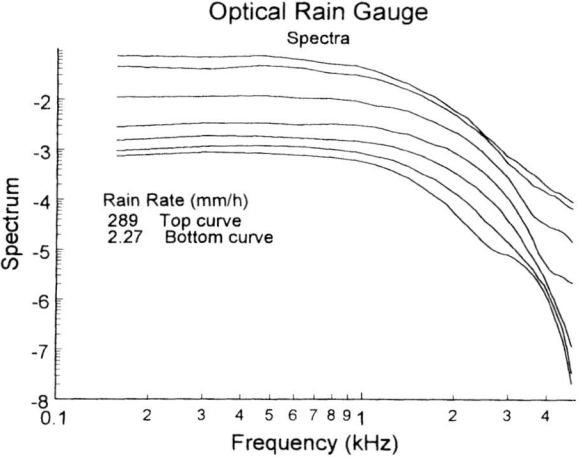

Fig. 9-9 Spectrum analysis of simulated ORG signal.

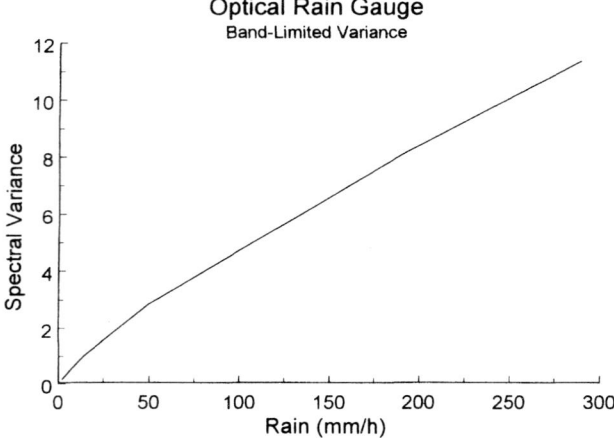

Fig. 9-10 Simulated spectral variance in the 1 to 4 kHz band.

Water is siphoned out of the flask at a rate controlled by the valve. The data logger counts the bucket tipping rate and the computer determines the mass loss of water from the flask per unit time, using the digital balance. From this, the equivalent rain rate can be determined. The rain rate will decrease as the head in the flask drops and, by using various valve settings, the gauge can be tested over a wide range of rain rates. Another advantage of this test is that erratic or noisy behavior of the gauge can be readily detected.

Optical rain gauges are calibrated by comparison with a high-quality tipping-bucket gauge.

9.2.1.5 Exposure

In turbulent flow, sometimes caused by flow around the gauge itself, small rain drops and snow may be deflected out of the gauge and, consequently, the gauge catch is reduced. Experimental measurements show that the reduction may be 20% for wind in the range 5 to $10\,\mathrm{m\,s^{-1}}$ and over 80% for winds above $10\,\mathrm{m\,s^{-1}}$. The ideal exposure for a precipitation gauge is an area free from obstructions that would create large eddies that deflect the flow, and where the winds are light to allow rain and snow to fall vertically. These conditions can be approximated reasonably well for large raindrops but are nearly impossible to achieve for snow. A gauge in the lee of an obstruction could over- or under-collect precipitation depending upon the obstruction, the wind speed and particle size. The usual exposure is to set the gauge orifice a few feet above ground well away from obstructions. Sometimes a special windscreen is placed around the gauge that creates turbulence that tends to minimize systematic wind bias. Gauges should never be placed around or on a building.

One way to improve the exposure of a standard rain gauge is to build a turf wall around it as shown in fig. 9-12. The inner area must be drained to prevent flooding. The turf wall is expensive and cumbersome and so is rarely used. A simpler and less

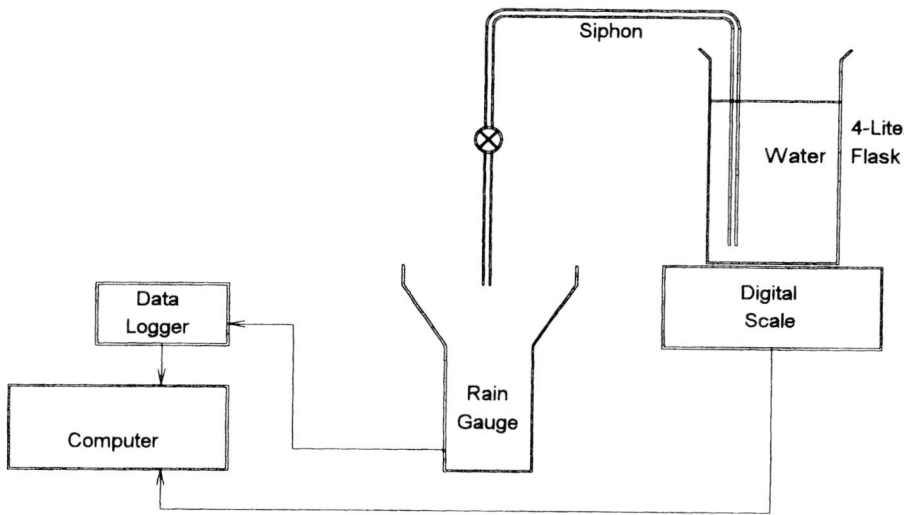

Fig. 9-11 Automated rain gauge calibration system.

expensive way to minimize the wind effect is to use an Alter windshield (see fig. 9-13). According to Groisman and Legates (1994) less than half the well-equipped weather stations have gauges equipped with Alter shields. Where much of the annual precipitation falls as snow, the Nipher gauge (see fig. 9-14) is preferred.

9.2.1.6 Sources of Error

Point measurement rain gauges are subject to many sources of error but, by far, the most important sources of error are exposure and representativeness.

Error sources common to all accumulation gauges

- *Representativeness.* When there is only one gauge for an area of 700–800 km^2 the fraction of the area covered by the gauge is on the order of 10^{-6}. This is poor sampling for any kind of rain and hopelessly inadequate for thunderstorms. Sampling error can be positive or negative. If a thunderstorm misses a gauge entirely, there is a large underestimation. But if the storm passes directly over the gauge, the gauge output will overestimate the rainfall for the area represented by the gauge.
- *Wind.* Wind flow around the gauge is perturbed by the gauge itself and deflects the smaller drops out of the gauge, thus leading to an underestimation of the catch. Snow will be deflected even more than raindrops. The amount of underestimation will be a function of the type of precipitation (rain or snow), size spectrum (large drops are deflected less than small drops), the wind speed at

180   Meteorological Measurement Systems

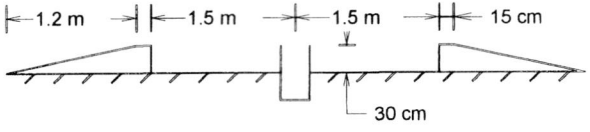

Fig. 9-12 Turf wall around a rain gauge to improve exposure.

the top of the gauge during the rain event, and the type of wind shield present, if any. Optimum exposure would be within some type of windshield and close to the ground where wind speeds are lower.
- *Wetting and evaporation.* When a light rain starts to fall into a dry gauge, some water will be held by the funnel surface and by the screens, thus delaying the apparent onset of precipitation. If the gauge is warm, some of this water will be lost by evaporation. Evaporation error is usually small, but delaying the apparent onset of precipitation may be a very significant error in some applications, particularly if short rain events with small amounts of precipitation are prevalent.
- *Splash out.* Large water drops hitting the top portion of the funnel may splash and part of the drop may be ejected.
- *Plugging.* The small end of the funnel can be plugged by bits of grass blown into the gauge and slipping past screens designed to prevent this. If the grass stems enter the screen end on, they pass through and sometimes jam the small end of the funnel, causing complete blockage. Ice and snow can accumulate in the funnel, and block it. Sometimes heaters are added to the gauge to prevent ice and snow blockage but they cause increased loss due to evaporation.
- *Dew accumulation.* Heavy dew formation can accumulate in a gauge to a measurable extent and be recorded by a data logger. Since a trace of rainfall may be indicated where there was none, this is overestimation. If the dew is not heavy enough to be recorded, it will usually evaporate during the day.

Error sources specific to gauge type

- *Tipping-bucket loss at high rain rates.* At high rain rates, the finite time required for a bucket to tip will cause some water to enter the bucket after it is full but before the next bucket is in place to receive water. This water is not counted, therefore a tipping-bucket gauge underestimates catch at high rain rates. It is possible to at least partially compensate for this loss with proper calibration.
- *Tipping-bucket jams.* Buckets can be jammed in place and then the output will be zero for all rain rates. This has been observed as a result of mechanical failure, spider webs strong enough to prevent bucket tips, and small frogs sitting in a bucket.

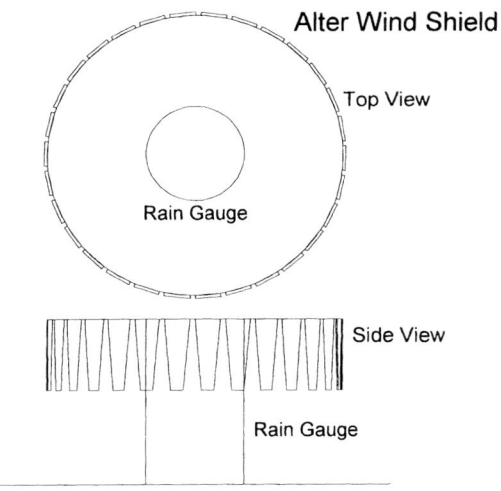

Fig. 9-13 Alter wind shield.

- *Pressure, siphon, and weighing gauges.* All gauges that present a continuous output signal (usually voltage) proportional to depth of water in the gauge may exhibit some temperature sensitivity that adds a component to the output that is proportional to the diurnal temperature cycle. These gauges may also be sensitive to wind flow over the gauge that can induce transients in the signal that may be recorded as rain.
- *Pressure, siphon, and weighing gauges.* These gauges fail to record precipitation while emptying.
- *Pressure, siphon, and weighing gauges.* As noted above, rain rate is estimated by taking the derivative of the output of these gauges. Humans who are very good at pattern recognition have analyzed the output trace of the classic weighing gauge with strip-chart output recording. They can readily detect and compensate for signal drop due to evaporation, sudden signal drop due to emptying the gauge, transients due to wind pumping, small perturbations due to residual temperature sensitivity, and so on. Some caution is required when automating this task as writing a program to recognize and handle these conditions is a non-trivial task.

Representativeness error can be positive or negative. Dew formation, temperature sensitivity, and wind pumping can cause a slight overestimation of rain that would probably be insignificant except that these events tend to occur during clear, dry weather and are, therefore, quite noticeable.

All other sources of error cause underestimation. These errors are mostly due to exposure so cannot be removed by calibration, but the net effect is to cause a negative bias. Evidently rain gauges are subject to various kinds of bias error as well as imprecision error. Calibration bias is usually very small or nonexistent in well-maintained instrumentation. Exposure bias is much larger and usually causes underestimation of rainfall. Unfortunately, it is difficult to estimate this bias except, perhaps, when dealing with many gauges over a fairly long period of time. The term "ground truth" has

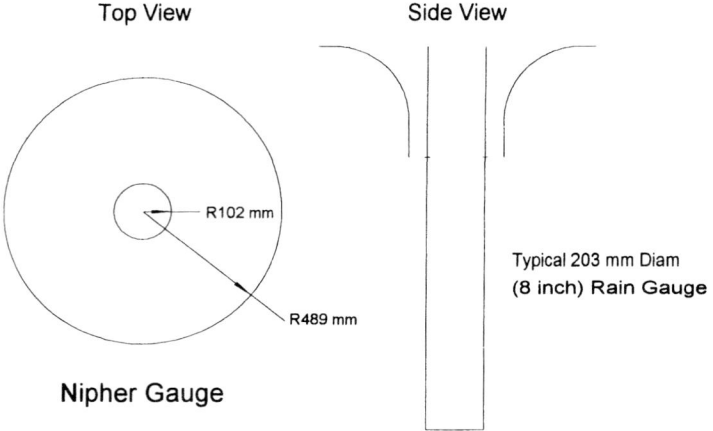

Fig. 9-14 Nipher wind gauge.

been used sometimes in reference to ground-based instrumentation including rain gauges. Since there will always be error in measurements, the word "truth" is simply inappropriate and, worse, may lead to yet another kind of bias error.

9.2.2 Radar Rain Measurement

Remote sensing is outside the scope of this text, but some mention of radar estimation of rainfall is appropriate because it complements ground or point measurement techniques that are severely compromised by the under-sampling problem. In areas where most rain gauges are reported monthly through the climatological network, these reports are useless for flood forecasting. Radar provides complete areal coverage and timely reporting. Gauges reporting through a real-time network are also timely but seldom provide adequate coverage.

A radar can be used to estimate raindrop concentration in the atmosphere from the strength of the returned signal. The rainfall rate can be inferred from this but it requires some assumptions about the raindrop size distribution and fall rate. The advantage of this method is that it provides an area estimate of rainfall in contrast to the point estimates from rain gauges. A single radar can measure rainfall over an area of at least 70 000 km² (assuming a 150 km range) and possibly more than 600 0 00 km² (with a range of 450 km). With the extensive coverage provided by the National Weather Service radar, the WSR-88D, it will be possible to estimate area rainfall over most of the US.

The weather radar equation is

$$P_r = \frac{\pi^3 P_t g^2 L^2 c\tau \theta_1^2 |K_w|^2 Z}{2^{10}(\ln 2) r^2 \lambda^2} \tag{9.8}$$

where P_r = received power in W, P_t = transmitted power in W (about 750 kW), g = antenna gain (e.g., 35 000), dimensionless, L = signal loss factor due to absorption along the path, c= speed of light, 3×10^8 m s^{-1}, τ = pulse width in s (e.g., 1.57 or

4.5 μs), θ_1 = antenna beam width in radians (e.g., 1° = 0.01745 radians), r = range to target in m (up to 450 km), λ = wavelength in m (from several mm to 10 cm), Z = effective reflectivity factor, and $|K_w|^2$ = parameter associated with the complex index of refraction of the scatterer; this is 0.93 for water droplets and 0.19 for ice particles.

The radar beam would propagate in a straight line but for the variation of atmospheric density with height that causes a gradient in the index of refraction which, in turn, causes the beam to bend down slightly even in normal atmospheric conditions. Figure 9-15 shows typical beam geometry under normal conditions. Even if the beam propagation were in a straight line, it would appear to bend up due to earth curvature. Normal atmospheric conditions cause the beam to bend down slightly; this partly compensates for the earth curvature. Figure 9-15 shows a beam elevation angle of 1° above the horizontal. The beam width is 1°, which causes it to spread as it propagates. The beam depth along the propagation path is fixed by the pulse width τ and is relatively narrow. The beam depth is drawn to scale in the figure.

The radar turns on its transmitter for a short period of time, the pulse width, and the peak power during this transmission period is P_t. An electromagnetic pulse, with a frequency of about 3 GHz is transmitted at intervals of 767 μs to 3067 μs. In between pulses, the radar listens for any signal reflected back from particles in the atmosphere that were "illuminated" by the transmitted beam, see Fig. 9-16. The return echo from a target at T1 (range = 90 km) is detected after 600 μs. If the pulse repetition interval is 1 ms, the maximum unambiguous range is 150 km. In this case, the echo from target T2 returns to the radar just as the next pulse is being transmitted and is masked by that transmission. The echo from target T3 at about 170 km returns after the next pulse has been transmitted and the radar cannot tell whether the target is T3 or T3′. It would assume that the target is T3′ at a range of about 20 km. If the pulse repetition interval were 3 ms, the maximum unambiguous range would be 450 km.

Modern radar has a beam width of 1° and a parabolic antenna (that provides the antenna gain of 35 000). The maximum unambiguous range of the radar is determined by the pulse repetition interval and by the speed of light.

Z is the effective reflectivity factor for particles of ice or water that fill the radar beam at the target range:

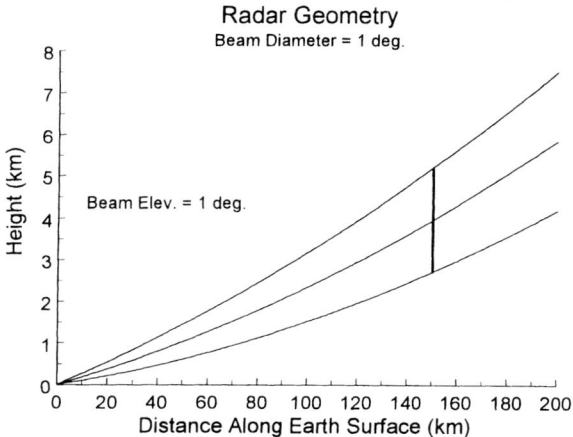

Fig. 9-15 Typical radar geometry showing beam spread and the effect of earth curvature.

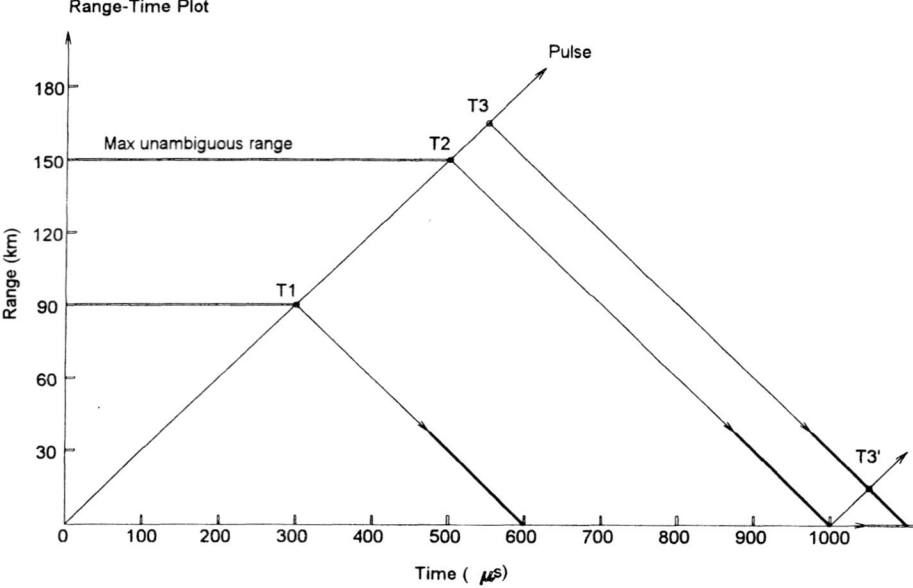

Fig. 9-16 Range–time diagram.

$$Z = \int_0^\infty N(D) D^6 \, dD \qquad (9.9)$$

where D is the droplet diameter and the units of Z are m^3. The rain rate can be defined as

$$R = \int_0^\infty \frac{\pi D^3}{6} N(D) w_t(D) dD \qquad (9.10)$$

in units of $m\,s^{-1}$. The first term in the integral represents the drop volume, the second the number density, drops of size D per unit volume per drop size interval, and the last one is the terminal fall speed of the drops. The terminal fall speed can be approximated by

$$w_t(D) = 9.65 - 10.3 \, e^{-600D} \qquad (9.11)$$

where drop size D is in units of m and the fall speed units are $m\,s^{-1}$.

Since the number density of drops in rain is not generally known, an empirical Marshall–Palmer formula relating rainfall rate to Z in units of $mm^6\,m^{-3}$ is used, such as $Z = 200R^{1.6}$ with R in $mm\,hr^{-1}$.

The radar receives a signal and measures the power returned, P_r. The reflectivity Z can be determined from P_r using the weather radar equation, and the rainfall rate R can be inferred from Z if some relation between Z and R is assumed to apply. The exact form of the applicable Z–R relation will depend upon the type of rain and varies with storm type and location within a storm.

A rain rate of 5 mm hr^{-1} (0.2 in/hr) would produce $Z = 2630$ mm^6 m^{-3} or 2.63×10^{-15} m^3 and, for the radar specified above, the return signal would be $P_r = 6.9 \times 10^{-11}$ W at a range of 150 km; that would be easily detected. The threshold for signal detection for the radar is on the order of 10^{-14} W.

The above equations have been derived under the following assumptions:

(1) the beam at the target range is uniformly filled with water or ice particles with a certain size distribution implied by the Z–R relationship; and
(2) there is nothing to absorb or scatter the beam between the radar and the target.

The difficulties associated with radar rain measurements are:

(1) radar calibration;
(2) unknown drop size distribution;
(3) horizontal and vertical winds;
(4) attenuation by atmospheric gases, rain, and a wetted radome (dome covering the dish);
(5) reflectivity enhancement (the bright band associated with melting hydrometeors);
(6) incomplete beam filling; and
(7) evaporation and rain rate gradients.

The solution to these problems is to use radar in conjunction with ground-based rain gauge networks to provide corrections to the radar measurements.

QUESTIONS

1. Sketch the static transfer plot of a tipping-bucket rain gauge. Label the axes and give appropriate units. What is the static sensitivity of the gauge?
2. Indicate what would happen to the power received by a weather radar if
   (a) the range doubles,
   (b) the pulse length doubles,
   (c) the antenna gain doubles.
3. If the rainfall rate doubles, will the reflectivity double?
4. List the two major sources of error for a rain gauge.
5. Does a tipping-bucket rain gauge have a threshold? Discuss.
6. Describe the ideal atmospheric exposure for a rain gauge.
7. Your assistant calibrates a weighing rain gauge (with voltage output) to determine the constants in the calibration equation. She/he then adds some nonvolatile antifreeze to the bucket to prevent freezing. He/she forgot to record the weight of the antifreeze before putting the gauge into service. Since then, it has rained and the gauge has collected data. Can you recover rain rate estimates from the raw voltage data? Explain.
8. Assuming that problems associated with radar such as signal attenuation, beam filling, winds, gradients, etc. are resolved, what is the primary difficulty in radar estimation of rain rate?
9. The scintillation power in a given wave band, measured by an optical rain gauge, is proportional to the rain rate, independent of the drop size distribution. Why?

10. What will cause a tipping-bucket rain gauge to underestimate the rain rate when the rain rate is very low and when it is very high?
11. What is the rain rate if N drops of diameter D_R fall into the gauge with an orifice diameter D_G in Δt seconds? If the gauge is a tipping-bucket gauge with bucket size Δh, what is the bucket tip frequency? What is the static sensitivity of the gauge? How would the static sensitivity change (a) if the bucket size were doubled? (b) if the catch area were doubled?
12. Figure 9.17 depicts the output from a weighing rain gauge showing rainfall accumulation versus time in hours.
    (a) What happened at 3:20?
    (b) What happened at 5:00?
    (c) When did it rain hardest?
    (d) What does the slope of this graph represent?
    (e) The output, after 5:00, might slowly decrease with time. Why?
13. A 30.5 cm (12 inch) diameter tipping-bucket rain gauge is to be calibrated using a digital scale as depicted in fig. 9-11. If the flask loses 0.1 kg of water per minute, what is the equivalent rain rate (in mm hr^{-1})? Assume the water density is 998 kg m^{-3}.
14. Define all of the terms in the weather radar equation and give the SI units.
15. Given a tipping-bucket rain gauge with a 30.5 cm (12 inch) diameter funnel and 0.2 mm buckets, what is the bucket volume and static sensitivity?
16. Write the calibration equation for a weighing rain gauge.
17. A tipping-bucket rain gauge exhibits catch errors in light rains and in heavy rains.
    (a) Light rain. Error is (pos/neg). Cause of error?
    (b) Heavy rain. Error is (pos/neg). Cause of error?
18. What is the largest source of error in *any* point-measurement type rain gauge.
19. What characteristic of precipitation does an optical rain gauge use to distinguish between rain and snow?

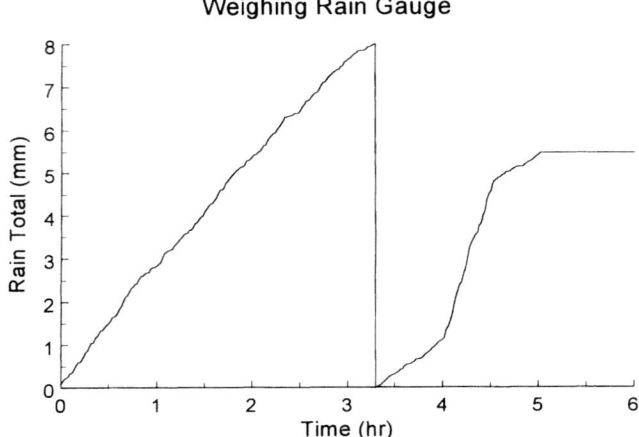

Fig. 9-17 Water depth versus time in a weighing rain gauge.

BIBLIOGRAPHY

Austin, P.M., and S.G. Geotic, 1980: Precipitation measurements over the ocean. In *Air-Sea Interaction: Instruments and Methods*, ed. F Dobson, L. Hasse, and R. Davis. Plenum Press, New York, 801 pp.

Barnston, A.G., 1991: An empirical method of estimating raingage and radar rainfall measurement bias and resolution. *J. Appl. Meteor.*, 30, 282–296.

Baumgardner, D., and A. Rodi, 1989: Laboratory and wind tunnel evaluations of the Rosemount icing detector. *J. Atmos. Oceanic Technol.*, 6, 971–979.

Bradley, J.T., R. Lewis, and P.A. Haas, 1993: Automated snow depth measurement. *Preprints 8th Symp. on Meteorological Observations and Instrumentation*, Anaheim, CA. American Meteorological Society, Boston, MA, pp. 74–79.

Costello, T.A., and H.J. Williams, Jr, 1991: Short duration rainfall intensity measured using calibrated time-of-tip data from a tipping bucket raingage. *Agric. and Forest Meteor.*, 57, 147–155.

Creutin, J.D., G. Delrieu, and T. Lebel, 1988: Rain measurement by raingage–radar combination: a geostatistical approach. *J. Atmos. Oceanic Technol.*, 5, 102–115.

Doviak, R.J., and D.S. Zrnic, 1993: *Doppler Radar and Weather Observations*, 2nd ed. Academic Press, San Diego, 562 pp.

Earnshaw, K.B., T.-I. Wang, R.S. Lawrence, and R.G. Greunke, 1978: A feasibility study of indentifying weather by laser forward scattering. *J. Appl. Meteor.*, 17, 1476–1481.

Gaumet, J.L., P. Salomon, and R. Paillisse, 1991: Present weather determination by an optical method. *Preprints 7th Symp. on Meteorological Observations and Instrumentation*, New Orleans, LA. American Meteorological Society, Boston, MA, pp. 327–331.

Golubev, V.S., P.Y. Groisman, and R.G. Quayle, 1992: An evaluation of the United States standard 8-in. nonrecording raingage at the Valdai Polygon, Russia. *J. Atmos. Oceanic Technol.*, 9, 624–629.

Goodison, B.E., 1978: Accuracy of Canadian snow gage measurements. *J. Appl. Meteor.*, 17, 1542–1548.

Groisman, P.Y., and D.R. Legates, 1994: The accuracy of United States precipitation data. *Bull. Amer. Meteor. Soc.*, 75, 215–227.

Hill G.E., 1991: Comments on "Laboratory and wind tunnel evaluations of the Rosemount icing detector." *J. Atmos. Oceanic Technol.*, 8, 305–306.

Lewis, R., 1993: Automating the observation of blowing snow. *Preprints 8th Symp. on Meteorological Observations and Instrumentation*, Anaheim, CA. American Meteorological Society, Boston, MA, pp. 86–90.

Lindroth, A., 1991: Reduced loss in precipitation measurements using a new wind shield for raingages. *J. Atmos. Oceanic Technol.*, 8, 444–451.

Messaoud, M., and Y.B. Pointin, 1990: Small time and space measurement of the mean rainfall rate made by a gage network and by a dual-polarization radar. *J. Appl. Meteor.* 29, 830–841.

Meteorological Office, 1981: *Measurement of Precipitation and Evaporation*, Vol. 5 of *Handbook of Meteorological Instruments*, 2nd ed. Meteorology Office 919e. Her Majesty's Stationery Office, London, 34 pp.

Peck, E.L., 1993: Biases in precipitation measurements: an American experience. *Preprints 8th Symp. on Meteorological Observations and Instrumentation*, Anaheim, CA. American Meteorological Society, Boston, MA, pp. 329–334.

Perrin, T., M. Cabane, A. Rigaud, and C. Pontikis, 1989: An optical device for the measurement of droplet size spectra in calm or low wind conditions. *J. Atmos. Oceanic Technol.* 6, 850–960.

Sauvageot, H., 1992: *Radar Meteorology*. Artech House, Boston, MA, 366 pp.

Segal, B., 1986: The influence of raingage integration time on measured rainfall-intensity distribution functions. *J. Atmos. Oceanic Technol.*, 3, 662–671.

Sevruk, B., and R. Tettamanti, 1993: General model of wind-induced error of precipitation measurement. *Preprints 8th Symp. on Meteorological Observations and Instrumentation*, Anaheim, CA. American Meteorological Society, Boston, MA, pp. 323–324.

Snow, J.T., and S.B. Harley, 1988: Basic meteorological observations for schools: rainfall. *Bull. Am. Meteor. Soc.*, 69, 497–507.

Sovine, K.E., and K.M. Starr, 1991: A development summary of meteorological sensors for future inclusion with automated surface observing systems. *Preprints 7th Symp. on Meteorological Observations and Instrumentation*, New Orleans, LA, American Meteorological Society, Boston, MA, pp. 202–206.

Starr, K. M., and R. Van Cauwenberghe, 1991: The development of a freezing rain sensor for automated surface observing systems. *Preprints 7th Symp. on Meteorological Observations and Instrumentation*, New Orleans, LA, American Meteorological Society, Boston, MA, pp. 338–343.

Wang, T.-I., G. Lerfald, R.S. Lawrence, and S.F. Clifford, 1977: Measurement of rain parameters by optical scintillation. *Appl. Optics*, 16, 2236–2241.

Wang, T.-I., K.B. Earnshaw, and R.S. Lawrence, 1979: Path-averaged measurements of rain rate and raindrop size distribution using a fast-response optical sensor. *J. Appl. Meteor.*, 18, 654–660.

Wang, T.-I., P.N. Kumar, and D.J. Fang, 1983: Laser rain gauge: near field effect. *Applied Optics*, 22(24), 4008–4012.

Wiggins, W.L., and B.E. Sheppard, 1991: Field test results on a precipitation occurrence and identification sensor. *Preprints 7th Symp. on Meteorological Observations and Instrumentation*, New Orleans, LA. American Meteorological Society, Boston, MA, pp. 348–351.

10

Solar and Earth Radiation

This chapter is concerned with the measurement of solar radiation that reaches the earth's surface and with the measurement of earth radiation, the long wave band of radiation emitted by the earth. The unit of radiation used in this chapter is the $W\,m^{-2}$. Table 10-1 lists some conversion factors.

10.1 Definitions

Radiant flux is the amount of radiation coming from a source per unit time in W.

Radiant intensity is the radiant flux leaving a point on the source, per unit solid angle of space surrounding the point, in $W\,sr^{-1}$ (sr is a steradian, a solid angle unit).

Radiance is the radiant flux emitted by a unit area of a source or scattered by a unit area of a surface in $W\,m^{-2}\,sr^{-1}$.

Irradiance is the radiant flux incident on a receiving surface from all directions, per unit area of surface, in $W\,m^{-2}$.

Absorptance, reflectance, and transmittance are the fractions of the incident flux that are absorbed, reflected, or transmitted by a medium.

Global solar radiation is the solar irradiance received on a horizontal surface, $W\,m^{-2}$. This is the sum of the direct solar beam plus the diffuse component of skylight, and is the physical quantity measured by a pyranometer.

Direct solar radiation is the radiation emitted from the solid angle of the sun's disc, received on a surface perpendicular to the axis of this cone, comprising mainly unscattered and unreflected solar radiation in $W\,m^{-2}$. At the top of the atmosphere this is usually taken to be $1367\,W\,m^{-2} \pm 3\%$ due to changes in the earth orbit and due to sunspots. The direct beam is attenuated by absorption and scattering in the atmo-

Table 10-1 Radiometric conversion factors.

$$1\,W\,m^{-2} = 1.433 \times 10^{-3}\,cal\,cm^{-2}\,min^{-1}$$
$$= 1.433 \times 10^{-3}\,langley\,min^{-1}$$
$$= 1.0\,J\,s^{-1}\,m^{-2}$$
$$= 1000\,erg\,s^{-1}\,cm^{-2}$$
$$= 0.3172\,BTU\,ft^{-2}\,h^{-1}$$
$$= 5.285 \times 10^{-3}\,BTU\,ft^{-2}\,min^{-1}$$

sphere. The direct solar radiation at the earth's surface is the physical quantity measured by a pyrheliometer.

Diffuse solar radiation (sky radiation) is the downward scattered and reflected radiation coming from the whole hemisphere, with the exception of the solid angle subtended by the sun's disc in $W\,m^{-2}$. Diffuse radiation can be measured by a pyranometer mounted in a shadow band, or it can be calculated using global solar radiation and direct solar radiation.

Visible radiation is the spectral range of the standard observer. Most of the visible radiation lies between 400 nm and 730 nm.

Ultraviolet radiation is the radiation with wavelengths in the range 100 to 400 nm. It is subdivided into three ranges: UVA is 315 to 400 nm, UVB is 280 to 315 nm, and UVC is 100 to 280 nm.

Infrared radiation is the radiation with wavelengths longer than 730 nm.

Photosynthetically active radiation (PAR) is the band of solar radiation between 400 nm and 700 nm that is used by plants in the photosynthesis process. PAR is usually measured in moles of photons, amole being Avogadro's number of photons, 6.022×10^{23} photons.

Solar and Earth Radiation. A black body is one that, at a given temperature, radiates as much or more, at every wavelength, than any other kind of object at the same temperature. The irradiance emitted by a black body per unit wavelength, in units of $W\,m^{-3}$, is described by Planck's law:

$$E_\lambda^* = \frac{2\pi hc^2 \lambda^{-5}}{\exp\left(\frac{hc}{k\lambda T}\right) - 1} \qquad (10.1)$$

where $T =$ temperature, K, $\lambda =$ wavelength, m, $h =$ Planck's constant, $k =$ Boltzmann's constant, and $c =$ speed of light. E_λ^* is the power emitted by a black body per unit area in the wavelength interval $d\lambda$. The superscript $*$ is used to indicate a black body. The maximum irradiance occurs at a wavelength given by Wien's displacement law (obtained by differentiating the Planck function with respect to wavelength and setting the result equal to zero):

$$\lambda_m T = 2.898 \times 10^{-3}\,m\,K. \qquad (10.2)$$

EXAMPLE

If the effective solar temperature is 5780 K, the wavelength of maximum emission is $0.501\,\mu m$. At this wavelength, the irradiance is $8.301 \times 10^{13}\,W\,m^{-2}\,m^{-1}$. To obtain the irradiance per μm of wavelength, multiply by 10^{-6} to obtain

$8.301 \times 10^7 \, \text{W m}^{-2} \, \mu\text{m}^{-1}$. This is the irradiance of the sun in the wavelength interval of 1 μm.

Let the effective earth temperature be 255 K; then the wavelength of maximum emission is 11.36 μm. The blackbody irradiance at this temperature is $1.387 \times 10^7 \, \text{W m}^{-2} \, \text{m}^{-1}$ and the irradiance per μm is $13.87 \, \text{W m}^{-2} \, \mu\text{m}^{-1}$.

The total power per unit area emitted by a black body is given by

$$E^* = \int_0^\infty E_\lambda^* \, d\lambda = \sigma T^4 \qquad (10.3)$$

where σ is the Stefan–Boltzmann constant, $5.6696 \times 10^{-8} \, \text{W m}^{-2} \, \text{K}^{-4}$. Real objects are not black bodies and emit $E = e\sigma T^4$, where the emissivity is in the range $0 < e < 1$. The absorptivity of a body, denoted by a, is equal to the emissivity and has the range $0 < a < 1$.

The sun has a wavelength of maximum emission 0.5 μm, assuming an equivalent black body temperature of 5780 K. Earth radiation has a maximum at a wavelength of 11.4 μm with an equivalent blackbody temperature of 255 K. All black bodies radiate at all wavelengths and the blackbody curve for a 5780 K radiator is greater than that of a 255 K radiator at all wavelengths. The solar (or shortwave) radiation is characterized as the wavelength band $0.3 \, \mu\text{m} \leq \lambda \leq 4 \, \mu\text{m}$ and earth or longwave radiation as the $4 \, \mu\text{m} \leq \lambda \leq 50 \, \mu\text{m}$ wavelength band. The crossover between solar and earth radiation is shown in fig. 10-1.

In this figure, the equivalent earth temperature is taken to be 255 K. That is the temperature required to achieve radiative equilibrium with the sun. The plot of earth radiation has not been adjusted but the plot for solar radiation, with an assumed temperature of 5780 K, has been adjusted. The adjustment is to compensate for the distance between the earth and the sun, to account for the earth's albedo, and to account for the fraction of the earth illuminated by the sun at any time (half of the earth is illuminated by the sun but the entire surface of the earth is radiating). The

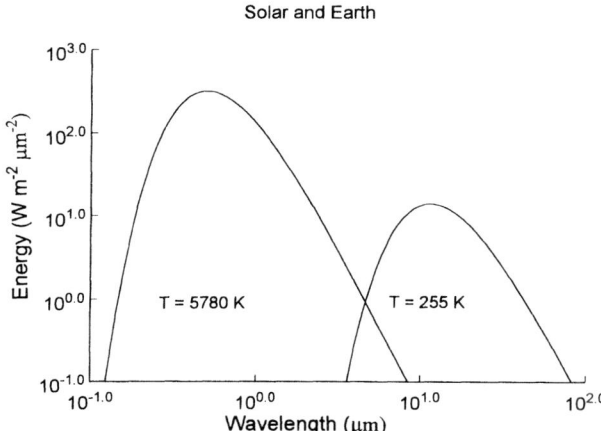

Fig. 10-1 Solar and earth radiation. Solar radiation has been corrected for the fraction intercepted by the earth (including the atmosphere), the earth surface area, and the earth albedo.

planetary albedo is the fraction of the total incident solar radiation that is reflected back into space without absorption. The solar correction factor, C_s, is

$$C_s = \left(\frac{R_s}{D_e}\right)^2 (1 - A)\left(\frac{\pi R_e^2}{4\pi R_e^2}\right) = 3.81 \times 10^{-6} \qquad (10.4)$$

where R_s = solar radius = 7×10^8 m, D_e = earth–sun distance = 1.5×10^{11} m, A = earth albedo = 0.3, and R_e = earth radius = 6.37×10^6 m. With this adjustment, the earth and solar curves cross between 4 and 5 µm and the crossover is usually taken to be 4 µm. Therefore, solar radiation is assigned the wavelength range $0 \leq \lambda \leq 4$ µm and earth radiation the range above 4 µm. The atmosphere attenuates radiation for $\lambda < 0.3$ µm, so, for practical reasons, solar radiation is taken to be $0.3 \leq \lambda \leq 4$ µm and earth radiation is assumed to be $4 \leq \lambda \leq 50$ µm

EXAMPLE
Use eqn. 10.3 to check the computation of the solar correction factor. If the sun and earth are in radiative equilibrium, then $C_s \sigma T_s^4 = \sigma T_e^4$. So $C_s = (255/5780)^4 = 3.81 \times 10^{-6}$.

10.2 Methods of Measurement

There are two primary methods used in the measurement of radiation: thermal detectors that respond to the heat gain or loss due to absorption of incoming or emission of outgoing radiation, and photovoltaic detectors that convert absorbed radiation to a voltage. As noted above, shortwave or solar radiation is defined to be the band from 0.3 µm to 4 µm but, since high-quality glass windows are transparent from 0.3 µm to 3 µm, an upper limit of 3 µm is generally accepted for solar radiation devices. Instruments for measuring longwave or earth radiation have windows that are transparent to radiation from 3 µm or 4 µm to at least 50 µm

Radiation measuring instruments can be classified according to their use. The generic term for all radiation measuring instruments is the radiometer.

- A pyrheliometer measures the direct solar beam. Its sensing element must be kept normal to the solar beam and therefore the pyrheliometer must be pointed at the sun.
- A pyranometer is used to measure global solar radiation, so it must respond to both the direct solar beam and to diffuse sky radiation from the whole hemisphere. The sensing element must be a horizontal flat surface.
- A pyrgeometer is used to measure global, long-wave earth radiation.
- A pyrradiometer is used to measure total global radiation, both shortwave and longwave.
- An instrument that measures the difference between incoming and outgoing radiation is called a net pyrradiometer or simply a net radiometer when the context makes the meaning clear.
- An albedometer measures the difference between incoming and reflected shortwave (solar) radiation.

10.2.1 Pyrheliometers

Direct solar radiation in the band $0.3\,\mu m \leq \lambda \leq 3\,\mu m$ is measured with a pyrheliometer. The viewing angle is about 5°, which encompasses the sun and some of the sky around it to measure just the direct solar beam, excluding nearly all scattered sky radiation. A pyrheliometer must be continuously oriented toward the sun during operation. Manual devices have a sighting device in which the solar image falls upon a mark in the center of a target when the receiving surface is exactly normal to the direct solar beam. Essential elements of a pyrheliometer include a cavity with a limited field of view, about 5°, a thermal detector at the other end, and a mechanism to keep it oriented toward the sun. A blackened plate with a temperature sensor (thermocouple or platinum RTD) is used for the thermal detector. Pyrheliometers are calibrated by comparison with an absolute pyrheliometer.

10.2.1.1 Primary Standard Pyrheliometers

An absolute pyrheliometer can measure irradiance without resorting to reference sources or radiatiors. In addition to the essential elements of a pyrheliometer, an absolute pyrheliometer contains an electrical heater positioned close to the thermal detector. The instrument compares heat received from the sun with heat obtained by forcing a measured quantity of current through a known resistor. Current is adjusted to yield approximately the same detector output as was obtained from the sun.

An absolute pyrheliometer is considered to be a primary standard if it fulfills the following conditions:

- At least one instrument of a series manufactured has to be fully characterized. This is accomplished by examination of the physical properties of the instrument and by calculation of the deviations from ideal behavior. For example, in determining heat generated in the resistor, one would take into consideration change of resistance with temperature of the resistor and I^2R power losses in the leads. The root-mean-square uncertainty should be less than 0.25% of full scale (1000 W m^{-2}).
- Each instrument must be compared with another that has been characterized.
- A detailed description of the results of such comparisons and of the instrument characterization must be available.
- Calibration must be traceable to the WRR, World Radiation Reference (WMO, 1983).

One type of absolute cavity radiometer has a cavity that can alternately be exposed to a radiation source, such as the sun, or heated by an electrical resistor in the cavity. A thermopile (multiple thermocouples) is used to measure the temperature difference between the cavity and a closed reference cavity. Then irradiance can be calculated using

$$E = KV_i \frac{P - C_2 I^2}{V_e} \tag{10.5}$$

where E = irradiance in W m^{-2}, K = a factor determined from the characterization, V_i = thermal detector output when exposed to radiation, V_e = thermal detector output when electrically heated, I = current into the resistor, P = electrical power supplied to the resistor, and C_2 = a factor to account for lead losses (about 0.065).

10.2.2 Pyranometers

Thermal detectors measure the temperature change induced by the heat gain (loss) due to absorption (emission) of radiation by a black surface. The temperature change is measured relative to a white surface or to the shell of the instrument. Figure 10-2 shows two versions of a thermopile pyranometer. A thermopile is a nested array of thermocouples, usually 10 to 100. The advantage of a thermopile over a single thermocouple is that the number of thermocouple pairs in the pile multiplies the output voltage. Following the terminology of chap. 1, the measurand is the incident radiation, the irradiance. The sensor is the thermal plate(s) and the raw output is the temperature difference between the absorbing plate and the non-absorbing plate or the instrument shell ($y_1 = \Delta T$). The thermopile converts the temperature difference to a voltage difference between the thermopile outputs ($y_2 = \Delta V$). This is a form of analog signal conditioning as a thermopile combines multiple individual thermocouple outputs to form the composite voltage output.

When the radiometer is a pyranometer, it can be shielded from the atmosphere with a glass dome because glass is transparent to radiation from 0.25 μm to 2.8 μm. With soda lime glass, used in the best instruments, the upper limit can be extended to 4.5 μm. The glass shield is necessary to keep ambient wind from affecting the heat gain or loss of the black plate and to keep atmospheric contamination from altering the spectral characteristics of the black surface. A surface that appears to be black in visible light may not be perfectly black and it may be gray in the infrared portion of the spectrum. The typical sensitivity of a thermopile pyranometer is 4 μV W^{-1} m^2 and the typical time constant is 5 s.

A black and white pyranometer measures the temperature difference between a black and white surface, both of which are exposed to the sun. These surfaces may

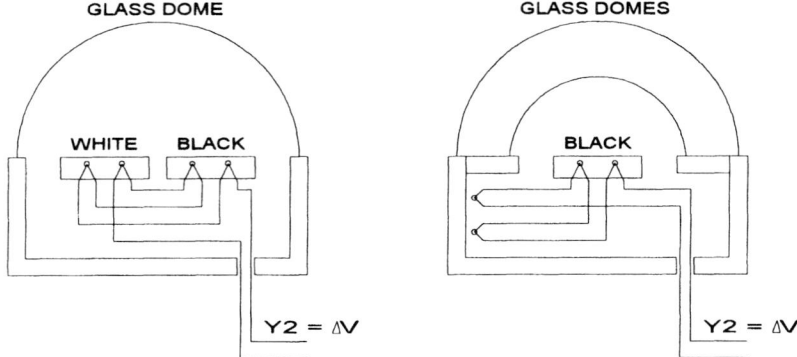

Fig. 10-2 Two styles of thermopile pyranometers showing only two pairs of thermocouples instead of the more usual 10–100 pairs. Not shown are the level and the mechanism for adjusting the level.

be concentric rings or segmented pie slices. Either way, both white and black surfaces are exposed to the same radiation and to the same ambient temperatures, so a simple glass dome suffices to isolate the sensor from wind and rain. The spectral response of the glass dome, shown in fig. 10-3, limits the pyranometer to shortwave or solar radiation.

When there is a single black surface, the reference temperature is the instrument body that is not exposed to incoming radiation in the same way as the white surface. The black surface pyranometer often uses two glass shields to provide more isolation from the longwave radiation emitted by the glass shield itself. The glass is opaque to long wave lengths thus making the sensor respond only to shortwave radiation and, by the same token, being itself nearly black for longwave radiation. As the glass dome heats up, due to convective heating from the air or by absorbing longwave radiation, it reradiates that energy to the black sensor plate. The second glass shield helps to isolate the black sensor.

Even with the use of two domes, errors in the longwave measurements can result when the glass dome heats up. To account for this, some radiometers have a temperature sensor installed at the base of the dome to measure the dome temperature. In addition, the temperature of the radiometer case can be measured to account for temperature differences between the case and the environment. The two measurements are used to correct the longwave measurement (Delany and Semmer, 1998):

$$R_{lw} = R_{pile} + \sigma T_{case}^4 - B\sigma(T_{dome}^4 - T_{case}^4) \tag{10.6}$$

where R_{lw} is the corrected longwave measurement, R_{pile} is the raw thermopile output multipled by the sensitivity, σ is the Stefan–Boltzmann constant (5.67×10^{-8} W m^2 K^{-4}), B is an empirically derived factor to account for dome transmissivity, and T_{dome} and T_{case} are the measured dome and case temperatures in degrees K. Another term is sometimes added to account for the transmissivity of the longwave domes to shortwave radiation.

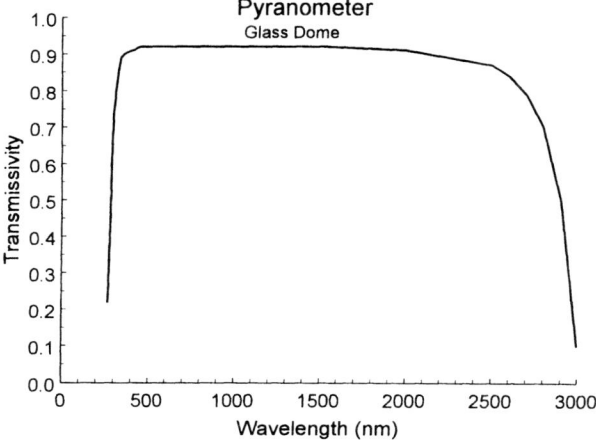

Fig. 10-3 Pyranometer glass transmissivity provides both shortwave and longwave length cutoffs.

Three classes of pyranometer can be defined on the basis of their accuracy and overall system performance, as listed in table 10-2.

10.2.2.1 Photovoltaic Detectors

A photovoltaic pyranometer uses a silicon detector whose response typically extends from 400 to 1100 nm, the response being far from uniform over this region. As shown in fig. 10-4, the response increases gradually from 400 nm, peaks at about 1000 nm, and then falls rapidly to zero at 1100 nm. Because of this large spectral response error (e.g., a perfect sensor would have a spectral response of 1.0 across the intended wavelength range), a photon detector must be calibrated against a high-quality thermopile-type sensor such as the Eppley Precision Spectral Pyranometer. In addition, the calibration must be performed under natural daylight conditions. The resulting calibration error, when used in natural solar radiation, can be quite small, typically ±3% to ±5%. The implicit assumption made in using a photon detector is that clouds, dust, and atmospheric water vapor uniformly attenuate all of the solar radiation in the band from 0.3 μm to 4 μm. This is a fairly good approximation. The advantage of photon detectors is that they are typically much less expensive than a thermopile-type detector. However, they cannot be used to measure ultraviolet or near-infrared radiation.

A shadow ring or shadow disk can be used with a pyranometer to block the direct solar beam so that just the diffuse sky radiation is measured. The shadow ring blocks direct solar radiation for the entire day and must be adjusted each day (or several days) for latitude and solar declination. The shadow disk blocks only the direct solar beam and requires a solar tracker that continuously keeps the shadow positioned over the radiometer. When a second, unshaded pyranometer is also used, the difference is the direct solar beam, which can be compared to the output of a pyrheliometer. The practice of measuring the diffuse solar using a shadow disk and adding to it the direct solar measurement from a pyrheliometer is considered one of the most accurate measurements of solar radiation because it avoids the effects of dome heating due to absorption of solar radiation.

Table 10-2 Classification of pyranometers.

Characteristic	Secondary standard	First class	Second class
Resolution (W m^{-2})	±1	±5	±10
Stability (% full scale per year)	±1	±2	±5
Cosine response (% deviation from ideal at 10° solar elevation on a clear day)	< ±3	< ±7	< ±15
Azimuth response (% deviation from the mean at 10° solar elevation on a clear day)	< ±3	< ±5	< ±10
Temperature response (% full scale)	±1	±2	±5
Non-linearity (% full scale)	±0.5	±2	±5
Spectral sensitivity (% deviation from mean absorptance)	±2	±5	±10
Response time (99% response)	< 25 s	< 60 s	< 4 min

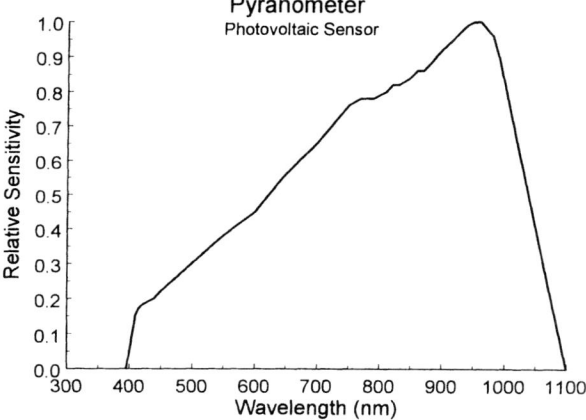

Fig. 10-4 Silicon detector relative response.

An albedometer or net pyranometer comprises two pyranometers, one facing up and the other pointed down. The one facing up measures global solar radiation and the other measures reflected solar radiation. The difference is the albedo.

10.2.3 Pyrgeometers

A pyrgeometer measures only earth or longwave radiation and therefore glass domes cannot be used (see fig. 10-3). Typically, a silicon window (flat, not domed) is used to measure from 3 µm to 50 µm and the field of view is thus limited to 150 degrees. Some sensors incorporate an electrical heater to prevent dew or frost formation. Silicon windows are more stable and reliable than polyethylene windows.

These instruments require a temperature correction. The simplest form involves a calibration with two terms: $E = a_1 V_1 + a_2 V_T$, where a_1 and a_2 are constants. V_1 is the raw sensor voltage output, and V_T, the voltage proportional to sensor temperature.

10.2.4 Pyrradiometers

Pyrradiometers measure the total global radiation, including shortwave and longwave. Glass domes cannot be used because glass is not transparent to wavelengths above about 2.8 µm (fig. 10-3). The glass dome has been replaced by a hemisphere (about 30 mm diameter) of silicon that is transparent to wavelengths above 5 µm (with an interference filter) and below 50 µm. However, silicon domes are delicate and degrade with exposure to sunlight (e.g., become brittle and can crack). Therefore, sensors that use silicon domes require maintenance and periodic dome replacement, as often as monthly, depending on atmospheric conditions.

Moisture can condense inside the domes, which can lead to errors. To avoid this, the domes can be pressurized with dry air or (nitrogen).

A new version of an old net radiometer design has recently become available which does not use domes. This sensor, if sufficiently accurate, may solve several problems associated with the long-term, unattended use of net radiometers (Brotzge et al., 1999).

Two pyrradiometers, one facing up and one facing down, are often combined into one sensor to measure the net radiation at the earth's surface. This sensor is called a net pyrradiometer, or net radiometer for short, and measures the total energy available at the earth's surface for atmospheric processes.

A pair of upward facing and downward facing pyrgeometers and pyranometers can be combined to measure the net radiation. This method is usually considered more accurate than a single net radiometer. For example, many field projects use an Eppley Precision Spectral Pyranometer (PSP) to measure shortwave radiation and a Precision Infrared Radiometer (PIR) to measure longwave radiation. A PSP and a PIR are mounted facing upward and another PSP and PIR are mounted facing the surface. The four measurements are combined into one measure of the net radiation. This method is considered very accurate; however, it is very expensive and not practical for many projects.

10.3 Measurement Errors

Absolute calibration error is the error due to the use of an imperfect reference sensor in the calibration of a sensor. The minimum calibration error is obtained by sending a sensor to NIST for calibration. All other calibrations involve the use of other standards which must be related to NIST and usually involve additional error.

Spectral response error is due to a sensor not conforming to the ideal spectral response. For example, the ideal pyranometer should absorb all incident radiation between 280 and 4000 nm. If a pyranometer, such as the photon-type sensor, deviates markedly from this ideal, it can still be used to measure global solar radiation with the appropriate correction factor.

Cosine error is sensor error due to inaccurate cosine correction that produces errors at low solar elevation angles. When a parallel beam of radiation of given cross-sectional area spreads over a flat surface, the area that it covers is inversely proportional to the cosine of the angle between the beam and a plane normal to the surface. Therefore, the irradiance due to the beam is proportional to the cosine of the angle. A radiometer whose response to beams coming from different directions follows the same relationship is said to be "cosine corrected."

Azimuth error is the change in the sensor output as the sensor is rotated about the normal axis at a particular angle of incident radiation. It is due to lack of symmetry. Many sensors are designed to be mounted in a particular direction relative to true North.

Linearity error exists when a sensor output is not linearly proportional to the input.

Hysteresis error exists when a sensor response to an increasing input differs from its response to a decreasing input at the same input value.

Temperature coefficient error exists when the sensor is sensitive to temperature as well as radiation.

Response time error exists when the input is changing rapidly and the sensor cannot respond.

Long-term stability error is caused when the sensor characteristics change with time.

User setup and application errors include, but are not limited to, reflections or obstruction from the supporting mast or adjacent buildings; dust and bird droppings on the sensor; shock, causing permanent damage to the sensor; use of an incorrect calibration constant; and failure to have the sensor recalibrated periodically.

Wind speed errors can be caused by wind heating or cooling the dome of the instrument.

10.4 Exposure

Some exposure requirements are common to all radiation instruments:

- Incidence of fog, smoke, and airborne pollution at the site should be typical of the surrounding area.
- Instrument windows should be kept clean. Dust, rain, dew, and bird droppings may adversely affect window transparency. A first-class station requires daily cleaning, and aspirators should be used to continually blow air on the radiometer domes to keep them free of dew and dry them quickly after rain.
- Condensation must never occur inside the instrument.
- The site must be free from shadows cast over the instrument for all sun angles for the entire year. A rooftop installation if preferable provided that it is free of antennas and guy wires that may cast shadows on the instrument. Obviously, rooftop mounting should not be used to measure upwelling longwave or reflected shortwave radiation.
- There must be no reflections of light toward the instrument for all sun angles for the whole year.
- The instrument must be kept level.

Requirements specific to a pyrheliometer:

- An equatorial mounting or an automatic tracker is required for continuous recording. It must be protected from environmental influences.
- The instrument must be kept aligned to the sun within $0.25°$.

Requirements specific to instruments that "look" down, including all net radiation instruments:

- Most pyranometers and pyrradiometers have a field of view of $180°$ so, even when mounted just 1 to 3 m above the surface, they have a large "footprint", the surface emitting radiation to the instrument.

$$E_T = \int_0^{2\pi} \int_0^{\pi/2} E_{\theta,\phi} \cos\theta \sin\theta d\theta \, d\phi$$

where ϕ sweeps from 0 to 2π around the footprint on the ground and $\theta (0 \leq \theta \leq \pi/2)$ is the angle between the nadir (point on the ground just below the sensor) to any point

200 Meteorological Measurement Systems

Table 10-3 Radius of footprint that receives fraction f of total radiation.

f	θ_f (degrees)	$\tan(\sin^{-1}\sqrt{f})$	R(m) (for $z = 3$ m)
0.90	71.6	3.00	9.0
0.95	77.1	4.36	13.1
0.98	81.9	7.00	21.0
0.99	84.3	9.95	29.8

in the footprint. Then $E_T = 2\pi E$ provided that $E_{\theta,\phi} = E$, a constant over the region. The footprint extends to infinity, but ground distant from the nadir contributes little to the total radiation received by the sensor. Let f be the fraction ($0 \leq f \leq 1$) of E_T received from a circle of radius R on the ground when the height of the sensor is z. Then

$$E_f = fE_T = 2\pi E \int_0^{\theta_f} \cos\theta \sin\theta d\theta = 2\pi E \sin^2\theta_f$$

and $f = E_f/E_T = \sin^2\theta_f$. The radius of the footprint that contributes a fraction f of the total radiation received by the sensor is $R = z\tan(\sin^{-1}\sqrt{f})$. Some values of R for $z = 3$ m are listed in table 10-3. This table shows that, if the sensor is mounted 3 m above the ground, 95% of the radiation received from the ground comes from a circle of radius 13.1 m about a point just below the sensor. This 95% footprint should contain ground cover representative of the area and should certainly not contain a tower leg.

QUESTIONS

1. If the sensitivity of a pyranometer is 10.35 mV W^{-1} m^2, the recorder sensitivity is 0.5 mV cm^{-1}, and the observed deflection is 17.6 cm, what is the irradiance incident on the sensor?
2. Name six factors that affect the accuracy of radiation sensors such as the Eppley pyranometer.
3. You wish to design a sensor to detect infrared radiation from people. The sensor should have a maximum sensitivity at what wavelength?
4. How could a properly calibrated pyranometer at the earth's surface be subjected to an irradiance greater than that received on a cloudless day when the atmosphere is dust free?
5. What is the justification for using a photovoltaic sensor in a pyranometer since it is subject to gross spectral errors? What assumption about the atmosphere is required to use a photon sensor in a pyranometer?
6. Let T_b be the temperature of the black plate and T_w the temperature of the white plate in a thermal-type pyranometer. Would you expect $T_b < T_w$ in any conditions? If so, what conditions?
7. In a thermal-type pyranometer, what kind of error would result if the white plate got slightly dirty?
8. Define the following terms: pyranometer, pyrheliometer, pyrradiometer and net pyrradiometer.
9. What is the boundary between shortwave and longwave radiation? How is it defined?
10. The glass cover over a pyranometer serves to protect the sensor from the environment. What other functions does it have?
11. What happens to the static sensitivity of a thermal pyranometer when

(a) the areas of the black and white receivers are doubled?
(b) the number of thermocouple pairs is doubled?
12. How would you minimize pyranometer exposure errors?
13. Why is it difficult to build a pyrradiometer as stable and rugged as a pyranometer?
14. What measurement do you get by combining a tracking pyrheliometer with a shaded pyranometer?
15. What measurement do you get by combining two pyranometers (one pointing up and one down) with two pyrgeometers (one up and one down)?

BIBLIOGRAPHY

Aceves-Navarro, L.A., K.G. Hubbard, and J. Schmidt, 1988: Group calibration of silicon cell pyranometers for use in an automated network. *J. Atmos. Oceanic Technol.*, 5, 875–879.

Alados-Arboledas, J., J. Vida, and J.I. Jimenez, 1988: Effects of solar radiation on the performance of pyrgeometers with silicon domes. *J. Atmos. Oceanic Technol.*, 5, 666–670.

Brotzge, J. A., S. J. Richardson, C. E. Duchon, S. E. Fredrickson, and D. L. Grimsley, 1999: Evaluation of a domeless net radiometer. *13th Symp on Boundary Layers and Turbulence*, AMS, Dallas, TX.

Delany, A. C., S. R. Semmer, 1998: An integrated surface radiation measurement system. *J. Atmos. and Oceanic Technol*, 15(1) 46–53.

Dichter, B.K., 1993: Instrumentation requirements for establishing UV climatology. *Preprints 8th Symp. on Meteorological Observations and Instrumentation*, Anaheim, CA. American Meteorological Society, Boston, MA, pp. 5–10.

Dixon, J., 1988: Radiation thermometry. *J. Phys. E.: Sci. Instr.*, 21, 425–436.

Fritschen, L.J., and C.L. Fritschen, 1991: Design and evaluation of net radiometers. *Preprints 7th Symp. on Meteorological Observatiosn and Instrumentation*, New Orleans, LA. American Meteorological Society, Boston, MA, pp. 113–117.

Halldin, S., and A. Lindroth, 1992: Errors in net radiometry: comparison and evaluation. *J. Atmos. Oceanic Technol.*, 9, 762–783.

Hinzpeter, H., 1980: Atmospheric radiation instruments. In *Air–Sea Interaction: Instruments and Methods*, ed. F. Dobson, L. Hasse, and R. Davis. Plenum Press, New York, 801 pp.

Meteorological Office, 1981: *Measurement of Solar and Terrestrial Radiation*, Vol. 6 of *Handbook of Meteorological Instruments*, 2nd ed. Meteorology Office 919f. Her Majesty's Stationery Office, London, 45 pp.

Wilk, G.E., and C.E. Duchon, 1993: Comparison of direct and component measurements of net radiation. *Preprints 8th Symp. on Meteorological Observations and Instrumentation*, Anaheim, CA. American Meteorological Society, Boston, MA, pp. 127–130.

World Meteorological Organization, 1983: *Guide to Meteorological Instruments and Methods of Observation*. WMO No. 8, 5th ed. Geneva, Switzerland.

11

Visibility and Cloud Height

Visibility measurement is the most human-oriented measurement discussed because the objective of such measurement is to determine the distance at which humans (pilots, seamen, etc.) can see objects. Thus we are concerned with light that can be seen by humans (0.4 to 0.7 µm), the way human eyes perceive such light, and then with the transparency of the atmosphere. Throughout this chapter, in the discussion of atmospheric transparency or absorption, the range of wavelengths from 0.4 (violet) to 0.7 µm (red light) will be assumed.

Cloud height is a remote sensing measurement but is included here because airport meteorological systems usually include a cloud height sensor.

11.1 Definitions

According to the WMO, meteorological visibility by day is defined as the greatest distance that a black object of suitable dimensions, situated near the ground, can be seen and recognized when observed against a background of fog, sky, etc. Visibility at night is defined as the greatest distance at which lights of moderate intensity can be seen and identified.

Attenuation of light in the atmosphere is described by Beer's law

$$E = E_0 e^{-k\rho x} \tag{11.1}$$

where E = irradiance of visible light in W m^{-2}, k = extinction coefficient in m^2 kg^{-1}, ρ = atmospheric density in kg m^{-3}, and x = path length, m. E_0 is the irradiance at the beginning of the path. Atmospheric extinction is due to absorption and scattering of light caused by hydrometeors (water and ice particles such as rain, snow, fog) and lithometeors (other kinds of particulates such as dust and smoke).

An observer determines visibility by the contrast of an object with respect to its background. This contrast can be defined as

$$C = \frac{B_o - B_b}{B_b} \qquad (11.2)$$

where B_o is the brightness of the object, B_b is the brightness of the background, and C is the contrast. The range of contrast is given by $-1 \leq C \leq \infty$ but typically $C \leq 10$. Critical contrast, C_c, is when the object is barely distinguishable against the background; this value is usually taken to be $C_c = 0.05$ (sometimes 0.02) but it is very much a function of the individual observer's eyesight as well as the angular dimensions of the object, glare, and general brightness. When the distance to the object increases until critical contrast is reached, that distance is the visual range, V.

Equations 11.1 and 11.2 have been related by Koschmieder (1925) who suggested that

$$B_o = B_b\left(1 - e^{-k\rho x}\right) \qquad (11.3)$$

then

$$C = \frac{B_0 - B_b}{B_b} = e^{-k\rho x} \qquad (11.4)$$

and, as the object–observer distance x increases, the contrast C approaches the critical value C_c, then

$$C_c = e^{-k\rho V} \qquad (11.5)$$

A human observer determines visibility at night by noting the extinction of lights of moderate intensity at known distances. Visibility determined by humans at night differs from daytime visibility in two important ways; see fig. 11-1. First, daytime visibility is adversely affected by light scattered into the observer's eyes that did not originate from the object viewed. Scattering particles in the atmosphere play a dual role: they scatter light out of the path from the target to the observer and they scatter

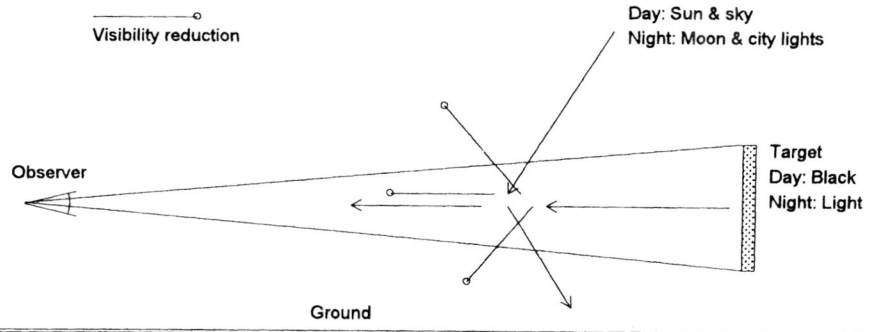

Fig. 11-1 Visibility reduction by scattering.

other light into the observer's eyes. Both effects reduce the contrast and definition of the target. The latter effect is virtually absent at night (fig. 11-2).

The other effect is that the peripheral parts of the retina, most useful for night vision, lack color perception but are more sensitive to low light levels (Wyszecki and Stiles, 1982). Night visual range is usually longer (as much as twice as long) as daytime visual range but it depends upon the observer and upon the presence of other illumination such as moonlight, twilight, and city lights.

Daytime visibility is affected by the

- observer's eyesight,
- value of the extinction coefficient in the atmosphere,
- brightness of object and background,
- reflective properties of the object and the background,
- elevation of the sun,
- angular separation of the sun and the object, and
- size of the object.

These effects complicate the problem of correlating visibility as measured by instruments with human visual range. Daytime visual range is taken as the standard. Conversion of nighttime visibility to daytime visibility is accomplished with standard tables or graphs.

11.2 Measurement of Visibility

To facilitate measurement of visibility with instruments, a new variable was defined, the meteorological optical range, MOR. MOR is the path length in the atmosphere required to reduce the light intensity in a collimated beam to 0.05 of its original value.

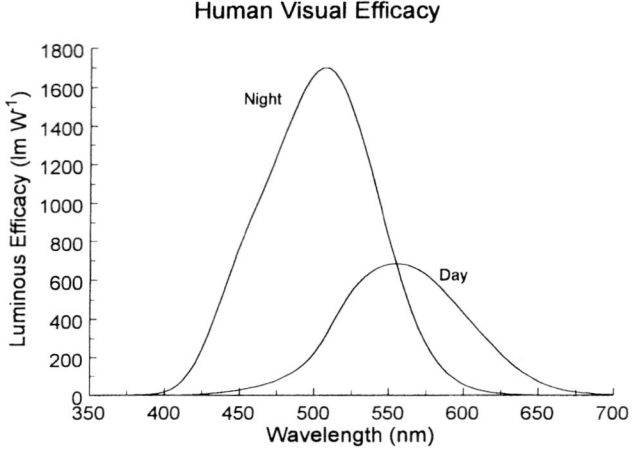

Fig. 11-2 Human visual efficacy by day and by night.

The standard light source is defined to be an incandescent lamp at a color temperature of 2700 K. A direct method of measuring MOR, the extinction of light over a known path, is with an instrument called a transmissometer. An indirect method is to measure the amount of light scattered, usually in a forward direction, out of a beam. The US National Weather Service has defined sensor equivalent visibility (SEV) as any equivalent of human visibility derived from instrumental measurements.

11.2.1 Transmissometer

A transmissometer comprises a light source and one or two light detectors at fixed distances from the source, usually 75 m and 450 m. To minimize sensitivity to ambient light, the source emits high-intensity light pulses of 0.1 μs duration at a repetition frequency of 1.5 Hz. The light detectors are designed to receive light only from the source direction. Because of the long path length, the transmissometer is located along and parallel to a runway, hence the term Runway Visual Range (RVR). This instrument directly measures E in eqn. 11.1. E_0 is the source intensity, assumed constant. The path length x, is known, so

$$k\rho = \frac{-\ln(E/E_0)}{x} \tag{11.6}$$

From eqn. 11.5, we know that

$$V = -\frac{\ln C_c}{k\rho} \tag{11.7}$$

and so we can solve for

$$V = x\frac{\ln C_c}{\ln(E/E_0)} \tag{11.8}$$

where C_c is the critical contrast ratio, usually taken to be 0.05, x is the known baseline of the transmissometer, E_0 is the source intensity, assumed constant, and E is the measured intensity, reduced by atmospheric extinction, at the receiver.

EXAMPLE

Suppose that the baseline of the transmissometer is 450 m and the ratio of received light signal to transmitted light was 0.72. Then the visibility was

$$V = 450\frac{\ln(0.05)}{\ln(0.72)} = 4100 \text{ m}$$

The quantity E/E_0 is called the transmissivity and is the raw output of the sensor as shown in fig. 11-3. Note that the sensitivity of the instrument is poor for very short and very long visual ranges. This shows the necessity for two receivers and for the long path length of 450 m. A transmissometer with a path length of only 75 m would be useless for visual ranges exceeding a few kilometers.

206 Meteorological Measurement Systems

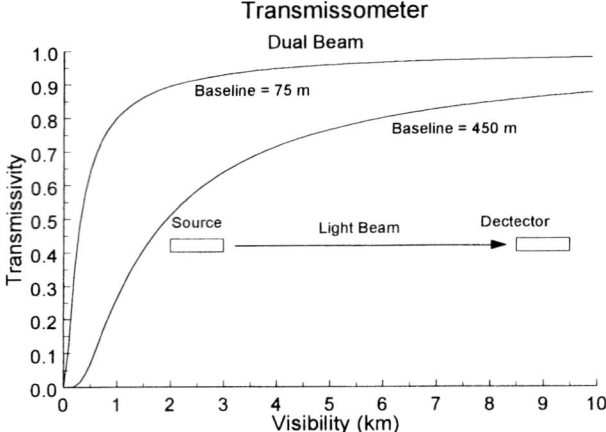

Fig. 11-3 Transfer function for a transmissometer.

11.2.2 Forward Scatter Meters

Light is scattered out of a beam in all directions by particulates that reduce visibility. Instruments have been built to detect the light scattered in the forward direction or in the backward direction (towards the source). The problem with measuring scattered light is that the amount scattered in a given direction is a function of the ratio of the particle diameter to the wavelength of the light in the beam. The wavelength is constant as we are concerned with visible light, but the particle size changes considerably from smoke to fog, for example. It seems that light scattered in a forward direction (20° to 50°) is less dependent on this ratio.

The one available visibility meter (fig. 11-4) has a sampling volume of 0.021 m^3 and uses a xenon flash bulb as the source. The reported visibility range is from 0.4 km to 16 km. The inaccuracy increases with the visibility; it is 0.4 km at the low end of the range and 0.8 km at about 4 km. A forward scatter meter must be calibrated against a transmissometer, the standard for MOR and SEV. The US Weather Service uses transmissometers with 150 m and 460 m base lines to calibrate the forward scatter meters. Visibility is calculated starting with 30 samples of the output. These are converted to visibility and averaged over one-minute intervals. A ten-minute harmonic mean is calculated from the one-minute averages using

$$\bar{V} = \frac{N}{\sum_{n=1}^{N} \frac{1}{V_n}} \qquad (11.9)$$

where the V_n are the one-minute average visibilities. A harmonic mean is used rather than the arithmetic mean because it is more responsive to rapidly decreasing visibility and yields a lower (more pessimistic) value.

EXAMPLE

If a visibility meter takes 10 samples over a period of a minute, the usual average, in km, would be $V_{\text{Avg}} = (3.2 + 3.1 + 3.1 + 2.9 + 3.3 + 3.2 + 2.8 + 2.3 + 1.8 + 1.3)/10$

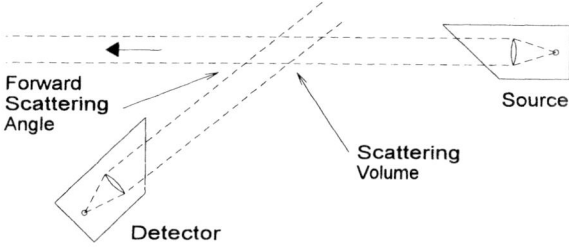

Fig. 11-4 A forward scatter visibility meter.

$= 2.7\,\text{km}$. But the data show that the visibility is degrading, part of the time, with only a small effect on the average. The same data using a harmonic average would be

$$V_{\text{Har}} = \frac{10}{\frac{1}{3.2} + \frac{1}{3.1} + \frac{1}{3.1} + \frac{1}{2.9} + \cdots + \frac{1}{1.3}} = 2.5\,\text{km}$$

and this is bit more pessimistic.

Visibility, as determined by the human observer, is integrated over a long path length, the visible range. But a forward scatter meter integrates over just a few centimeters (or less) so there is a fundamental difference between the two measurements. Signals from a forward scatter meter are integrated over time to compensate for the very short path length. Integration over time is not always comparable to integration over distance, especially when there are local patches of fog.

11.3 Measurement of Cloud Height

The human observer can estimate cloud heights by scanning the sky whereas instruments measure cloud height at a single point. The instrumental technique uses time averaging instead of the observer's space averaging to detect the presence of multiple cloud layers and to estimate the fraction of sky cover. The instruments used to measure cloud height are the rotating beam ceilometer (RBC), developed in the 1940s, and the more recent laser ceilometer.

11.3.1 Rotating Beam Ceilometer

A rotating beam ceilometer (RBC) contains a powerful light source which is rotated from a nearly horizontal position to the vertical. Another variation is the rotating receiver ceilometer (RRC). The RRC keeps the light source pointing vertically and rotates the receiver to detect the cloud height. In either case, the cloud height $z = X \tan \theta$, where X is the baseline distance between the beam and the receiver and θ is the elevation angle. Both the RBC and the RRC are depicted in fig. 11-5. The rotating element is the light source for the RBC or the light detector for the RRC. In either case, the rotating element is swept from the horizontal to the vertical until

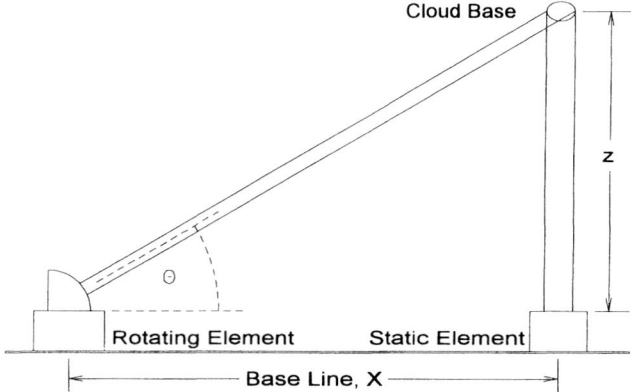

Fig. 11-5 Rotating beam or rotating receiver ceilometer.

the detector "sees" the light reflected from the cloud base. One version of this instrument uses a xenon lamp that is pulsed to produce a high intensity, focused flash of light. The baseline in that instrument is 76 m.

The static transfer plot of a rotating beam ceilometer is shown in fig. 11-6. Note the decreasing sensitivity as the cloud height increases.

11.3.2 Laser Ceilometer

The laser ceilometer emits a pulse of laser light aimed straight up and measures the time required for the "echo" or reflected light to return. The cloud height $z = ct/2$, where $c = $ speed of light $(3 \times 10^8 \text{ m s}^{-1})$ and t is the range time, or time for the pulse to travel to the cloud layer and be reflected back to the receiver. The range limit of the laser ceilometer could be determined by the pulse repetition frequency or by atmospheric attenuation and the receiver sensitivity. In practice, it is the latter that sets the maximum cloud height limit. As the light pulse travels upward, it spreads out, so the power received by a point target is proportional to z^{-2}. The light scattered by the target back toward the receiver also spreads proportionally to z^{-2} so the intensity of the signal received is proportional to z^{-4}. However, if the cloud deck completely fills the beam, which is a reasonable assumption given the narrow beam width, then the cross-sectional area of the target is proportional to z^2 so the intensity of the signal returned from a cloud deck is proportional to z^{-2}. Eye safety is a critical issue with laser ceilometers. The laser wavelength and pulse power are selected to be safe for eyes.

Precipitation creates a special problem for a cloud height indicator. Precipitation particles attenuate both the upward beam and the downward echo. This can completely obscure the cloud base. One response to this problem is to compute a vertical visibility during this condition, similar to that of the visibility sensor.

EXAMPLE

The ceilometer in table 11-1 has a range of 7.5 km divided into 500 range gates. Each gate represents a height interval of 15 m and, since the returning signal is

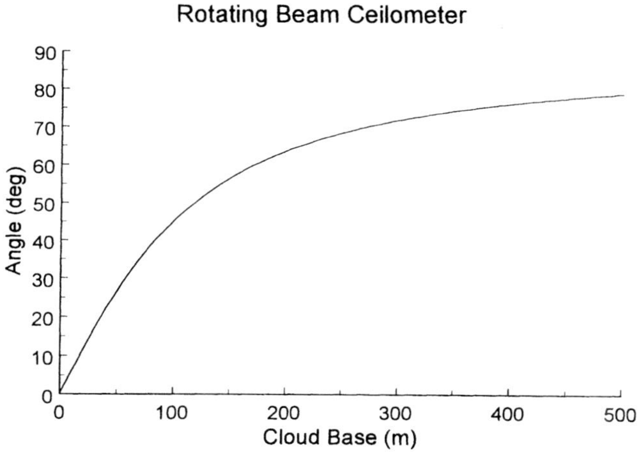

Rotating Beam Ceilometer

Fig. 11-6 Transfer function of a rotating beam ceilometer with a 100 m baseline.

averaged (filtered) over that interval, the resolution is 15 m. The signal is propagated at the speed of light, so the travel from the source to the range limit is 25 µs. Pulses are transmitted at about 2 ms intervals.

The signal for each range gate is averaged across all the pulses during the acquisition time (15 to 60 s). If the averaging is 15 s, each range gate would be the average of 7500 samples.

With a beam width of 5 mrad, the diameter of the beam width at 7.5 km would be $2 \times 7500 \times \tan(0.005/2) = 37.5$ m. The assumption is that the beam width is defined to be the angular distance between points where the power is 1/2 of the power on the center line.

Each return detected within its range is accumulated in bins (or range gates). Each bin represents a height interval of range resolution. The distribution of counts in each height bin over a time averaging is analyzed to detect multiple cloud layers. The ceilometer also notes the frequency of occurrence of each of these layers so that it can differentiate scattered, broken, and overcast layers.

Some laser ceilometers are designed to tilt at variable angles. They can measure cloud height (given the tilt angle) along the beam path and measure visibility along

Table 11-1 One possible implementation of a laser ceilometer.

Specifications	Units
Range	0...7.5 km (0...25 000 ft)
Range gates	500
Resolution	15 m (50 ft)
Laser type	Pulsed, 905 nm
Pulse repetition frequency	500 pulse/s
Beam width	5 mrad (0.29°)

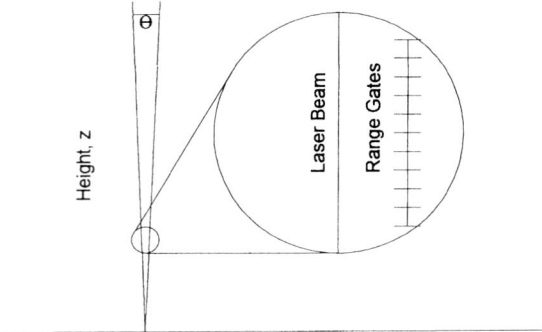

Fig. 11-7 Laser ceilometer showing some the range gates in detail. The horizontal scale has been exaggerated to show the beam width angle.

this slant path. Streicher et al. (1993) have built a lidar (light detection and ranging) to measure the slant range. This kind of visibility is useful since the slant path can be similar to the aircraft approach path. Then the measured visibility is more like the pilot's perspective than visibility along a horizontal path.

QUESTIONS

1. What is the apparent brightness of a black object against a sky background
   (a) when the contrast $= -1$.
   (b) when the contrast $= 0$.
2. A laser ceilometer detects a return signal 6.7 µs after the transmitted pulse. What is the cloud height?
3. If a laser ceilometer maximum detectable cloud height were limited by its pulse repetition frequency, what would be the maximum for a laser with a pulse repetition frequency of 770 Hz? What does set the maximum range?
4. If a laser ceilometer emits a 1 J pulse, what would be the energy in the pulse returned from a cloud deck at 3000 m if the cloud deck were a perfect reflector and there were no atmospheric attenuation?
5. Why use a harmonic mean instead of an arithmetic mean for visibility?
6. Lower limit of contrast $= C_C$, what is the equivalent brightness?
7. List some ways in which nighttime visibility differs from daytime visibility.
8. Aside from dust, smoke, hydrometeors, etc., daytime visibility is affected by what?
9. In an absolutely clear atmosphere, what would E be for
   (a) a transmissometer?
   (b) a forward scatter meter?
10. Define:

    critical contrast
    extinction
    eye-safe laser
    forward-scatter visibility
    harmonic mean
    laser ceilometer

meteorological optical range, MOR
rotating beam ceilometer, RBC
rotating receiver ceilometer, RRC
runway visual range, RVR
sensor equivalent visibility, SEV
slant visual range
transmissometer

BIBLIOGRAPHY

Bradley, J.T., R. Lewis, and J. Nilsen, 1991: Visibility measurements for the automated surface observing system (ASOS). *Preprints 7th Symp. on Meteorological Observations and Instrumentation*, New Orleans, LA. American Meteorological Society, Boston, MA, pp. 344–347.

Brown, H.A., and D.C. Burnham, 1993: Evaluation of a prevailing visibility sensor based on a scanning solid state video camera. *Preprints 8th Symp. on Meteorological Observations and Instrumentation*, Anaheim, CA. American Meteorological Society, Boston, MA, pp. 80–85.

Burnham, D.C., 1991: High visibility measurements and reference standards. *Preprints 7th Symp. on Meteorological Observations and Instrumentation*, New Orleans, LA. American Meteorological Society, Boston, MA, pp. 332–337.

Burnham, D.C., 1993: Fog, snow and rain calibrations for forward-scatter visibility sensors. *Preprints 8th Symp. on Meteorological Observations and Instrumentation*, Anaheim, CA. American Meteorological Society, Boston, MA, pp. 66–71.

Koschmieder, H., 1925: Theorie der horizontalen Sichweite. *Beitr. Phys. Atmos.*, 12, 35–53.

Meteorological Office, 1981: *Measurement of Visibility and Cloud Height*, Vol. 7 of *Handbook of Meteorological Instruments*, 2nd ed. Meteorology Office 919g. Her Majesty's Stationery Office, London, 38 pp.

Shipley, S.T., and I.A. Graffman, 1991: ASOS ceilometer measuremnts of airosols and mixed layer height in the lower atmosphere. *Preprints 7th Symp. on Meteorological Observations and Instrumentation*, New Orleans, LA. American Meteorological Society, Boston, MA, pp. J228–J229.

Streicher, J., C. Münkel, and H. Borchardt, 1993: Trial of a slant visual range measuring device. *J. Atmos. Oceanic Technol.*, 10, 718–724.

Tonna, G., and K.S. Shifrin, 1992: Reliability of the polar nephelometer for the measurement of visibility in fog. *Appl. Optics*, 31, 2932–2941.

Viezee, W., and R. Lewis, 1986: Visibility measurement techniques. In *Probing the Atmospheric Boundary Layer*, ed. and D.H. Lenschow. American Meteorological Society, Boston, MA, 269 pp.

Wyszecki, G., and W.S. Stiles, 1982: *Color Science: Concepts and Methods, Quantitative Data and Formulae*. 2nd ed. John Wiley & Sons, New York, 950 pp.

12

Upper Air Measurements

Measurements of atmospheric properties become progressively more difficult with altitude above the surface of the earth, and even surface measurements are difficult over the oceans. First balloons, then airplanes and rockets, were used to carry instruments aloft to make in-situ measurements. Now remote sensors, both ground-based and satellite-borne, are used to monitor the atmosphere. In this context, upper air means all of the troposphere above the first hundred meters or so and, in some cases, the stratosphere.

There are many uncertainties associated with remote sensing, so there is a demand for in-situ sensors to verify remote measurements. In addition, the balloon-borne instrument package is relatively inexpensive. However, it should be noted that cost is a matter of perspective; a satellite with its instrumentation, ground station, etc. may be cost-effective when the mission is to make measurements all over the world with good space and time resolution, as synoptic meteorology demands.

12.1 Methods for Making Upper Air Measurements

Upper air measurements of pressure, temperature, water vapor, and winds can be made using in-situ instrument packages (carried aloft by balloons, rockets, or airplanes) and by remote sensors. Remote sensors can be classified as active (energy emitters like radar or lidar) or passive (receiving only, like microwave radiometers), and by whether they "look" up from the ground or down from a satellite. Remote sensors are surveyed briefly before discussing in-situ instruments.

12.1.1 Remote Sensing

Profiles of temperature, humidity, density, etc. can be estimated from satellites using multiple narrow-band radiometers. These are passive sensors that measure longwave radiation upwelling from the atmosphere. For example, temperature profiles can be estimated from satellites by measuring infrared radiation emitted by CO_2 (bands around 5000 μm) and O_2 (bands around 3.4 μm and 15 μm) in the atmosphere. Winds can be estimated from cloud movements or by using the Doppler frequency shift due to some component of the atmosphere being carried along with the wind. An active sensor (radar) is used to estimate precipitation and, if it is a Doppler radar, determine winds.

The great advantage of satellite-borne instruments is that they can cover the whole earth with excellent spatial resolution. Time resolution can be good, depending upon the number of satellites and the orbit used. Typically, vertical resolution is the limiting factor.

All of the GPS satellites transmit radio signals that are used by ground receivers to determine receiver location. One of the sources of position error is the transit delay induced by atmospheric water vapor, which increases the index of refraction. This signal delay, which is a problem for position measurement, can be measured to obtain total precipitable water vapor.

A wind profiler is a highly specialized Doppler radar designed to measure wind profiles. They are relatively low-power, highly sensitive clear-air radars that operate with wavelengths of 33 cm to 6 meters (in contrast to 10 cm used by most meteorological radars). They detect fluctuations in atmospheric density that are used as a tracer of the mean wind. Their antennas are mechanically fixed, pointing straight up, but can be electronically steered a few degrees (10 to 15) from the vertical in at least two directions (e.g., N and E). A typical 74-cm wavelength wind profiler provides hourly averaged wind profiles to 16 km. It measures winds almost directly above each site. Temporal resolution is very good but spatial resolution is limited to the number and location of sites.

Wind profilers can be supplemented with RASS (Radio Acoustic Sounding System) to get a virtual temperature profile. Four vertically oriented acoustic transmitters, one on each side of the profiler antenna, emit sound waves at various frequencies. The wind profiler can detect the density fluctuations caused by the sound waves and obtains a maximum return when the acoustic wavelength is equal to one half the radar wavelength. Since the acoustic frequency at that moment is known, the speed of sound can be determined. From this measurement, the virtual temperature can be calculated, as it is proportional to the square of the speed of sound. The virtual temperature profile can be measured routinely to 4 km and sometimes to 7 km.

12.1.2 In-Situ Platforms

Instrumented aircraft can make measurements along the flight path with a wide variety of instrument packages. Research aircraft have made specialized measurements in support of research programs but these flights are not available for routine measurements. Since there are few research aircraft available for meteorological soundings, they may be equipped with very sophisticated instrument packages. Rockets have also been used to carry special instrument packages to high altitudes.

Unmanned aircraft have been used to make atmospheric measurements. Some are able to fly above the tropopause (18 km in the tropical Pacific) for extended periods (49–72 hours) at a reasonable cost.

Commercial airplanes have been equipped with simple instrument packages to make measurements of pressure, temperature, and humidity along the standard flight path. Data are transmitted to ground via satellite communication links. This is especially useful for trans-oceanic flights. Disadvantages of this method are that vertical profile data are not obtained from these flights, data can be obtained only along standard commercial flight paths, and the airline controls the flight schedule.

A simple balloon can be tracked with radar or a theodolite to obtain winds along the flight path of the balloon as it ascends. When equipped with an instrument package or sonde, it can measure pressure, temperature, and humidity in addition to the wind vector. Balloon technology changes very slowly, but sonde technology has improved rapidly to include advanced sensor designs and GPS receivers for navigation. There is a continuing demand for sounding balloons, despite advances in remote sensors, because the sounding balloon is the least expensive way to make in-situ measurements. The great disadvantage of the sounding balloon is its very poor temporal (typically one or two flights per day from each station) and spatial resolution.

Specialized versions of sondes are used to measure other atmospheric variables such as ozone concentration or electric field. Another variation of the sonde is dropped from aircraft (the drop-sonde) instead of being carried by a balloon. Its fall speed is controlled with a parachute.

12.2 Balloons

Meteorological balloons can be categorized according to their size, color and extensibility. Most balloons are extensible; that is, they stretch. The volume of the balloon increases as the balloon ascends until it reaches the bursting altitude. An inextensible balloon does not stretch. It ascends until the buoyancy force moving it up is balanced by the weight of the balloon and payload, and then it maintains constant altitude. These are called constant level balloons. The color of the balloon matters only when the position of the balloon is to be observed visually. A light color is preferred on a clear, sunny day and a dark color on a cloudy day.

Pilot balloons are tracked visually to determine the horizontal winds. They do not carry a payload (no instruments) so they are smaller, weighing 10 to 100 g. Sounding balloons, on the other hand, carry a payload and are larger, weighing 600 to 1800 g. Even larger balloons are available for special payloads. A few balloons that are available are listed in table 12-1. Balloons are inflated with either hydrogen or helium. Helium is an inert gas, so there is no explosion or fire hazard, but it is relatively rare and has a slightly higher density than hydrogen. Hydrogen is readily available and can be easily generated as needed. However, hydrogen can be very explosive.

The rate of ascent of a balloon is a function of the balloon lift, weight, and payload. The total lift force on a balloon is

$$F_T = \frac{\pi}{6} D^3 (\rho - \rho_B) g \qquad (12.1)$$

Table 12-1 Some meteorological balloons.

Weight, g	Volume, m³	Burst Alt., km	Burst Diam., m
10	0.05	7	0.6
30	0.14	12.5	1.02
100	0.21	15	1.33
300	1.9	21	3.9
1000	3.5	30.5	7.7
3000	5.3	39	13.5

where D = balloon diameter, m, ρ = air density, kg m^{-3}, ρ_B = density of gas in the balloon (hydrogen or helium), kg m^{-3}, g = acceleration due to gravity, m s^{-2}; the total lift mass is $m_T = F_T/g$. The free lift force is the total lift force minus the mass of the balloon and the payload,

$$F_L = \frac{\pi}{6}D^3(\rho - \rho_B)g - (m_B + m_P)g \qquad (12.2)$$

where m_B = mass of the balloon, kg, and m_P = mass of the payload, kg. As before, the free lift mass is $m_L = m_T - m_B - m_P$. Balloon drag is given by

$$F_D = \tfrac{1}{2}C_D\rho A w_B^2 \qquad (12.3)$$

where C_D = drag coefficient, a function of the Reynolds number, $R_e = w_B \rho D/\mu$ (non-dimensional), μ = dynamic viscosity, N s m^{-2}, A = balloon cross-sectional area = $\pi D^2/4$, m^2, and w_B = balloon vertical speed, m s^{-1}.

If the balloon altitude is known, the density and dynamic viscosity can be calculated from the observed data or from the US Standard Atmosphere listed in table 12-2. Then the Reynolds number can be determined and one can look up the drag coefficient in fig. 12-1. Steady-state ascent is achieved when $F_D = F_L$. We can solve for the balloon vertical speed,

$$w_B = \sqrt{\frac{\pi D^3(\rho - \rho_B)g/6 - (m_B + m_P)g}{C_D \rho \pi D^2/8}} \qquad (12.4)$$

which is valid at any altitude. At standard temperature and pressure (STP) (273.15 K, 1013.25 hPa), the density of air is $\rho_0 = 1.293$ kg m^{-3}, the density of hydrogen is $\rho_{H0} = 0.0901$ kg m^{-3}, and the density of helium is $\rho_{He0} = 0.1785$ kg m^{-3}. The zero subscript is used to indicate STP conditions.

EXAMPLE

Calculate the vertical speed of a 30 g pilot balloon (with no payload) from table 12-1. Assume it will be launched at $z = 0$ using the Standard Atmosphere, table 12-2. Assume that we are using hydrogen and that $C_D = 0.4$ and $g = 9.81$ m s^{-2}

SOLUTION

The density of hydrogen $\rho_B = \rho_{B0}\rho/\rho_0 = 0.0901 \times 1.293/1.2250 = 0.0854$ kg m^{-3}.

$$w_B = \sqrt{\frac{9.81(3.141593(0.644)^3(1.2250 - 0.0854)/6 - (0.03 + 0))}{0.4(1.2250)3.14159(0.644)^2/8}} = 4.05 \text{ ms}^{-1}$$

Table 12-2 US Standard Atmosphere.

z (km)	T (K)	p (hPa)	ρ (kg m^{-3})
0.0	288.1	1013.25	1.2250
1.0	281.7	898.76	1.1117
2.0	275.2	795.01	1.0066
3.0	268.7	701.21	0.9093
4.0	262.2	616.60	0.8194
5.0	255.7	540.48	0.7364
6.0	249.2	472.17	0.6601
7.0	242.7	411.05	0.5900
8.0	236.2	356.51	0.5258
9.0	229.7	308.00	0.4671
10.0	223.3	264.99	0.4135
12.0	216.6	193.99	0.3119
14.0	216.6	141.70	0.2279
16.0	216.6	103.52	0.1665
18.0	216.6	75.65	0.1217
20.0	216.6	55.29	0.0889
22.0	218.6	40.47	0.0645
24.0	220.6	29.72	0.0469
26.0	222.5	21.88	0.0343
28.0	224.5	16.16	0.0251
30.0	226.5	11.97	0.0184
31.0	227.5	10.31	0.0158
32.0	228.5	8.89	0.0136

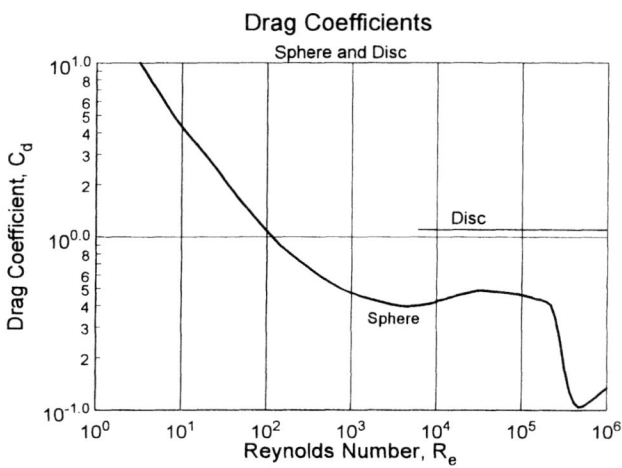

Fig. 12-1 Drag coefficients as a function of the Reynolds number for a disc and for a sphere.

EXAMPLE

Use the same balloon as in the example above but calculate the ascent rate at $z = 5$ km with conditions as given in table 12-2 for the Standard Atmosphere. Use the same drag coefficient and acceleration due to gravity.

SOLUTION

Since the density is 0.7364 kg m^{-3}, the density of hydrogen decreases to 0.05131 kg m^{-3} and balloon diameter increases to 0.777 m. Then

$$w_B = \sqrt{\frac{9.81\left(3.141593(0.777)^3(0.7364 - 0.05131)/6 - (0.03 + 0)\right)}{0.4(0.7364)3.141593(0.777)^2/8}} = 4.41 \text{ ms}^{-1}$$

The ascent rate has increased slightly due to the decreased density.

To calculate the vertical speed throughout the ascent, one must consider the change of these variables with height. As the atmospheric density changes, so will the density of the balloon gas, the balloon diameter, and the volume. As the speed and diameter change, so will the drag coefficient. As a first approximation, we can assume that the balloon volume will be inversely proportional to the density; that is, $V_B \rho = V_{B0} \rho_0$. This will hold up to the burst altitude. The balloon gas density is given by $\rho_B \rho_0 = \rho_{B0} \rho$. If the air density is known, perhaps from the US Standard Atmosphere (an excerpt is listed in table 12-2), the ascent rate can be calculated from

$$\left. \begin{array}{l} \rho_B = \rho_{B0} \dfrac{\rho}{\rho_O} \\[8pt] D = D_0 \left(\dfrac{\rho_0}{\rho}\right)^{1/3} \\[8pt] w_B = \sqrt{\dfrac{g(\pi D^3 (\rho - \rho_B)/6 - (m_B + m_P))}{C_D \rho \pi D^2 / 8}} \end{array} \right\} \quad (12.5)$$

This calculation requires some assumptions about the drag coefficient. It is possible to use a graph (fig. 12-1) with the Reynolds number. It is convenient to assume that $C_D = 0.4$, but balloons sometimes enter the transition region ($R_e > 3 \times 10^5$) where the drag coefficient quickly drops. The diameter and ascent speed are plotted in fig. 12-2 for a 200 g balloon with a 190 g payload.

The vertical speed deviates from this model for a number of reasons, as listed below.

- The balloon material is not perfectly elastic, so the balloon volume will not be inversely proportional to atmospheric density, as assumed. At some point the balloon will stop stretching and eventually burst.
- Hydrogen (or helium) will leak though the balloon.
- As the balloon ascends into cooler air, thermal lag may cause the gas inside to be warmer than the outside air. Reynolds (1966) found that only 9% of the variation in ascent rate could be attributed to the lapse rate effect.
- The balloon shape is not always spherical. This will change the diameter-to-volume relation and will affect the drag coefficient.
- Vertical wind components will add to or subtract from the buoyancy speed.

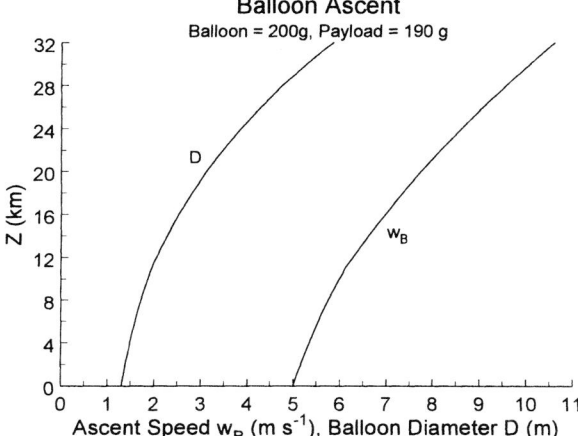

Fig. 12-2 Balloon ascent using a simple model.

- For the ascent shown in fig. 12-2, the drag coefficient was assumed to be a constant 0.4. The drag coefficient will not be constant but varies as the product of the density, balloon diameter, and ascent speed (the dynamic viscosity does not change very rapidly).

The typical initial ascent rate is 5 to 6 m s^{-1} and, due to the problems noted above, the ascent rate is usually nearly constant with altitude.

12.3 Wind Measurement

Horizontal wind speed can be determined by tracking a balloon. When a balloon is used without an instrument package, it is called a pilot balloon. It can be tracked by a single theodolite, double theodolites, or by radar. If the balloon has an instrument package, some additional methods are available such as the radio theodolite and the use of navigation aids including GPS.

Since balloons are characterized by large surface area and small mass, they usually follow the mean wind flow. Even when incumbered with an instrument package, they follow the flow because the package usually has a fairly small surface area compared to that of the balloon.

The force exerted on a balloon by the wind is

$$F = m_B \frac{dV_B}{dt} = C_D \rho A (V - V_B)^2 \tag{12.6}$$

where m_B = balloon mass, kg, V = horizontal wind speed, m s^{-1}, V_B = horizontal balloon speed, m s^{-1}, C_D = balloon drag coefficient, ρ = air density, kg m^{-3}, and A = balloon cross-sectional area. If we assume that the balloon speed is approximately equal to the wind speed, we can simplify this equation to

220 Meteorological Measurement Systems

$$\tau \frac{dV_B}{dt} + V_B = V \qquad (12.7)$$

where

$$\tau = \frac{m_B}{C_D \rho A V} \qquad (12.8)$$

and the time constant τ is inversely proportional to the wind speed. For balloons, m_B is small and A is large so τ is small. Thus the balloon accelerates and decelerates rapidly to follow the wind. Therefore, the assumption that the balloon follows the wind seems to be well justified.

EXAMPLE
Calculate the time constant of a 100 g balloon at sea level.

SOLUTION
From table 12-1 we find the balloon volume to be $0.21\,\text{m}^3$ so the cross sectional area is $0.427\,\text{m}^2$. Let $C_D = 0.4$ and assume that the density is $1.293\,\text{kg}\,\text{m}^{-3}$. If the wind speed is $10\,\text{m}\,\text{s}^{-1}$, then

$$\tau = \frac{0.100}{0.4 \times 1.293 \times 0.427 \times 10} = 0.045\,\text{s}$$

Even if the wind speed were only $1\,\text{m}\,\text{s}^{-1}$, the time constant would be 0.45 s.

12.3.1 Theodolites

A pilot balloon can be tracked visually with a single theodolite that measures the azimuth, relative to true North, and the elevation angle. If the balloon ascent rate is known (assumed from balloon and gas specifications), the position of the balloon can be determined and the wind velocity inferred from successive balloon positions. With a second theodolite, the balloon position can be determined more accurately. This method is still limited to visual range and is unsatisfactory when there is cloud cover.

Figure 12-3 shows the geometry for the single theodolite method.

The procedure is to track the balloon optically with the theodolite and measure the azimuth and elevation angles at, typically, one-minute intervals. Height z can be determined from the known (assumed) ascent rate. From fig. 12-3, it can be shown that the following relations

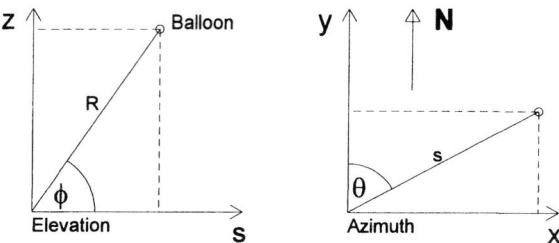

Fig. 12-3 Single theodolite geometry showing both the elevation and azimuth angles.

$$\left.\begin{array}{l} x = S\sin\theta \\ y = S\cos\theta \\ z = R\sin\phi \\ s = R\cos\phi \end{array}\right\} \quad (12.9)$$

apply and that one can solve for x and y using

$$\left.\begin{array}{l} x = z\cot\phi\sin\theta \\ y = z\cot\phi\cos\theta \end{array}\right\} \quad (12.10)$$

where θ and ϕ are the azimuth and elevation angles, x and y define the horizontal position of the balloon relative to the theodolite, S is the distance from the theodolite to the balloon along the earth's surface (ignoring earth curvature), and R is the slant range to the balloon. If measurements are made at one-minute intervals, the u- and v-components of the horizontal wind vector are given by $u = (x_n - x_{n-1})/60$ and $v = (y_n - y_{n-1})/60$. At low elevation angles, below 10°, this method is very sensitive to errors in the measurement of the elevation angle (see fig. 12-4). Note that a bias error in the calculated x, y position would not affect the calculated wind speed since the speed is determined from the difference in successive balloon positions. All

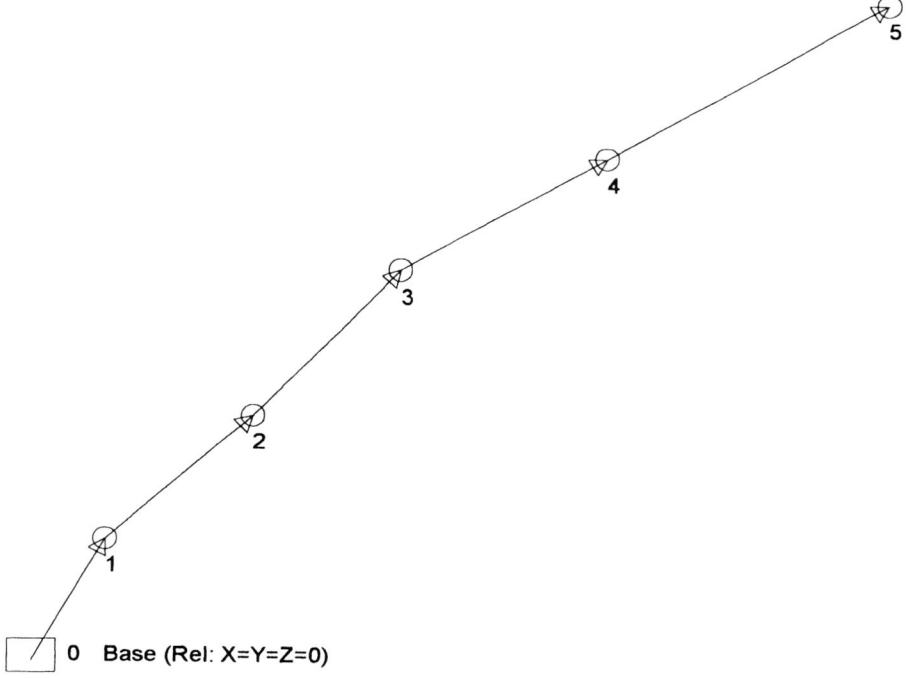

Fig. 12-4 Wind-finding using balloon positions at regular time intervals.

balloon or radiosonde wind-finding techniques require only differential location methods.

With a radiosonde, a balloon equipped with an instrument package that transmits measured data to ground using a small radio transmitter, position can be determined by a radio theodolite, or radio direction finder, that tracks the radio signal from the radiosonde. Combined with the pressure reported by a sonde, the complete position, and thus the winds, can be determined. The size of the radio theodolite antenna is directly related to the transmission frequency of the sonde. The higher the frequency, the smaller the antenna needed. Therefore, radio theodolites work best with higher-frequency sondes. The usual frequency is 1680 MHz.

EXAMPLE: SINGLE THEODOLITE COMPUTATION

Assume that a balloon ascends at a constant rate of 6 m s^{-1}. You are given data at 8, 9, and 10 minutes into the flight. The measured azimuth and elevation angles at these times are (187.5, 71.2), (205.8, 70.2), and (221.2, 68.4), respectively. Compute the balloon position and the wind speed and direction.

SOLUTION

The balloon elevation is given by the balloon ascent rate times the flight time. At a flight time of 8 minutes the elevation is 2880 m (8 min ×60 s/ min × 6 m s^{-1} = 2880 m), 3240 m at minute 9, and 3600 m at minute 10.

Calculate the x, y position using eqn. 12.10. (At minute 8, $x = z \cot\phi \sin\theta =$ 2880 cot(71.2) sin(187.5) = −128.0 m and $y = z \cot\phi \cos\theta =$ 2880 cot(71.2) cos (187.5) = −972.0 m.) We cannot calculate the winds at minute 8 because we need to know how far the balloon has moved during the previous minute and we have no data for minute 7. The x, y, z position at minute 9, from a similar calculation, is (−507.7, −1050.2, 3240). Next, calculate the horizontal wind components, u and v.

$$u = (x_n - x_{n-1})/\Delta t = (-507.7 + 128.0)/60 = -6.33 \text{ m s}^{-1}$$

$$v = (y_n - y_{n-1})/\Delta t = (-1050.2 + 972.0)/60 = -1.30 \text{ m s}^{-1}$$

Then the horizontal speed = $(u^2 + v^2)^{1/2} = ((-6.33)^2 + (-1.30)^2)^{1/2} = 6.46 \text{ m s}^{-1}$. Wind direction is given by conversion from rectangular to polar coordinates, to obtain 78.4° [See Appendix D]. The wind components indicate that the wind is blowing from the first quadrant at an angle approaching 90°. The results of these calculations are summarized in the following table.

Computation of winds from a fragment of a balloon flight.

Time (min)	Az (deg)	El (deg)	x (m)	y (m)	z (m)	Wind Dir.	Wind Spd.
8	187.5	71.2	−128.0	−972.0	2880	–	–
9	205.8	70.2	−507.7	−1050.2	3240	78.4	6.5
10	221.2	68.4	−938.9	−1072.5	3600	87.1	7.2

12.3.2 Radar

If the balloon carries a corner reflector, it can be tracked with radar that measures the slant range to the balloon and the azimuth and elevation angles. From these data, the position, including height, can be calculated. The radar measures angles to within $0.1°$ and range to within 30 m. The balloon height can be corrected for earth curvature. Due to refraction of the beam in the atmosphere, the beam path is not a straight line but is curved. Many radar installations routinely correct for earth curvature and for typical refraction effects.

12.3.3 Navigation Aids

Some radiosondes carry special radio receivers that detect navigational signals and relay them to a ground station where a computer calculates the balloon position. This scheme requires a more expensive sonde but a much less expensive ground station because neither radar nor a radio direction finder is required. This arrangement is ideal for portable ground stations, especially those mounted on ships or land vehicles. The navigation aids available are LORAN C, OMEGA, and GPS.

12.3.3.1 LORAN C

LORAN C is a high-accuracy, medium-range navigational aid operating in the low-frequency band centered on 200 kHz. Since its primary use is for marine navigation, particularly in coastal and continental shelf areas where high accuracy is demanded, LORAN C coverage has been provided mostly in certain parts of the world, mainly in or close to maritime areas in the northern hemisphere. This coverage has been extended even to some continental areas by radiosondes. LORAN C transmission consists of groups of about 8 pulses of 100 kHz carrier, each some $150\,\mu s$ in duration. Each chain of transmitters consists of one master station and two or more slaves. In principle, chain coherence is established by references to the master transmission. Each slave transmits its groups of pulses a fixed interval after the master and at a rate which is specific to a given chain. Typically, this is once every 100 ms. The signal received at a given location is a composite of that received as a ground wave and that as sky waves after one or more reflections by the ionosphere. The pulsed nature of the transmissions allows the possibility of selecting the ground wave preferentially by detecting the phase or time of arrival of the leading section of the pulse.

12.3.3.2 OMEGA

OMEGA is a network of eight atomic-clock-controlled transmitters operating in the very-low-frequency band, designed to provide global coverage. Each station transmits sequentially for 0.9 to 1.2 s on the three assigned frequencies of 10.2, 11.3, and 13.6 kHz. No two stations transmit simultaneously on one frequency, nor does any individual station transmit on more than one frequency at a time. The cycle is repeated every 10 s. At the chosen frequencies the ionosphere and the Earth's surface act as a waveguide. The OMEGA transmitters excite various modes of propagation whose amplitudes and phase velocities vary with the height of the ionosphere, direction of propagation, and range from the transmitter. As a result of the presence of many high order modes, the signal phase is difficult to predict within about 1000 km of a transmitter. Beyond this range the phase is sensitive to diurnal and sporadic fluctuation caused by changes in the height and nature of the ionosphere. To the extent that such disturbances have a uniform effect on the signal phase over an area, a correction can be applied. The correction can be inferred by subtracting from the inferred motion of the balloon the fictitious motion of a local ground station monitoring the same OMEGA transmissions as those received at the balloon. This is known as differential OMEGA. Of course, if the ground station, such as a ship, has a real motion, this must be known and a further correction applied.

12.3.3.3 GPS

The Global Positioning System (GPS) is an array of 24 satellites in orbit 20 200 km above the earth. The orbital period is 12 hours and the orbital plane is inclined 55° with respect to the equatorial plane. Each satellite has an atomic clock and transmits a signal at precise intervals on two frequencies: 1.226 Ghz and 1.575 GHz. If a receiver can hear four satellites simultaneously (i.e., if there are at least four satellites above the receiver's horizon), the position of the receiver can be determined to within 100 m (worst case). This system works anywhere on the earth's surface. Sondes have been built that incorporate a portion of a GPS receiver. The sonde transmits the received GPS signals to its ground station which also receives the GPS signals directly from the satellites. The ground station computes the position of the sonde and of itself and obtains the position of the sonde relative to the ground station. This procedure, called differential GPS, compensates for many of the system errors including selective availability and propagation delays.

The basic principle of GPS is triangulation from satellites based on accurate knowledge of satellite positions in space. The distance from a satellite to a receiver is measured using the travel time of a radio signal and, as the GPS signal travels from the satellite to the receiver, the ionosphere and the earth's atmosphere delay it. This delay is converted into a distance with the knowledge that the speed of the radio signal is (very close to) the speed of light.

To examine how GPS is used for location finding in three-dimensions, we will examine the analogous two-dimensional case in which one less satellite is required. The principles to be discussed are the same but the required diagrams and interpretation are simpler.

Consider the situation in which you are trying to pinpoint your location using only one satellite. If you know precisely your distance from a satellite (found from the signal travel time), call it X km, then your location must be on a circle (a sphere is 3-D) X km in diameter centered on the satellite. If the distance to a second satellite is known to be Y km, then the desired location is where the two circles intersect, as shown in fig. 12-5. As shown in the figure, the two circles actually intersect in two different places so there are two possible locations. However, one can usually be ruled out on the basis of its position or speed that is usually unreasonable. In 3-D, two spheres intersect to form a circle, the desired location lying somewhere on that circle, and a third satellite is required to give the exact location.

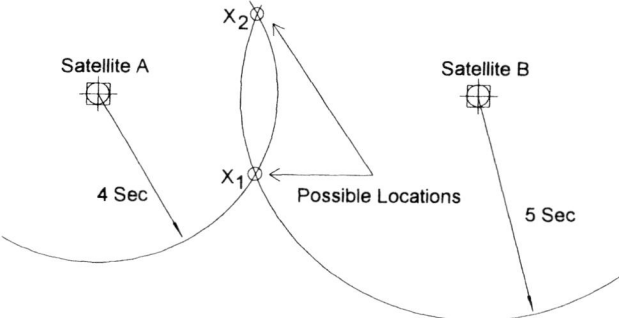

Fig. 12-5 GPS navigation in 2D with precise clocks. It requires just two satellites.

The method described above works if the exact signal travel times are known, which would require both the satellite and the receiver to have extremely accurate clocks (e.g., atomic clocks). If this were the case, then the position could be determined in 3-D with three satellites. Unfortunately, atomic clocks are far too expensive to be used in low-cost GPS receivers and so an alternative solution had to be developed. Trigonometry dictates that if three perfect measurements locate a point in space, then four imperfect measurements can be used to account for timing offsets, assuming the offset is consistent. Consider the 2-D example shown in fig. 12-6. The travel times from satellites A and B are 4 s and 5 s respectively and the correct location is X_1. Now if the GPS receiver has a 1-second offset, then the travel times will be read as 5 s and 6 s and the incorrect position will be given as X_3. Note that there is no way to know the position is wrong with only two satellites. If a third satellite (see fig. 12-7) is used with a travel time of 6 s, then there is only one location, X_1, that can simultaneously be 4 s, 5 s, and 6 s from all three satellites. In 3-D, four satellites are required.

This is the basic principle involved in GPS wind-finding, although the actual mathematics, trigonometry, and computations involved are much more complicated.

Table 12-3 summarizes the navigational aids (Navaid) techniques with approximate speed and vertical displacement errors. One special advantage of these Navaid techniques is that they give the height of the sonde directly. Radio direction finder (RDF) techniques do not give sonde height. It is inferred from the pressure measurement.

12.4 Radiosondes

Sounding balloons carry a payload, usually a radiosonde, to heights of 30 km or more. This instrument package measures atmospheric variables, usually pressure, temperature, and humidity, and sends these data to a ground station via a radio transmitter. Available radio frequency bands are 27.5 to 28 MHz, 400 to 406 MHz, 1668 to 1700 MHz, and 35.2 to 36 GHz. Special instrument packages have been designed to measure solar and terrestrial radiation, ozone concentration, electrical potential gradient, air conductivity, and radioactivity of the atmosphere. Horizontal winds are measured

226 Meteorological Measurement Systems

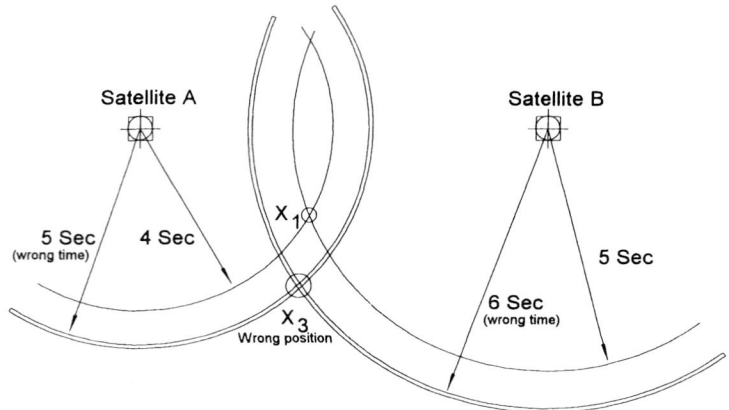

Fig. 12-6 Navigation with imprecise clocks.

by tracking the instrument package, or the package can determine and report its own location using one of several navigation aids listed above.

Since a radiosonde is generally used only once, it must be designed for mass production at low cost. It must be capable of operating on an internal battery for about three hours, and transmitting its radio signal at least 200 km. For aircraft safety, its mass density must be low enough to not damage a jet engine if ingested, nor must it be capable of breaking a windscreen. The sensors have only a short active service life but may have a long shelf life before use. Either drift must be carefully controlled or the sensors must be recalibrated with simple ground equipment before the flight.

Signals from each sensor in turn are multiplexed to a modulator and then sent to the radio transmitter. In older designs a mechanical multiplexer, advanced by the aneroid cell, switched the signals. Newer designs sometimes incorporate a microprocessor to multiplex the signals on a timed basis and digitally encode the data. To obtain readings representative of layers about 100 m thick, a timed switch must complete a cycle in 15 to 20 seconds if the ascent rate is $5\,\mathrm{m\,s^{-1}}$.

The pressure sensor is usually a single metallic diaphragm aneroid or a silicon diaphragm sensor. They are usually designed to respond to the logarithm of pressure rather than linearly with pressure, to enhance their high-altitude performance. Even so, metallic diaphragm sensors lose sensitivity at low pressures (< 50 hPa) so they are sometimes augmented with hypsometers. As noted in chap. 2, the sensitivity of a hypsometer increases as the pressure decreases.

Temperature sensors may be small rod thermistors or temperature-sensitive capacitors. Since the radiosonde moves through a wide range of pressures (and densities),

Table 12-3 Wind vector error and vertical position error.

Navaid	Speed $(\mathrm{m\,s^{-1}})$	Vertical (m)
LORAN C	0.7	150
OMEGA	1.5	300
GPS	0.1	30

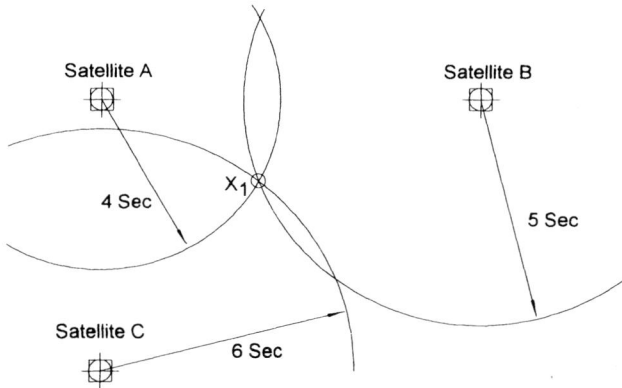

Fig. 12-7 GPS locations found in 2D with three satellites.

the thermal lag of the sensor must be small. The typical time constant for a rod thermistor 1.27 mm in diameter is 4.5 s at 1000 hPa, increasing to 10.6 s at 100 hPa and to 30 s at 10 hPa. With a time constant of 30 s, the sensor is not capable of resolving 15 to 20 s reporting layers at the top of the flight. Radiation error is another major concern. The sensor is ventilated at about 5 m s^{-1} by the ascent of the sonde but must not be directly exposed to the sun. The radiation shield is part of the sensor itself; for example, the "shield" consists of a highly reflective coating and a shield like that shown in fig. 4-14 is not used. Thus the shield used for the radiosonde temperature sensor does protect the sensing element from radiation heating, although not in the traditional sense.

With few exceptions, sorption humidity sensors are used in radiosondes. Psychrometric (wet- and dry-bulb) sensors have been used in at least one sonde and, recently, small chilled-mirror devices are being flown on sondes for field projects.

It is characteristc of sorption sensors that the response time increases as the temperature decreases since there is less water vapor present at lower temperatures. Some examples are listed in table 12-4.

The carbon element is temperature sensitive and is subject to drift due to contamination of the resistive surface. It is kept is a sealed container until just before launch

Table 12-4 Approximate time constant, in seconds, of humidity sensors.

Sensor	Temperature (°C)					
	20	10	0	−10	−20	−30
Hair	30	40	55	175	400	800
Hair, rolled flat	10	10	12	15	20	30
Goldbeater's skin	6	10	20	50	100	200
Carbon hygristor	0.3	0.6	1.2	3	7	15
Capacitive thin-film	0.3	0.7	1.5	4	9	20

when it must be calibrated and then inserted into the sonde. Pre-launch calibration consists of measuring the resistance when exposed to a humidity of 33% using a saturated salt solution.

12.5 Exposure Error

Radiosonde exposure errors are accentuated by the extreme variations of temperature, pressure, and density that the package must be able to withstand. Sensors are exposed to extremes of solar radiation, temperature, electric field gradients, environmental shock in handling, wetting and icing by cloud droplets, etc. To counter sonde environmental problems, it is generally best to make all of the sensors as small as possible.

Radiosondes are tethered to balloons by long strings to remove the sonde package from the environment of the balloon and, therefore, the sonde package swings below the balloon like a pendulum. For wind-finding systems with update cycles small compared to the period of the pendulum motion (which is approximately 0.1 Hz), the swinging motion can be filtered out using digital filters in the ground station.

Specific sensor errors are addressed below.

Barometer: small size means smaller temperature gradients across the sensor and faster response to temperature changes. A silicon integrated circuit device only a few mm in diameter and thickness can be used.

Temperature sensor: small size to reduce radiational heating error and dynamic lag error. The Vaisala RS-90 sonde uses an unshielded capacitive sensor 0.1 mm in diameter. The radiational heating error is approximately 0.5 K at 10 hPa and, after correction, residual error is on the order of 0.1 K. The time constant of this sensor is approximately 0.5 s at 1000 hPa.

Humidity sensor: some manufacturers use capacitive sorption sensors that are specially designed to eliminate condensation in supercooled clouds. The Vaisala RS-90 sonde uses a dual sensor to measure RH. In this design, one sensor is heated to remove potential condensate while the other is used to measure RH. Then, heat is switched to the other sensor and, after the first one cools, it is used to measure RH. The heating cycle is 40 s. Overall sensor dimensions are about 4 mm by 1.5 mm by 0.4 mm.

Both temperature and relative humidity sensors are located on an arm projecting from the main sonde housing to isolate them from heat from the sonde (e.g., the battery) and provide better atmospheric exposure.

QUESTIONS

1. Show why the model stated in eqn. 12.5 predicts that the balloon ascent speed will increase with altitude.
2. Does the expression in the numerator of eqn. 12.4 change as a balloon ascends? How about the denominator?
3. A simple model of balloon ascent showed ascent speed (*increasing or decreasing?*) with height. Actual ascent speeds are almost constant. Why?
4. A radiosonde navigation system finds the sonde position at one time as (x_n, y_n) and as (x_{n+1}, y_{n+1}) at time Δt later. What is the horizontal wind vector in this layer?

BIBLIOGRAPHY

Ahnert, P.R., 1991: Precision and comparability of national weather service upper air measurements. *Preprints 7th Symp. on Meteorological Observations and Intrumentation*, New Orleans, LA. American Meteorological Society, Boston, MA, pp. 221–226.

Boire, G., D.C. Sautter, S.P. Pryor, and A.K. Brown, 1993: A low cost GPS rawinsonde system. *Preprints 8th Symp. on Meteorological Observations and Instrumentation*, Anaheim, CA. American Meteorological Society, Boston, MA, pp. 23–24.

Doviak, R.J., and D.S. Zrnic, 1993: *Doppler Radar and Weather Observations*. 2nd ed. Academic Press, San Diego, 562 pp.

Elachi, C., 1987: *Introduction to the Physics and Techniques of Remote Sensing*. John Wiley & Sons, New York, 413 pp.

Finger, F.G., and F.J. Schmidlin, 1991: Upper-air measurements and instrumentation workshop. *Bull. Am. Meteor. Soc.*, 72, 50–55.

Hasse, L., and D. Schriever, 1980: Pilot balloon techniques. In *Air–Sea Interaction: Instruments and Methods*, ed. F. Dobson, L. Hasse, and R. Davis. Plenum Press, New York, 801 pp.

Hock, T.F. and H.L. Cole, 1991: A new aircraft universal lightweight digital dropsonde. *Preprints 7th Symp. on Meteorological Observations and Instrumentation*, New Orleans, LA. American Meteorological Society, Boston, MA, pp. 291–296.

Huovila, S., and A. Tuominen, 1991: Influence of radiosonde lag errors on upper-air climatological data. *Preprints 7th Symp. on Meteorological Observations and Instrumentation*, New Orleans, LA. American Meteorological Society, Boston, MA, pp. 237–242.

Jain, M., M. Eilts, and K. Hondl, 1993: Observed differences of the horizontal wind derived from Doppler radar and a balloon-borne atmospheric sounding system. *Preprints 8th Symp. on Meteorological Observaitons and Instrumentation*, Anaheim, CA. American Meteorological Society, Boston, MA, pp. 189–194.

Lauritsen, D.K., 1991: A review of the CLASS sounding system and an overview of its successor: NEXUS. *Preprints 7th Symp. on Meteorological Observations and Instrumentation*, New Orleans, LA. American Meteorological Society, Boston, MA, pp. 265–269.

Leroy, M., 1993: Test procedures and acceptance tests of radiosonde. *Preprints 8th Symp. on Meteorological Observations and Instrumentation*, Anaheim, CA. American Meteorological Society, Boston, MA, pp. 40–43.

Luers, J.K., 1990: Estimating the temperature error of the radiosonde rod thermistor under different environments. *J. Atmos. Oceanic Technol.*, 7, 882–895.

McMillin, L., M. Uddstrom, and A. Coletti, 1992: A procedure for correcting radiosonde reports for radiation errors. *J. Atmos. Oceanic Technol.*, 9, 801–811.

Nash, J., 1991: Implementation of the Vaisala PC-CORA upper air sounding system at operational radiosonde stations and test ranges in the United Kingdom. *Preprints 7th Symp. on Meteorological Observations and Instrumentation*, New Orleans, LA. American Meteorological Society, Boston, MA, pp. 270–275.

Nash, J., 1993: Characteristic errors in radiosonde temperature observations identified in the WMO radiosonde comparisons. *Preprints 8th Symp. on Meteorological Observations and Instrumentation*, Anaheim, CA. American Meteorological Society, Boston, MA, pp. 98–103

Olsen, R.O., R.J. Okrasinski, and F.J. Schmidlin, 1991: Intercomparison of upper air data derived from various radiosonde systems. *Preprints 7th Symp. on Meteorological Observations and Instrumentation*, New Orleans, LA. American Meteorological Society, Boston, MA, pp. 232–236

Parsons, C.L., G.A. Norcross and R.L. Brooks, 1984: Radiosonde pressure sensor performance: evaluation using tracking radars. *J. Atmos. Oceanic Technol.*, 1, 321–327.

Pratt, R.W., 1985: Review of radiosonde humidity and temperature errors. *J. Atmos. Oceanic Technol.*, 2, 404–407.

Reynolds, R.D., 1966: The effect of atmospheric lapse rates on balloon ascent rates. *J. Appl. Meteor.*, 5, 537–541.

Sauvageot, H., 1992: *Radar Meteorology*. Artech House, Boston, MA, 366 pp.

Schmidlin, F.J., 1991: Derivation and application of temperature corrections for the United States radiosonde. *Preprints 7th Symp. on Meteorological Observations and Instrumentation*, New Orleans, LA. American Meteorological Society, Boston, MA, pp. 227–231.

Schroeder, J.A., 1990: A comparison of temperature soundings obtained from simultaneous radiometric, radioacoustic, and rawinsonde measurements. *J. Atmos. Oceanic Technol.*, 7, 495–503.

Schwartz, B.E., and C.A. Doswell III, 1991: North American rawinsonde observations: problems, concerns and a call to action. *Bull. Am. Meteor. Soc.*, 72, 1885–1896.

Stephens, G.L., 1994: *Remote Sensing of the Lower Atmosphere*. Oxford University Press, New York, 523 pp.

Thompson, N., 1980: Tethered balloons. In *Air-Sea Interaction: Instruments and Methods*, ed. F. Dobson, L. Hasse, and R. Davis, Plenum Press, New York, 801 pp.

Tiefenau, H.K.E., and A. Gebbeken, 1989: Influence of meteorological balloons on temperature measurements with radiosondes: nighttime cooling and daylight heating. *J. Atmos. Oceanic Technol.*, 6, 36–42.

Uddstrom, M.J., 1989: A comparison of Philips RS4 and Vaisala RS80 radiosonde data. *J. Atmos. Oceanic Technol.*, 6, 201–207.

Wade, C.G., 1994: An evaluation of problems affecting the measurement of low relative humidity on the United States radiosonde. *J. Atmos. Oceanic Technol.*, 11, 687–700.

Weber, B. L., and D.B. Wuertz, 1990: Comparison of rawinsonde and wind profiler radar measurements. *J. Atmos. Oceanic Technol.*, 7, 157–174.

Williams, S.F., C.G. Wade and C. Morel, 1993: A comparison of high resolution radiosonde winds: 6-second MICRO-ART winds versus 10-second CLASS LORAN winds. *Preprints 8th Symp. on Meteorological Observations and Instrumentation*, Anaheim, CA. American Meteorological Society, Boston, MA, pp. 60–65.

13

Sampling and Analog-to-Digital Conversion

Along the signal path from the atmosphere, through the sensors and the data logger to the final archive, the signal quality may be irreversibly comprised. These faults include aliasing caused by poor sampling practice and quantization in an analog-to-digital converter. Aliasing and quantization will be defined in this chapter. Drift in some of the system parameters, such as temperature sensitivity, is generally preventable but is not always reversible.

Sampling of a signal occurs in the time domain and, frequently, in the space domain with one, two, or three dimensions. In the time domain, the time interval between successive points is called the sampling interval and the data logger controls this interval. When two or more sensors are distributed, vertically, along a mast then the system is sampling both in the time domain and in the space domain. When multiple measurements are arrayed along the surface of the earth, the sampling is occurring in time and in two or three space dimensions. Most meteorological systems are undersampled both in time and space. Space undersampling is an economic necessity. The consequence of undersampling is that frequencies above a certain limit, called the Nyquist frequency, will appear at lower frequencies and this is an irreversible effect.

Quantization occurs when the signal is converted from analog to digital in the analog-to-digital converter. Since the range of the converter is expressed in a finite number of digital states, signal amplitudes smaller than this quantity will be lost. This is another irreversible effect.

These are not the only irreversible effects. For example, drift is caused by physical changes in a sensor or other component of the measurement system. Drift may have a causal component, such as undocumented temperature sensitivity, and a random component such as wearing of an anemometer bearing. The former is theoretically preventable and reversible, whereas the latter is irreversible.

13.1 Signal Path

Each element of the system (see fig. 13-1) may include some signal averaging, and each element may add bias and gain. As noted in earlier chapters, a sensor is a transducer, a device that changes energy from one form to another. Every sensor has one or more energy reservoirs; therefore every sensor does some averaging of the signal. This point has been made in chaps 6 and 8.

Analog signal conditioning may include a transducer to convert resistance to voltage, for example, to add bias or to increase the gain of the signal, or to filter the signal. A filter could be used as a low-pass filter, which is a form of signal averaging.

The multiplexer in the data logger is just a switch in that only one signal at a time is to be connected to the amplifier and the analog-to-digital converter. The timing of the multiplexer and of the ADC is controlled by the computer in the data logger. At time intervals, the computer in the data logger selects the first channel of the multiplexer then triggers the ADC. Then the next channel is selected and the ADC is started again. In this way, each analog signal is sampled and converted to digital form.

The amplifier increases the gain of the signal and is designed to minimize bias. The gain is usually set at one of a few fixed gains, say 1, 10, 100, and 1000.

An analog-to-digital converter is used to convert an analog signal to an accurate digital number proportional to its amplitude. That digital number can then be used in the data logger computer for formatting, filtering, transmission, and more exotic functions like spectrum analysis.

Digital signal conditioning has many more functions than analog conditioning. But, of course, DSCs are more versatile and should not have any inadvertent bias or gain.

The sensor raw output signal is always an analog signal. It may take many forms including voltage or current, resistance, etc., and may be converted to other forms of energy by transducers or analog signal conditioning as shown in table 13-1.

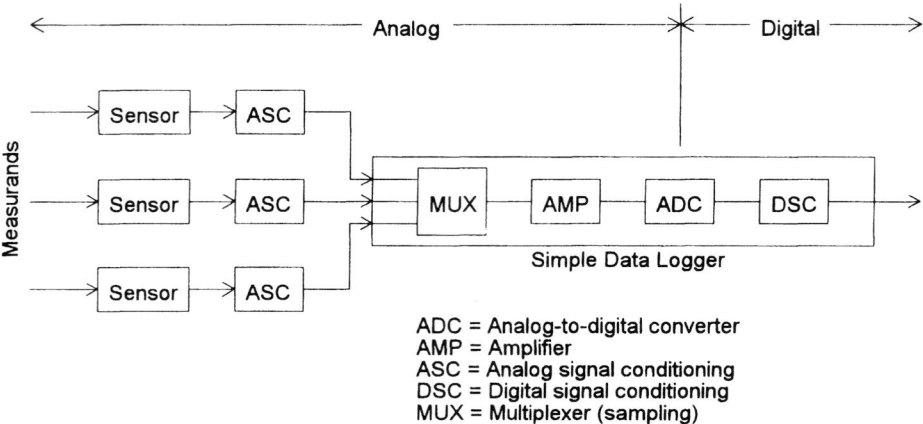

Fig. 13-1 Signal path from the sensors through the data logger.

Table 13-1 Some sensors and associated signal types.

Sensor	Raw Sensor Output	Transducer or Analog Signal Conditioning	Resulting Analog Signal
Thermocouple	Voltage	Amplifier	Voltage
RTD, thermistor	Resistance	Bridge circuit	Voltage
Rain gauge	Frequency or period	–	Frequency or period
Wind vane	Angular position	Potentiometer	Voltage
Cup and propeller anemometers	Rotation rate	DC generator	DC voltage
Cup and propeller anemometers	Rotation rate	Relay or light chopper	Frequency or period
Aneroid barometers	Deflection or position	Capacitive circuit	Frequency or period
Liquid-in-glass thermometers	Position	–	Position

13.2 Drift

Every component of a measurement system, electronic or mechanical, could be subject to drift. In a well-designed system the probability of a large drift is very small. Even small drift can be significant at a crucial point in a measurement circuit. The word "drift" implies that the changes are slow and gradual, but sometimes the changes can be relatively fast. And then the sensor or data logger will exhibit an unusual change from one calibration to another. The user then cannot determine when the change occurs nor can he correct the measurements.

In sensitive applications the system must provide redundancy or a self-testing mode. Redundancy with sensors can work when the sensor cost is fairly low or when the data are critical. Self-testing is not applicable in many sensors, including cup anemometers and barometers, but redundancy would work. A data logger is an example where self-testing will work. The simple data logger in fig. 13-1, the multiplexer, could provide another channel devoted to checking the multiplexer, the amplifier, and the analog-to-digital converter. The computer in the data logger could verify the temperature sensitivity of the electronics and correct any errors, within limits. It is difficult to reduce the residual errors to below the noise level.

13.3 Sampling

An analog signal may be sampled in space or time. For the following discussion, it is perhaps easier to think of time sampling. The resulting signal is still an analog signal; it has a continuous range of amplitudes. But the sampling process may cause a special kind of signal distortion called frequency folding if the sampled signal contains frequencies above a certain threshold called the Nyquist frequency which is defined to be $f_N = 1/(2\Delta T)$, where ΔT is the sampling interval. This is illustrated in fig. 13-2.

Signal Sampling

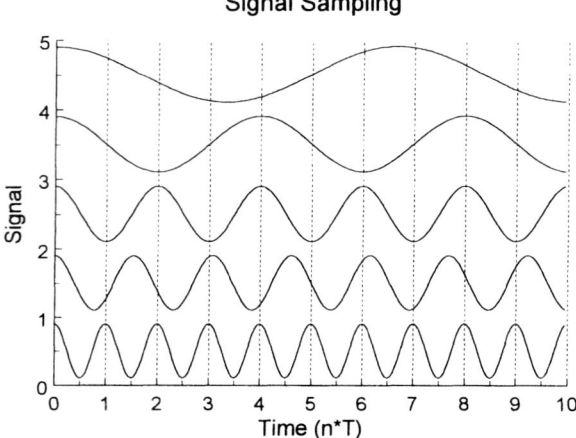

Fig. 13-2 Cosine waves of increasing frequency (from top to bottom) sampled at interval T.

When the frequency of the input signal is less than the Nyquist frequency, the sampled signal has the same frequency as the input, but when the absolute magnitude of the input frequency is $\geq f_n$, the apparent frequency of the sampled signal is $< f_N$. When $f_i = 2f_N$, the apparent output frequency is zero; a straight line.

The frequency-folding diagram is shown in fig. 13-3. The horizontal dashed line represents an observed frequency in the sampled data and the vertical dashed lines show that the observed frequency may have been any number of frequencies in the original data. In general, after the data have been sampled, it is impossible to determine the input frequency without some other information. The actual input frequency is said to be aliased with the apparent sampled frequency. The same phenomena occur with sampled digital data.

The concept of negative frequency applies whenever the sign of the frequency is considered. For example, a wheel can turn in either direction, so we could assign positive frequency for clockwise rotation and negative frequency for counter-clockwise rotation. This suggests that as the wheel rotation rate increases to exceed the Nyquist frequency, the wheel will appear to rotate in the opposite direction and at a lower frequency. This is commonly observed in motion pictures of rotating wheels. This is also an example of a sampled signal that now occurs in discrete time but with continuous amplitude, and so the amplitude is still an analog value.

In another example, a Doppler radar detects a frequency shift relative to the base transmitter frequency due to the velocity of the target. Target motion away from the radar causes a negative frequency shift, the received signal will have a lower frequency than the transmitted signal. Motion towards the radar causes a positive frequency shift. The radar samples the signal at the pulse repetition interval, so the perceived frequency (or target velocity) will fold about the Nyquist frequency and there will be an apparent change in the sign of the target velocity when the alias occurs.

More often the sign of the frequency is lost and we consider only positive frequencies, as illustrated in fig. 13-4. There, frequency folding occurs at $f = f_N$ and $f = 2f_N$.

Frequency Folding

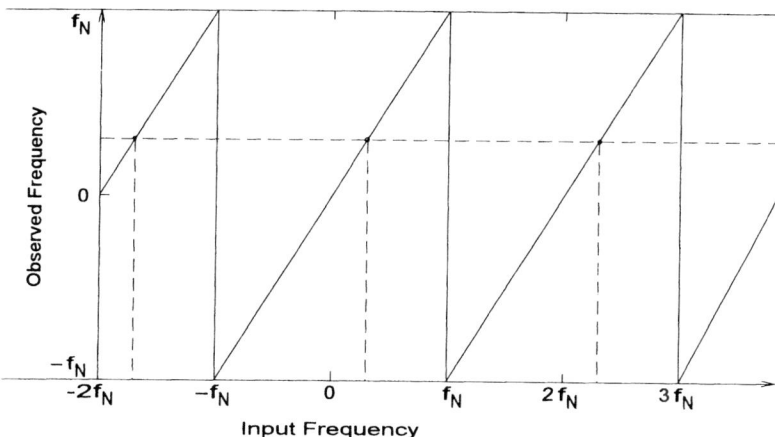

Fig. 13-3 Frequency folding. The abscissa is the actual frequency from $-\infty$ to ∞.

Sampling and averaging are illustrated in fig. 13-5. The signal is in analog form up to the analog-to-digital converter. Some averaging will occur in the sensor and, perhaps, also in the analog signal conditioning. The analog signal, with some averaging, is labeled as X_A. The sampled and digitized signal is X_D and is shown with an offset, for clarity. The conversion from X_A to X_D is critical; that is where aliasing can occur. And it is where quantization occurs. Then the computer can perform a final average, X_{Avg} in the figure. This is shown again with another offset. Without the offset, the signal X_A would coincide with X_D. But the signal X_{Avg} would not coincide with X_D because averaging would have reduced the amplitude of the X_{Avg} and induced some phase shifting.

An example of sampling and averaging (fig. 13-6) is practiced in the Oklahoma Mesonet. In that system, sampling occurs at 3-second intervals. The data are averaged over 5 minutes in the small computer in the data logger. Data are reported at 15-minute intervals. These reports comprise three sets of 5-minute data, so all of the 5-minute data are collected and transmitted via radio.

The stations were located at about 30 km spacing in an irregular grid. Since the effective length of all of the sensors was much smaller than the station spacing, the system was under sampled in space. Clearly it would have been impossible to make the station grid spacing smaller than the dimensions of the sensors. Most meteorological systems are undersampled in space.

13.4 Analog-to-Digital Conversion

The term analog-to-digital converter is usually associated with conversion of an analog voltage to digital form, but we could postulate a generic analog-to-digital converter (ADC) that would convert an analog signal in any form to a digital one. Such a

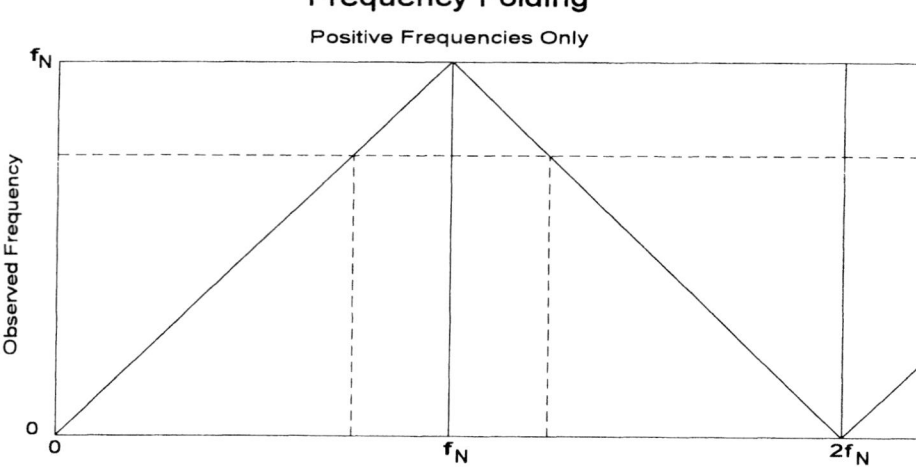

Fig. 13-4 Frequency folding where the actual frequencies are restricted to $f \geq 0$ Hz.

machine does not exist, but analog signals of all forms are converted to digital so it is useful to explore the concept of a generic ADC. The essential elements of a generic ADC, and of any real ADC, are (1) an analog input signal, (2) a defined range of input, (3) a reference quantity, and (4) output rendered in discrete quanta over a finite number of states. In addition to these required elements, an ADC may perform conversions at discrete time intervals or may be continuous in time.

The input must be an analog signal in any form, some are listed in table 13-1. As noted, the conventional ADC is a device that converts voltage, and sometimes current, to digital. A device that converts voltage or frequency to digital is usually called a counter, something of a misnomer as it does more than just count.

The input range of the ADC may be a voltage range (−5 to 5 V), a frequency range (10 Hz to 1 MHz), an angular range (0 to 360 degrees), etc. It must be defined to allow output representation with a finite number of states.

The reference quantity is required to allow the converter to determine the magnitude of the input signal relative to the possible assigned states. The reference quantity may be generated within the ADC or may be external to it.

The output must be represented in a finite number of states that represent the range of the ADC input. This is the essence of a digital signal. The output may be expressed in any number system. Binary is used when the ADC output is to be sent to a computer, decimal if the output directly drives a display as in a digital voltmeter.

EXAMPLE

Consider the simple voltage comparator which generates an output of "1" if the input is greater than some reference voltage and an output of "0" if the input is less than the reference. This is a "idiot" indicator, frequently used as an oil or temperature warning signal. While this is not usually called an ADC, it does meet all of the requirements.

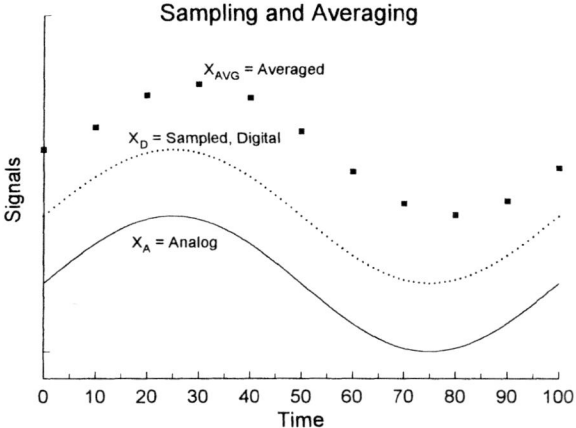

Fig. 13-5 The effect of sampling and averaging. The averaged signal X_{Avg} and the sampled signal X_D are offset above the analog signal X_A.

To explore the nature of an ADC further, let us make some definitions

- A = analog input signal
- D = digital output signal
- A_L = lower limit of the ADC input range
- A_H = upper limit of the range
- S_P = span = $A_H - A_L$
- N_S = number of states available in the output D
- Q = input quantum represented by an output state, uniformly distributed over the input range; $Q = S_P/N_S$.

An ADC used with a computer usually uses the binary number system and, for convenience, let us assume the output is a positive, binary integer. Then N_B = number of bits used and the range of D is $(0, N_S - 1)$ where

$$Q = \frac{S_P}{N_S} = \frac{S_P}{2^{N_B}} \tag{13.1}$$

The output is given by

$$D = \text{integer}\left[\frac{A - A_L}{Q} + 0.5\right] \tag{13.2}$$

subject to the constraint that $0 \leq D \leq N_S - 1$.

Fig. 13-7 shows the transfer plot (input–output plot) for a 3-bit binary converter with an input range of 0 to 12 V. The error is shown below. For most of the range, error is bounded by $-Q/2$ and $+Q/2$ but near the top of the range the error goes to $-Q$.

EXAMPLE

Consider a 12-bit binary ADC with an input range of -5 to 5 volts. $A_L = -5\,\text{V}$, $A_H = 5\,\text{V}$, $N_B = 12$, $N_S = 4096$, $Q = 2.441\,\text{mV}$.

Input A (V)	Output D (binary)	Output D (decimal)
−5.000	0000 0000 0000	0
0.000	0111 1111 1111	2047
3.142	1101 0000 0111	3335
4.996	1111 1111 1110	4094
4.997	1111 1111 1111	4095
5.000	1111 1111 1111	4095

We can express the output directly in the decimal number system even though the ADC works in binary. Computers store integers internally as binary numbers and represent them on output as decimal numbers, so this is a natural thing to do.

EXAMPLE

A simple wind vane system might have a magnet fastened to the vane shaft and four reed relays positioned around the shaft such that when the magnet is close to a relay it closes but is otherwise open; see fig. 13-8. The four relays are positioned at the cardinal points of the compass (N, E, S, and W). When the wind is from the North the magnet is close to the N relay and closes it. When the wind is from the NE the magnet is positioned between the N and E relays and closes both of them. The relays are connected to a display with four lights labeled N, E, S, and W. When a relay closes, the associated light is turned on. When lights N and E are on we interpret the display as indicating a wind from the NE.

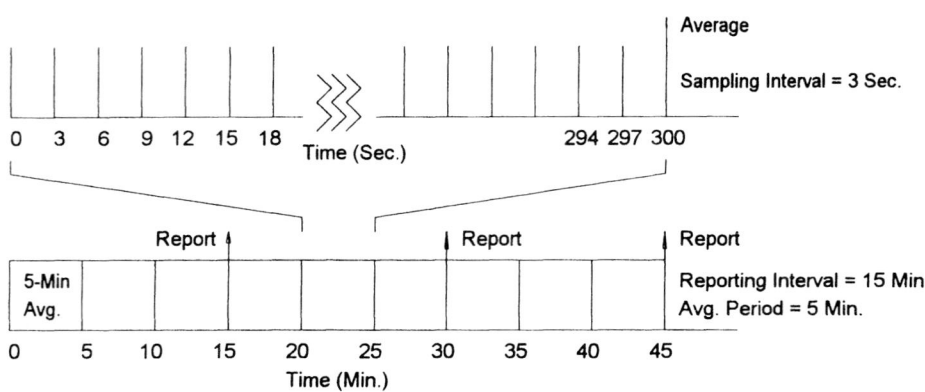

Fig. 13-6 An example of sampling, averaging, and reporting in the Oklahoma Mesonet.

Sampling and Analog-to-Digital Conversion 239

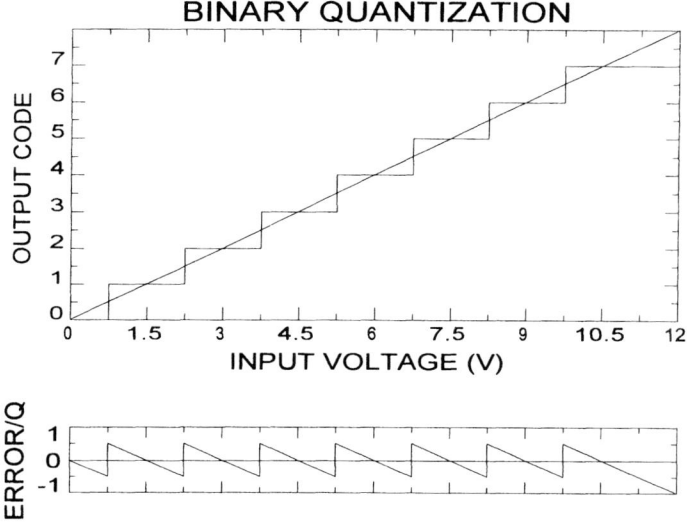

Fig. 13-7 A 3-bit binary quantizer (analog-to-digital converter).

This is an example of a mechanical analog-to-digital converter. The analog signal is the shaft angular position. This meets the requirements for an ADC; it has an analog input with a limited range (0 to 360°), there is a reference quantity (orientation to North) and the output is expressed as a digital signal with discrete quanta. The system output has eight possible states: N, NE, E, SE, S, SW, W, and NW. $A_L = 0$, $A_H = 360°$ and $N_S = 8$ so $Q = 45°$.

13.5 Information Content of a Signal

Consider a measurement system as discussed in chap. 1 comprising one or more sensors, transducers, a analog-to-digital converter, and other modules. If the sensor range is mapped into the ADC range and the ADC is a binary converter, we can estimate the probability of any possible ADC output state. If we have no *a priori* information about the state of the input signal, then all ADC output states are equally probable: $p_1 = p_2 = \cdots = p_i = 1/n$, which can be normalized to

$$\sum_{i=1}^{n} p_i = 1 \tag{13.3}$$

The average information content or information entropy of a digital signal is

$$H = -\sum_{i=1}^{n} p_i \ln p_i \tag{13.4}$$

where p_i = probability of occurrence of each digital state. If the logarithm in eqn. 13.4 is to base 2, then the information content value is the number of binary bits required

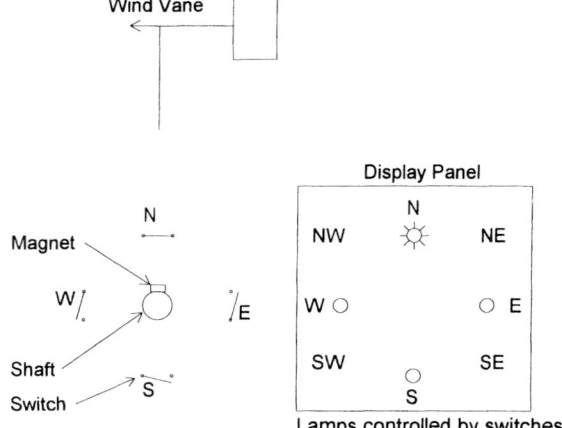

Fig. 13-8 A mechanical analog-to-digital converter.

to represent that value. However, since it is easier to use natural logarithms (base e), natural logarithms will be used except when referring explicitly to the number of binary bits required to represent a value. Note that the binary logarithm can be obtained from the natural logarithm by $\log_2(x) = 1.4427 \log_e(x)$. All digital signals have a finite number of states.

By definition, an analog signal is continuous; thus it has an infinite number of states. The ideas of information theory can still be applied to analog signals by assigning a finite number of states over the defined range of the sensor. The size of the finite states can be set equal to the imprecision. Because of the noise content of all analog signals, there is some uncertainty as to the exact value of the signal, so we set $\Delta x = $ imprecision. Then the average information content of an analog signal is

$$H = -\sum_{i=1}^{n} p_i \Delta x \ln(p_i \Delta x) \tag{13.5}$$

and

$$\sum_{i=1}^{n} p_i \Delta x = 1 \tag{13.6}$$

the span is equal to $(2n+1)\Delta x$. If we apply this concept to the measurand, we get

$$H_M = \lim_{\Delta x \to 0} \left[-\sum_{i=1}^{n} p_i \Delta x \ln(p_i \Delta x) \right] \tag{13.7}$$

where H_M is the information content of the measurand and $\Delta x \to 0$ since, by definition, the imprecision of the measurand is zero. Taken to the limit, H_M becomes

$$H_M(x) = -\int_{-\infty}^{\infty} p(x) \ln(p(x)) dx + \lim_{\Delta x \to 0} \ln \frac{1}{\Delta x} \tag{13.8}$$

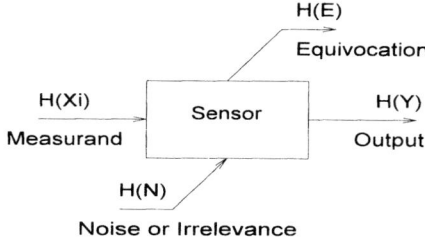

Fig. 13-9 Flow of information into a sensor from the measurand to the output.

but the last term goes to infinity. It is reasonable for the information content of the measurand to be infinite, or at least very large, since we assume that the imprecision of the measurand is zero. When the measurand is air temperature, complete information would involve knowing the velocities of all of the molecules in a parcel of air.

The measurand has very high information content; every molecule of air contains some information about the air temperature. But it should be evident that the output of a sensor has finite information content, sometimes zero information. What happens to the information that flows into a sensor? Some must be lost and some noise is always added, as shown in fig. 13-9.

In the process, some information is lost; this is called equivocation. The loss occurs since some input energy or information is lost as flow perturbation by an anemometer, hysteresis by an aneroid barometer, and so on. Some noise is always added to the signal; this is called noise or irrelevance. It may be due to heat loss in a resistor, rough bearings, self-heating in a thermistor, and similar phenomena.

This leads to a definition of noise in a signal. It is common to use the term noise without any kind of a definition, or, in some cases, to simply define noise as that part of a signal that is not wanted. This includes the legitimate high-frequency content of the measurand. From the above figure, we can define noise as any contribution to the sensor output signal that arises from unwanted inputs. In this sense, high frequency components of the measurand are not noise but still may be unwanted.

QUESTIONS

1. What is the minimum number of binary bits required in an ADC to digitize the output of a barometer, whose range is 800 to 1100 hPa, with a desired resolution of 0.05 hPa?

2. A wind vane turns a pot which has 1 volt applied across it. The arm of the pot is connected to a 14-bit ADC with an input range of -2.5 to 2.5 V. Express the ADC output resolution in terms of volts and in degrees. How does the resulting vane angle resolution compare with the probable inaccuracy of the vane? Is there any justification for this system?

3. The input signal has a frequency of 10 Hz and is sampled 8 times per second. What will be the apparent output frequency?

4. Given a system comprising a first-order, low-pass analog filter and an ADC where the filter time constant is 31.83 ms and the sampling rate is $10\,\mathrm{s}^{-1}$, find the output when the input is a finite series:

$$X_i(t) = A_{i0} + A_{i1}\sin(2\pi f_1 t) + A_{i2}\sin(2\pi f_2 t) + A_{i3}\sin(2\pi f_3 t) + A_{i4}\sin(2\pi f_4 t)$$

What is the Nyquist frequency?
Complete the following table:

Input		Output	
Amplitude	Frequency	Amplitude	Frequency
2.0	0 Hz		
1.0	1 Hz		
0.8	3 Hz		
0.6	8 Hz		
0.4	14 Hz		

5. Is it possible for a realizable analog filter to entirely prevent aliasing? Why?
6. Is it possible to correct the sampled output for the effects of the filter? For the effects of sampling?
7. The variance of the signal in question 4 above, is $\sigma^2 = (A_{i1}^2 + A_{i2}^2 + A_{i3}^2 + A_{i4}^2)/4 = $ _____. What is the variance of the above output signal?
8. Is the variance of the sampled signal ever affected by the sampling rate? Consider what happens when one of the signal frequency components is twice the Nyquist frequency.
9. A certain ADC has an input range of -10 to 10 V. What are the span, the number of states, the quantization, and the maximum quantization error?
10. Suppose that the signal is $3\sin(0.2\omega t)$ and the noise extends to 5 Hz. The signal is sampled at 1 s intervals. Is this OK? Will there be any aliasing?
11. Indicate which of the following are digital and which are analog:
    (a) the temperature indicated by a mercury-in-glass thermometer;
    (b) the temperature you write down on an observation sheet;
    (c) the position of a pointer on an aneroid barometer dial;
    (d) output from an analog signal conditioning device;
    (e) output of a tipping bucket rain gauge;
    (f) output of a cup anemometer equipped with a light-chopping transducer.
12. Would a measurement system be free from random noise if the system resolution granularity were greater than the imprecision? For example, the imprecision might be 4 mV at the ADC input and the LSB = 8 mV.
13. List noise sources and equivocation losses for a cup anemometer, an aneroid barometer, and a temperature sensor.
14. In order to analyze the information content of an analog signal, it was necessary to define analog granularity. How did we define analog granularity in this context? What happens to the information content of the measurand? Why?
15. What is the instrument definition of noise?
16. Consider the information content of an analog signal. What happens to the information content when
    (a) the imprecision is zero?
    (b) the bias is zero?
17. If the information content of a signal is zero when the measurand is known, what is the justification for performing a laboratory calibration?
18. If p_i is constant over the span of a sensor output and $\Delta x > 0$ and constant over the span, simplify eqns. 13.5 and 13.6.
19. A humidity sensor produces a 0 to 1 volt output, representing 0 to 100% RH. Its imprecision is 2%. The signal is input to a binary ADC that has an input range of 0 to

1 V. How many bits are required to make the quantization less than or equal to the imprecision? When the RH = 25%, what is the digital output?

20. If we record the output of a mercury-in-glass thermometer as 296.3 K, is this a digital output? If so, where did the analog-to-digital conversion take place?

BIBLIOGRAPHY

Jaeger, R.C., 1982a: Tutorial: analog data acquisition technology: Part I – Digital-to-analog-conversion. *IEEE Micro*, May, pp. 20–37.

Jaeger, R.C., 1982b: Tutorial: analog data acquisition technology: Part II – Analog-to-digital. *IEEE Micro*, August, pp. 46–56.

Jaeger, R.C., 1982c: Tutorial: analog data acquisition technology: Part III – Sample-and-holds, instrumentation amplifiers, and analog multiplexers. *IEEE Micro*, Nov., pp. 20–35.

Jaeger, R.C., 1983: Tutorial: analog data acquisition technology: Part IV – System design, analysis, and performance. *IEEE Micro*, Feb., pp. 52–61.

Zuch, E.L., 1982: Signal data conversion. In *Handbook of Measurement Science*, Vol. 1, *Theoretic Fundamentals,* ed. P. Sydenham. Wiley Interscience, Chichester, pp. 489–538.

Appendix A

Units and Constants

International System of Units (SI)

Base SI Units

Length	meter	m
Mass	kilogram	kg
Time	second	s
Thermodynamic temperature	kelvin	K
Amount of substance	mole	mol
Electric current	ampere	A
Luminous intensity	candela	cd

Supplementary Units

Plane angle	radian	rad
Solid angle	steradian	sr

Derived Units

Pressure, stress	pascal	Pa	$kg\,m^{-1}\,s^{-2}$	$N\,m^{-2}$
Energy, work, quantity of heat	joule	J	$kg\,m^2\,s^{-2}$	$N\,m$
Force	newton	N	$kg\,m\,s^{-2}$	

Power, radiant flux	watt	W	$kg\,m^2\,s^{-3}$	$J\,s^{-1}$
Frequency	hertz	Hz	s^{-1}	
Magnetic flux	weber	Wb	$kg\,m^2\,A^{-1}\,s^{-2}$	$V\,s$
Inductance	henry	H	$kg\,m^2\,A^{-2}\,s^{-3}$	$Wb\,A^{-1}$
Magnetic flux density	tesla	T	$kg\,A^{-1}\,s^{-2}$	$Wb\,m^{-2}$
Electric charge	coulomb	C	$A\,s$	
Capacitance	farad	F	$A^2\,s^4\,kg^{-1}\,m^{-2}$	$C\,V^{-1}$
Electric potential	volt	V	$kg\,m^2\,A^{-1}\,s^{-3}$	$W\,A^{-1}$
Electric resistance	ohm	Ω	$kg\,m^2\,A^{-2}\,s^{-3}$	$V\,A^{-1}$
Conductance	siemens	S	$A^2\,s^3\,kg^{-1}\,m^{-2}$	$A\,V^{-1}$
Luminous flux	lumen	lm	$cd\,sr$	
Illuminance	lux	lx	$cd\,sr\,m^{-2}$	

Units in Use with SI Indefinitely

Minute	min	=	60 s
Hour	h	=	3600 s
Day	d	=	86 400 s
Degree	°	=	$(\pi/180)$ rad
Minute	′	=	$(1/60)°$
Second	″	=	$(1/60)′$
Liter	L	=	$10^{-3}\,m^3$
Metric ton	t	=	$10^3\,kg$
Hectare	ha	=	$10^4\,m^2$

Factors, prefixes and symbols for forming multiples in the SI system of units

Factor	Prefix	Symbol	Factor	Prefix	Symbol
10^{18}	exa	E	10^{-1}	deci	d
10^{15}	peta	P	10^{-2}	centi	c
10^{12}	tera	T	10^{-3}	milli	m
10^{9}	giga	G	10^{-6}	micro	μ
10^{6}	mega	M	10^{-9}	nano	n
10^{3}	kilo	k	10^{-12}	pico	p
10^{2}	hecto	h	10^{-15}	femto	f
10^{1}	deka	da	10^{-18}	atto	a

Numerical Values

Universal Constants

| Boltzmann's constant | k | $1.380\,658(12) \times 10^{-23}\,J\,K^{-1}$ |
| Stefan–Boltzmann constant | σ | $5.669\,6 \times 10^{-8}\,W\,m^{-2}\,K^{-4}$ |

Planck's constant	h	$6.626\,075(40) \times 10^{-34}$ J s
Speed of light in vacuum	c	$2.997\,924\,58 \times 10^{8}$ m s^{-1}

The Earth

Average radius	R_E	6.37×10^{6} m
Acceleration due to gravity	g	9.81 m s^{-2}
Solar constant		1380 W m^{-2}

Dry Air

Density (0°C, 100 kPa)	ρ	1.275 kg m^{-3}

Water

Density (liquid, 0°C)	1000 kg m^{-3}
Latent heat of vaporization (100°C)	2.25×10^{6} J kg^{-1}

BIBLIOGRAPHY

Hillger, D.W., and L.F. Sokol, 1987: Guidelines for use of SI units in technical writing and presentations. *Bull. Am. Meteor. Soc.*, 68, 36–39.

Appendix B

Thermistor Circuit Analysis

A simple thermistor is frequently used in a voltage-divider circuit for temperature measurement. This circuit is simpler than a bridge circuit.

B.1 A Thermistor

Thermistors are characterized by their resistance at a standard temperature, 25°C, the slope of the resistance–temperature curve, and their tolerance, that is, how interchangeable they are. Typical 25°C resistance values range from $10\,\Omega$ to $400\,\mathrm{k}\Omega$. Available tolerances range from 0.5% to 20% of resistance. Thermistors with tight group tolerances are designated as precision thermistors, meaning that the manufacturer guarantees that any thermistor of a specified model will fit the standard resistance–temperature curve within a certain tolerance. The precision thermistor considered here has a resistance of $100\,000\,\Omega$ at 25°C. We will use the designation 103 for the $10\,\mathrm{k}\Omega$ thermistor and 104 for the $100\,\mathrm{k}\Omega$ thermistor; see fig. B-1.

In addition to the wide variety of thermistor resistance versus temperature characteristics noted above, thermistors are available in an enormous range of package sizes and configurations. The thermistor chip may be enclosed in epoxy or glass and sometimes further protected in a stainless steel jacket. Minimal protection is usually indicated for air temperature measurement since it is necessary to enhance heat flow between the thermistor chip and the air; protective configurations tend to interfere with that heat flow. For most meteorological purposes, the temperature range of -30 to 50°C is adequate. Precision thermistors are typically matched in the range 0°C to 70°C or to 100°C. But the additional matching error incurred in extending the range down to $-30°C$ is small. The time constant of a typical glass- or epoxy-encapsulated

Thermistor Circuit Analysis

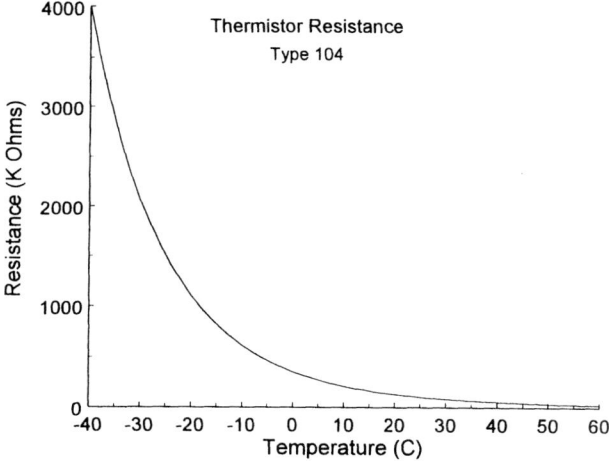

Fig. B-1 Nominal resistance versus temperature curve for a simple thermistor.

thermistor is 10 s in still air. For that configuration, the heat dissipation constant is $1 \text{ mW} \, ^\circ\text{C}^{-1}$, also in still air.

Thermistor resistance–temperature data can be modeled using the transfer function

$$R_T = \exp\left[a_0 + \frac{a_1}{T} + \frac{a_3}{T^3}\right] \quad (B.1)$$

where T is the temperature in Kelvin and R_T is in kΩ and the a's are coefficients. (There is no a_2 coefficient because there was no need for a term in T^2.)

The equivalent calibration equation is

$$\frac{1}{T} = c_0 + c_1 \ln R_T + c_3 (\ln R_T)^3 \quad (B.2)$$

where T, as before, is in Kelvin. The coefficients for eqns (B.1) and (B.2) are listed in table B-1 for a type 103 thermistor and a type 104. The transfer function model, eqn. B.1, is an excellent fit; the residual error is shown in fig. B-2.

Table B-1 Characteristics of two selected thermistors and associated circuit.

Type	103	104
R @ $25°C$	$10\,000 \, \Omega$	$100\,000 \, \Omega$
R_1	$30\,100 \, \Omega$	$249\,000 \, \Omega$
R_2	$1000 \, \Omega$	$1000 \, \Omega$
a_0	$-4.894\,03$	$-3.921\,65$
a_1	4358.14	4825.87
a_3	$-1.359\,83 \times 10^7$	$-1.991\,60 \times 10^7$
c_0	$1.126\,68 \times 10^{-3}$	$8.495\,79 \times 10^{-4}$
c_1	$2.345\,04 \times 10^{-4}$	$2.058\,93 \times 10^{-4}$
c_3	$8.633\,82 \times 10^{-8}$	$8.766\,13 \times 10^{-8}$

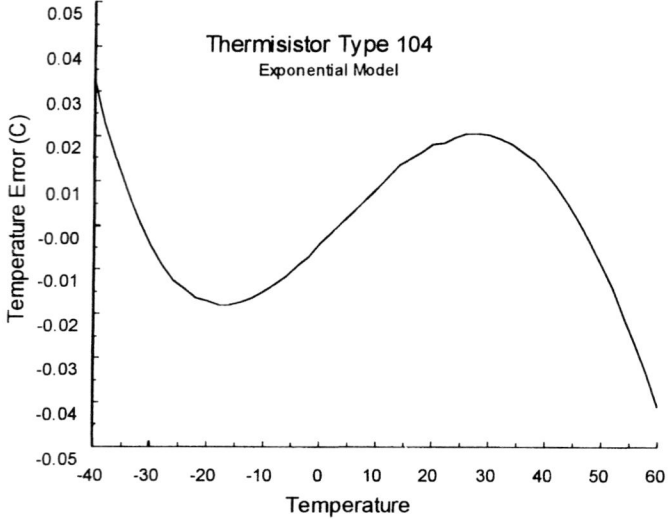

Fig. B-2 Residual error of the exponential model, expressed as equivalent temperature error.

B.2 A Circuit

A commonly used circuit is a variation of the simple voltage divider, depicted in fig. B-3. The component values are listed in table B-1. The circuit output is given by

$$V_1 = \frac{V_R R_2}{R_1 + R_2 + R_T} \tag{B.3}$$

where V_R is the reference voltage and R_1 and R_2 are fixed, precision resistors. Their values must be chosen with consideration given to heat dissipation in the thermistor, the available range of values for V_R, the input range of an analog-to-digital converter (ADC), if used, and the linearity of the circuit.

We can arbitrarily set the reference voltage to 1000 mV for now and obtain the final value later. The selection of R_1 has some impact on the signal output range and upon the linearity, as shown in fig. B-4. But the circuit is not very sensitive to the value of R_1, so one could select an arbitrary value (from the values available) such as 249 kΩ.

If the available input range for the ADC is ± 5 V (as would be the case for the generic data logger described in Appendix C), we could select $V_R = 1.4$ V and set the amplifier gain $= 1000$. Then all the values in the V_1 column of table B-2 should be multiplied by 1400; thus the range of V_1 becomes 0.605 V to 4.937 V.

If the ADC is a 13-bit binary converter, its resolution will be 1.22 mV. At the low end of the range, from -30 to $-20°$C, the circuit generates 51 mV° C^{-1} or 41 counts of the ADC per °C. At the high end of the range, from 40 to 50°C, the circuit generates 28.7 mV° C^{-1} or 23 counts per °C. It is important that the ADC resolution should not impact the overall system accuracy and it does not in this case. It is always best to utilize as much of the range of an ADC as possible. Given that the characteristics of the generic data logger are fixed, the only way we could improve the performance of this circuit would be to increase R_2, which could be readily done. Note that we have

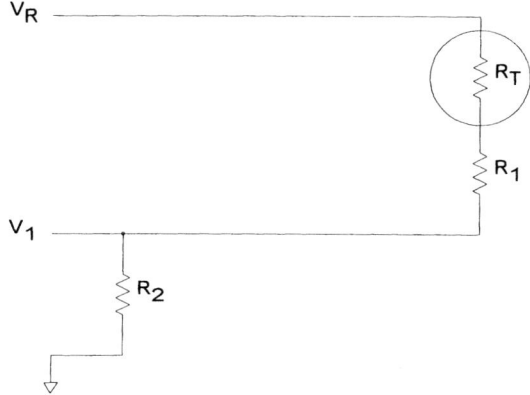

Fig. B-3 A voltage divider in a thermistor in the top half of the circuit.

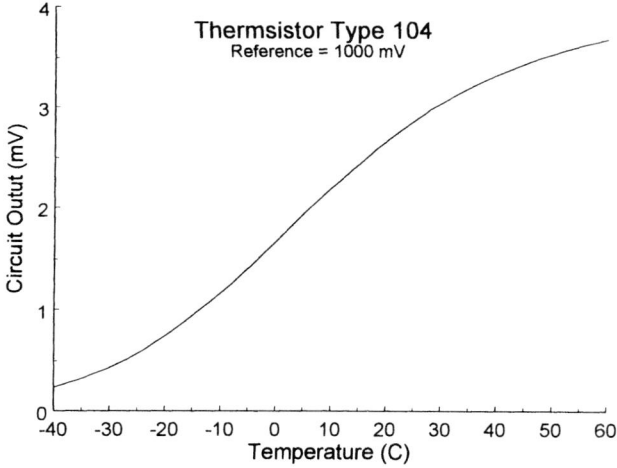

Fig. B-4 Output of the voltage divider circuit.

Table B-2 Tabulated circuit values with $V_R = 1\,V$, $R_1 = 249\,k\Omega$, and $R_2 = 1\,k\Omega$.

$T°$ (C)	R_T (kΩ)	V_1 (mV)
−30	2064.00	0.432
−20	1103.40	0.739
−10	611.87	1.160
0	351.02	1.664
10	207.85	2.184
20	126.74	2.654
30	79.422	3.036
40	51.048	3.322
50	33.591	3.526

said nothing, so far, about the impact of ADC inaccuracy. We have merely established that there is no problem with the resolution imposed by the ADC.

Assuming that nominal circuit component values are used and that there is no ADC error, the only error in calculating the temperature from the digitized and recorded voltage V_1 is the use of the model, eqn. B.2, which, as noted, is not an exact fit to the manufacturer's temperature–resistance curve. However, this is a small error source and results in bias $= 0.00°C$ and imprecision $= 0.028°C$.

The power dissipated in the thermistor is $I^2 R_T$, which must be less than $100\,\mu\text{W}$ for all T to keep the circuit self-heating to less than $0.1°C$. Remember, the heat dissipation constant given by the manufacturer is $1\,\text{mW}°\text{C}^{-1}$ in still air. It can be readily verified that the power dissipated in the thermistor is always less than $2\,\mu\text{W}$ with $V_R = 1.4\,\text{V}$.

B.3 An Alternative Calibration Equation

It is always possible to use a general polynomial for the calibration equation instead of eqn (B.2). The 5th-order calibration polynomial is

$$T_c = d_0 + d_1 V + d_2 V^2 + d_3 V^3 + d_4 V^4 + d_5 V^5 \tag{B.4}$$

For the same thermistor, designated as 104, the residual error, with nominal component values is plotted in fig. B-5. The coefficients, in this case, are $d_0 = -55.738$, $d_1 = 82.349$, $d_2 = -67.331$, $d_3 = 35.697$, $d_4 = -9.4119$, and $d_5 = 0.995\,68$. The error ranges from $-0.22°C$ to $0.11°C$ and the RMS error is $0.06°C$. The error due to this form of the calibration equation is larger than when using eqn. B.2. There seems to be no advantage in using B.5 as long as the data logger has a log instruction and can implement B.2 directly.

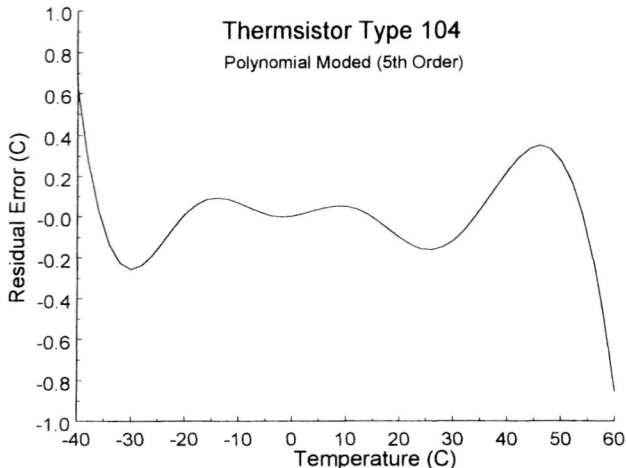

Fig. B-5 Residual error when using a 5th-order polynomial to model the voltage divider.

Appendix C

A Data Logger

A minimal data acquisition system was shown in chap. 1. A more typical, although still somewhat simplified, data logger is described in these pages. The purpose is to show some of the available features of data loggers and to show how they fit into a complete measurement system that incorporates many kinds of sensors.

C.1 The Data Logger

A data logger has to incorporate many functions, only hinted at in chap. 1, to handle the number and variety of sensors commonly used in field data acquisition systems. The data logger shown in fig. C-1 has a minimal subset of necessary functional modules built in to it. No data display is shown, as that function can be readily provided by an external device via the digital data communications port.

The functional modules in the "generic data logger" will be described starting from the upper left corner. The performance specifications are generic, that is, not specific to any commercially available data logger. Real data loggers are more complex, provide many more functions, and provide far more flexibility.

Excitation output. The box labeled "EXC" generates a voltage output that can be set to any level between −5 and 5 V within the resolution of the digital-to-analog converter (DAC), which is 13 bits or about 1 mV. The DAC provides the complementary function of an analog-to-digital converter (ADC) described in chap. 13. It converts a digital signal sent from the microprocessor into an analog voltage and holds that voltage until provided with another digital value. The purpose of the EXC is to provide the voltage reference or excitation required by many sensors such as the pot in a wind vane or the voltage divider or bridge in a temperature sensor.

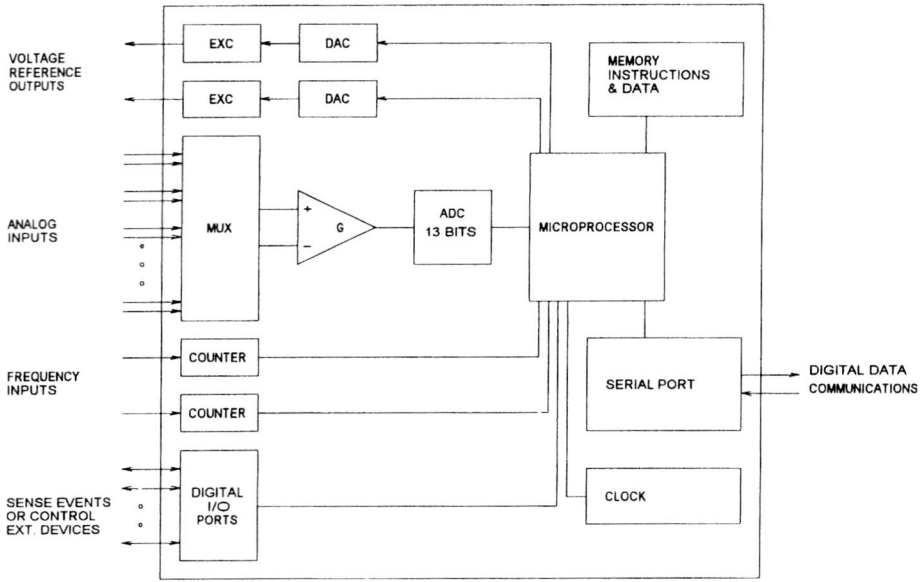

Fig. C-1 A generic data logger with minimal functionality.

Multiplexer input. The multiplexer or "'MUX'" switches eight analog voltage input pairs, one at a time, into a programmable gain instrumentation amplifier. Since many sensors provide a differential output, this MUX switches the analog signal in pairs, a high and a low voltage. Each voltage must be between -5 and 5 V. The difference between the high and low voltages is amplified (multiplied by a constant, G) in the instrumentation amplifier. The possible gains are $G = 1$, 10, 100, or 1000. The microprocessor controls the multiplexer, the amplifier, and the ADC. It selects the channel (1 to 8), selects the amplifier gain, and then directs the ADC to perform the conversion. The ADC is a 13-bit binary converter with a -5 to 5 V range. The effective analog input range and resolution is a function of the amplifier gain, as shown in table C-1.

Counter input. Analog signals whose information content is the frequency or period of the signal can be measured using the counter modules. These count the number of signal transitions (low to high) per unit time or measure the frequency of the signal. Sensors such as a tipping-bucket rain gauge or a cup or propeller anemometer with frequency-type output can be connected to the counter input.

Table C-1 Effective analog input range and resolution of the generic data logger.

Gain	Range	Resolution
1	-5 V to $+5$ V	1.22 mV
10	-500 to $+500$ mV	122 µV
100	-50 to $+50$ mV	12.2 µV
1000	-5000 to $+5000$ µV	1.22 µV

Digital I/O ports. The generic data logger has eight digital lines, each of which can be either an input line or an output line as selected by the microprocessor. As inputs, each line can sense the state of an input, whether the input is high (5 V) or low (0 V) at the time the microprocessor reads that line. An alternative input mode is for the microprocessor to sense when a line makes a low-to-high transition and enter some user subroutine when that happens. In the state mode, the microprocessor can detect whether an external switch is open or closed.

In the transition mode, the microprocessor can count external events provided they do not happen too fast. This is similar to the counter input and could accommodate a rain gauge, for example.

The digital I/O ports can be used to control and read data from external digital sensors such as a digital barometer which transmits data as a serial bit stream when commanded to do so. As an alternative, the data logger might be equipped with several serial ports, one of which could be used to read digital sensors.

Memory. The electronic memory of the data logger is designed to store the user's program instructions and to provide temporary data storage. The latter is required for intermediate storage of data and for storage of data awaiting transmission. Sometimes the data communications link fails, perhaps due to component failure, and it is desirable for the data logger to provide sufficient storage so that data can be stored for one or more days until communications are restored and the data can be recovered.

Microprocessor. The microprocessor is the heart of the data logger as it reads the user's instructions, or program, from memory, controls the other modules, and temporarily stores data in memory. It is the user's program that makes the generic data logger accommodate specific sensors and perform a specific task such as to collect data at user-specified time intervals, average the data over a user-specified period, and output the data in a particular format.

Serial port. The serial port is the interface between the microprocessor and the data communications hardware: modems, radio transceivers, hard wire links to other computers, and so on. Serial data transmission is sufficiently standardized that a wide variety of devices can be linked to the data logger.

Clock. The clock is a stable oscillator that controls the rate at which the microprocessor functions and enables the microprocessor to keep track of time. This time base is essential for many functions, including determining the sensor sampling interval and the averaging time. The combination of a counter input, the microprocessor, and the clock comprise an analog-to-digital converter for frequency-type analog signals.

C.2 Application in a Measurement System

The generic data logger could be applied to a typical meteorological measurement system as shown in fig. C-2. Excitation output #1 of the data logger provides the reference voltage for two sensors, a resistance temperature detector (RTD) in a bridge circuit, and to the pot of a wind vane. The wind vane is mechanically linked to a variable resistor (pot). One end of the pot is tied to the voltage reference (excitation #1) and the other end to ground. The arm, driven by the vane, selects a voltage proportional to the wind direction and this voltage is input to the ADC via the MUX and the amplifier.

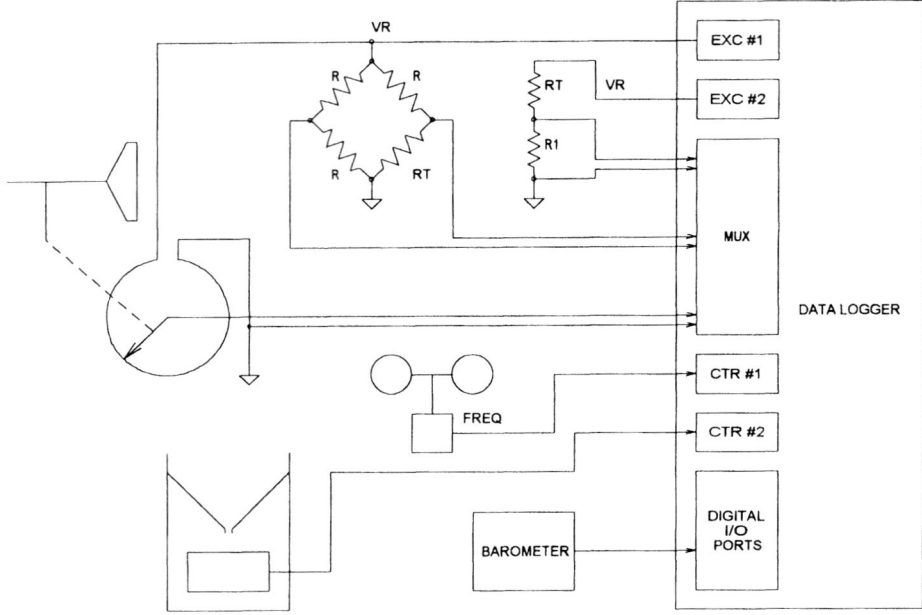

Fig. C-2 Generic data logger in a simple measurement system.

Excitation #1 also drives the bridge circuit. Differential voltage from the two arms of the bridge are input to the multiplexer and thence to the amplifier and the ADC. Excitation #2 drives a voltage divider circuit that includes a temperature sensor. Voltage signals are input to the multiplexer.

A cup anemometer drives a signal conditioner circuit that converts the shaft rotation rate to a frequency that is input to counter channel #1. The relay contact closure of a tipping-bucket rain gauge provides the input to counter channel #2.

Table C-2 Program to read three analog input signals.

Time Step	Program Action
1	Set EXC #2 = 1.000 V
2	Select MUX channel #1 and amplifier gain = 100
3	Start ADC
4	Store data read from ADC and set EXC #1 = 0.000 V
5	Set EXC #1 = 1.500 V
6	Select MUX channel #2 and amplifier gain = 1000
7	Start ADC
8	Store data read from bridge via the ADC and set EXC #1 = 0.0 V
9	Set EXC #1 = 5.000 V
10	Select MUX channel #3 and amplifier gain = 1
11	Start ADC
12	Store data read from vane via the ADC and set EXC #1 = 0.000 V

Note that when EXC #1 is used for reading the bridge the voltage reference is 1.5 V and that it is 5.00 V when used to read the vane pot.

The signal from a digital barometer, with a serial digital output, is input to the digital I/O port.

This data logger is shown accommodating six sensors. A real data logger could accommodate many more. Notice that some sensors could be hooked directly to the data logger without any intervening electronics while others require, at most, some resistors to complete a voltage divider or a bridge circuit.

Since the excitation ports are driven with a voltage specified by the user's program and only for the time period required for analog-to-digital conversion, power is saved. The voltage divider, the bridge, and the wind vane pot do not have a reference voltage applied continuously. While the vane pot and the bridge are driven by the same excitation port, they need not have the same value of reference voltage when active. Consider the sample program shown in table C-2 to read three analog channels.

Appendix D

Circuits

Some simple circuits and circuit elements used in earlier chapters are explained in this appendix.

D.1 Fundamentals

Some simple circuits with only resistor elements are shown in fig. D-1. In this figure, electrical resistance is designated R and the appropriate unit is the ohm, Ω. The performance of resistors can be expressed in terms of resistance value and precision (e.g. $1000\,\Omega \pm 5\%$), stability, and temperature coefficient. All resistors have some temperature coefficient.

The voltage symbol is V (E is sometimes used) and the measurement unit is the volt, V. Voltage is sometimes called the potential difference or electromotive force (EMF). The source of the voltage applied to a circuit may be a battery, a generator, or solar cells.

The current symbol is I (sometimes i). The unit is the ampere or amp (A). Current is said to flow from high voltage (positive) to low voltage (negative) even though electrons actually flow from negative to positive.

Ohm's law. The voltage across a resistor is related to the resistance and current by $V = IR$. In fig. D-1(a) this becomes $V_1 - V_2 = IR$.

Kirchhoff's current law. The sum of all currents into a node or point equals the sum of currents out; this is a statement of conservation of charge. In fig D.1(b), the current through R_1 is I_1 and must be equal to the current through R_2 which is I_2, since the current flowing from V_1 to V_2 must also flow from V_2 to V_3. We can replace the two resistors with an equivalent resistor, R_e and the current through it must be I.

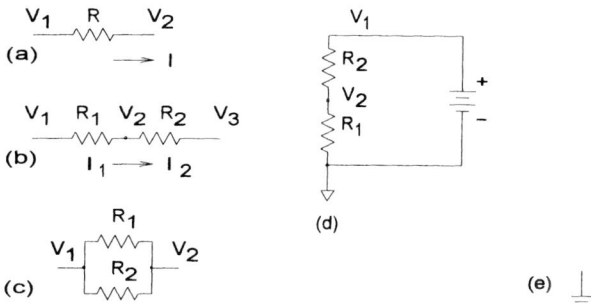

Fig. D-1 Illustration of circuit fundamentals.

Application of Ohm's law to this circuit yields $IR_e = V_1 - V_3$, $I_1 R_1 = V_1 - V_2$, and $I_2 R_2 = V_2 - V_3$. From these equations we can determine that $R_e = R_1 + R_2$ or that resistors in a serial circuit can be added.

Two resistors in parallel are shown in fig. D-1(c) and, again, they can be replaced by a single equivalent resistor R_e. According to the current law, the current through R_e, $I = I_1 + I_2$. Using $I = (V_1 - V_2)/R_e$ and the equations for I_1 and I_2, we can determine that $1/R_e = (1/R_1) + (1/R_2)$, which is the rule for resistors in parallel.

Kirchhoff's voltage law. The sum of voltage drops around any closed circuit is zero. The circuit in fig. D-1(d) has a battery to generate a voltage across the resistors. We usually measure voltage with respect to ground, an arbitrary reference voltage, which is simply one side of the battery in this case. The ground point is designated by the symbol below R_1. An alternative ground symbol is shown in fig. D.1(e). According to the voltage law, the voltage generated by the battery must be exactly equal to the voltage drop across resistors R_1 and R_2.

Power dissipated. The power dissipated as current flow through a resistor is $P = VI = I^2 R$ in units of watts, W.

Resistors. The resistance of a piece of material is a function of its cross-sectional area and length. So we can define the resistivity as $\rho = RA/L$, where R is the resistance as before, A = area, and L = length. The unit of resistivity is the ohm meter (Ω m) and some typical values of resistivity are shown in table D-1. The semiconductors are of special importance because of the way the resistivity is affected by temperature and by small amounts of impurities.

Table D-1 Resistivity of some metals, semiconductors, and insulators.

Metals	Resistivity (Ω m)	Semi-conductors	Resistivity (Ω m)	Insulators	Resistivity (Ω m)
Copper	1.72×10^{-8}	Carbon	3.5×10^{-5}	Glass	10^{10} to 10^{14}
Aluminium	2.63×10^{-8}	Germanium	0.60	Lucite	$> 10^{13}$
Tungsten	5.51×10^{-8}	Silicon	2300	Mica	10^{11} to 10^{15}
Steel	20×10^{-8}			Quartz	75×10^{16}
Constantan	49×10^{-8}			Teflon	$> 10^{13}$
Nichrome	100×10^{-8}			Wood	10^8 to 10^{11}

The resistivity and therefore the resistance of a material varies with temperature and this variation may be approximately linear over a small range:

$$R_T = R_0[1 + \alpha(T - T_0)] \qquad (D.1)$$

where R_T is the resistance at temperature T, and R_0 is the resistance at temperature T_0. Some values of the temperature coefficient α are given in table D-2.

The variation of resistance with temperature is not even approximately close to linear for semiconductors (except over very small temperature ranges) or for any material close to its superconducting temperature.

The most common resistors are of carbon composition. They are available in standard values ranging from $1\,\Omega$ to $100\,\text{M}\Omega$, with power ratings of 1/4 or 1/2 W, and tolerances of 5% or 10%. A better grade resistor is the 1% metal film with a temperature coefficient of 50 to 100 ppm (100 ppm = 100 parts per million = 0.01%). Metal film resistors are available with tolerances down to 0.025% and temperature coefficients to 5 ppm. It is impractical to make resistors much smaller than $1\,\Omega$ as its resistance would approach that of the connecting wire, nor is it practical to make ordinary resistors much greater than $100\,\text{M}\Omega$ because surface contamination, perhaps even a fingerprint, could provide a leakage path of less resistance than the nominal value of the resistor.

A resistor dissipates energy (converts it to heat) by the I^2R loss; it cannot store energy. It is impossible to design an electrical circuit without some resistance, so there is always some loss of energy. The function of a resistor in a circuit is to control current flow.

Capacitors. The current through a capacitor is proportional to the rate of change of voltage and the capacitance is the proportionality constant:

$$I = C \frac{dV}{dt} \qquad (D.2)$$

The symbol for capacitance is C, and the appropriate unit is the farad. A farad of capacitance is very large, so one usually deals with microfarads (µF) or picofarads (pF). If the voltage across a $1\,\mu\text{F}$ capacitor is rising at 1000 volts per second, the current will be 1 mA. The simplest capacitor is just two conductors in close proximity but not touching. The capacitance is proportional to the area of the two conductors and inversely proportional to the distance between them. A large capacitance is often obtained by plating a conductor onto some insulating material and rolling this sandwich into a cylinder. Several capacitors in parallel are like resistors in series, the equivalent capacitance is the sum of the capacitances, whereas for capacitors in series,

Table D-2 Temperature coefficient of resistance (near room temperature).

Material	$\alpha(°C^{-1})$	Material	$\alpha(°C^{-1})$
Aluminium	0.0039	Lead	0.0043
Carbon	−0.0005	Nichrome	0.0004
Constantan	0.000 002	Silver	0.0038
Copper	0.003 93	Tungsten	0.0045

$$C_{eq} = \frac{1}{\frac{1}{C_1} + \frac{1}{C_2} + \frac{1}{C_3} + \cdots} \tag{D.3}$$

If the circuit to a capacitor is opened, current cannot flow and the capacitor stores energy in its internal electric field. When the circuit is restored, that field produces a current that releases the energy. Therefore a capacitor is an energy storage reservoir; the electrical equivalent of mechanical momentum or position in a gravitational field. A capacitor would be a perfect energy storage reservoir if there were no internal losses or leakage. Of course, all real capacitors leak and therefore cannot retain stored energy indefinitely. But if you have ever poked around in high voltage circuits, such as in a TV, you know that capacitors can store energy, after the power has been switched off, long enough to produce tangible results. The function of a capacitor in a circuit is to store energy to smooth (filter) voltage fluctuations or to integrate current flow (see next section).

D.2 Simple Circuits

We can apply Ohm's law to resistor R_1 in fig. D-2(a) where $V_1 - 0 = IR_1$. The voltage across R_1 is taken to be V_1 since the other side of the resistor is tied to ground (0 V), as is indicated by the small triangle symbol.

Applying Ohm's law to resistor R_3, we get $V_R - V_1 = IR_3$ and the current through R_3 is equal to that through R_1 if no current flows in the branch to V_1, which we will assume here. Then

$$I = \frac{V_1}{R_1} = \frac{V_R - V_1}{R_3} \tag{D.4}$$

and, solving for V_1,

$$V_1 = \frac{V_R R_1}{R_1 + R_3} \tag{D.5}$$

where V_R is assumed to be a constant reference voltage. Since $V_1 < V_R$, this circuit is called a voltage divider. It is used in temperature measurement, where R_1 might be a resistance temperature detector (RTD) while R_3 is a fixed resistor. In another imple-

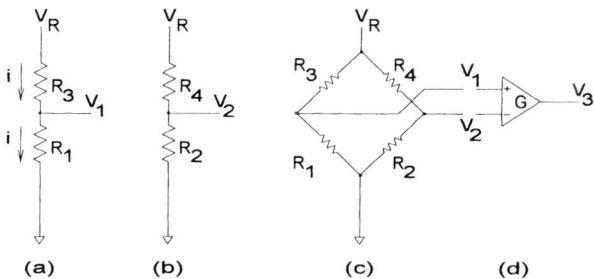

Fig. D-2 A voltage divider is shown in (a) and in (b), a bridge circuit in (c), and an instrumentation amplifier in (d).

mentation, R_3 is a thermistor, another type of temperature-sensitive resistor, and R_1 is the fixed resistor.

Another example of a voltage divider is shown in fig. D-2(b) where, following the above procedure, we find that

$$V_2 = \frac{V_R R_2}{R_2 + R_4} \tag{D.6}$$

The circuit in fig. D-2(c) is called a bridge circuit and careful inspection reveals that it is identical to the two voltage dividers in (a) and (b) taken together. If we take the difference between voltages V_1 and V_2, using equations (D.5) and (D.6), we get

$$V_1 - V_2 = V_R \left(\frac{R_1}{R_1 + R_3} - \frac{R_2}{R_2 + R_4} \right) \tag{D.7}$$

and, in the special case where $R_3 = R_4 = R_2$, this becomes

$$V_1 - V_2 = V_R \left(\frac{R_1}{R_1 + R_2} - \frac{1}{2} \right). \tag{D.8}$$

A bridge circuit is easy to analyze if we take it apart and see that it is just a pair of voltage divider circuits.

The last circuit element, fig. D-2(d), is called an instrumentation amplifier. It is sufficient to our purposes to consider that it multiplies the difference between its two inputs by a constant. This constant is called the gain, G. The transfer function for the instrumentation amplifier is

$$V_3 = G(V_1 - V_2) \tag{D.9}$$

and, if we connect V_1 of circuit (c) to V_1 of (d) and V_2 of (c) to V_2 of (d), we get

$$V_3 = G V_R \left(\frac{R_1}{R_1 + R_2} - \frac{1}{2} \right) \tag{D.10}$$

An instrumentation amplifier can have fairly large gains, ranging from 1 to 1000, so it is used to amplify the output of circuits, such as bridge circuits, to useful levels. The input impedance or resistance of an instrumentation amplifier is very high, on the order of $10^9 \, \Omega$. Therefore, when the voltage divider or bridge circuit output is connected to the instrumentation amplifier input, some current must flow into the amplifier. But the assumption that no current flows into the V_1 or V_2 branches of the circuit is justified because the resistors used in voltage dividers or bridge circuits are typically less than $10^3 \, \Omega$ and certainly less than $10^5 \, \Omega$. The current flowing into the amplifier must be at least four orders of magnitude less than that flowing in the voltage divider or bridge resistors.

When a wind vane is connected to a pot (potentiometer, a variable resistor), R_P, as shown by the dashed line in fig. D-3, the voltage output V_D should be proportional to the wind direction if V_R is a constant. The voltage V_D must be read by some device, usually a data logger, which presents some load resistance R_L. This load resistance is the input impedance of the data logger. Let $V_R = 1 \, \text{V}$ and $R_p = 10 \, \text{k}\Omega$. Assume that the vane moves the pot arm such that $V_D = 0$ when the wind direction Dir $= 0°$ and $V_D = V_R$ when Dir $= 360°$. Actually, the pot is formed in a circular arc

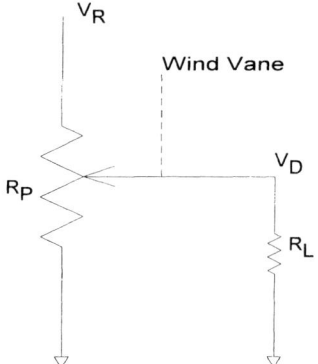

Fig. D-3 A variable resistor (pot) driven by a wind vane.

such that the ends nearly meet and the vane shaft is coaxial with the pot. The vane rotates the pot arm (a wiper that contacts the pot resistance). In this arrangement, it is impossible to have the pot ends meet or even be very close together. If this happened, the pot arm would bridge the two ends when the wind direction was from the North and short V_R to ground. Therefore there is usually a 5° gap in the pot, usually oriented to North. Then the circuit output V_D will be 0 V for $0° <$ Dir $<$ 2.5° and for $357.5° <$ Dir $< 360°$. This is called the dead zone of the pot and the data logger interprets 0 V as North. Since a finite load resistance causes an error (see question 7), it is useful to have an instrumentation amplifier incorporated into the data logger.

Application of Ohm's law and eqn. D.2 to fig. D-4 yields

$$C\frac{dV}{dt} = I = \frac{V_B - V}{R} \qquad (D.11)$$

If the switch is initially open and is closed at time $t = 0$, the initial condition of this differential equation is $V(0) = 0$. The solution is

$$V(t) = V_B(1 - e^{-t/RC}) \qquad (D.12)$$

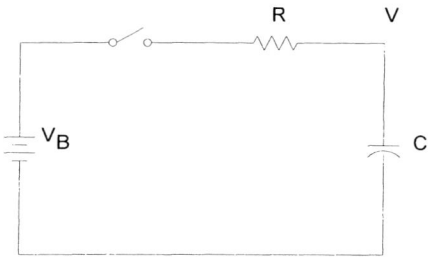

Fig. D-4 A simple resistor–capacitor circuit.

where V_B is the battery voltage. In this circuit, the quantity $RC = \tau$, the time constant. The units of RC are ohms-farads and are equivalent to seconds.

QUESTIONS

1. Three resistors (100 Ω, 200 Ω, and 300 Ω) are connected in series to a 9-V battery (ignore its internal resistance).
   (a) Find the equivalent resistance of the combination.
   (b) Find the current in each resistor.
   (c) Find the total current through the battery.
   (d) Find the voltage across each resistor.
   (e) Find the power dissipated in each resistor.
   Answers: (a) 600 Ω. (b) 15, 15, 15 mA. (c) 15 mA. (d) 1.5, 3, 4.5 V. (e) 22.5, 45, 67.5 mW.
2. The three resistors, mentioned in the above problem, are connected in parallel to the same battery. Answer the five questions for this situation.
3. Three equal resistors are connected in series. When a voltage V is applied across the combination, the total power dissipated is 1 W. What power would be dissipated if the three resistors were connected in parallel across the same voltage?
4. The power rating of a 10 kΩ resistor is 1/4 W. This is the maximum power the resistor can safely dissipate. What is the maximum allowable voltage across the resistor?
5. With respect to the circuit in fig. D-5, $R_1 = 800$ Ω, $R_2 = R_3 = 1600$ Ω, $R_4 = 2000$ Ω, $R_5 = 900$ Ω, $R_6 = 1800$ Ω, and $R_7 = 600$ Ω.
   (a) Calculate the equivalent resistance of the circuit between V_1 and V_3.
   (b) What is $V_1 - V_2$ if the current in $R_1 = 0.5$ mA?
   (c) If the current in $R_1 = 0.5$ mA, how much power is dissipated in this circuit?
   Answers: (a) 800 Ω. (b) 1.2 V. (c) 7.2 mW.
6. With respect to the circuit in fig. D-2(a), let $V_R = 2$ V, $R_1 = 24$ kΩ, and $R_3 = 10$ kΩ.
   (a) What is the reading of a voltmeter, with resistance 10^6 Ω, connected between V_1 and ground?
   (b) What would the voltmeter reading be if its resistance were 10^8 Ω?
   (c) What if the voltmeter had infinite resistance?

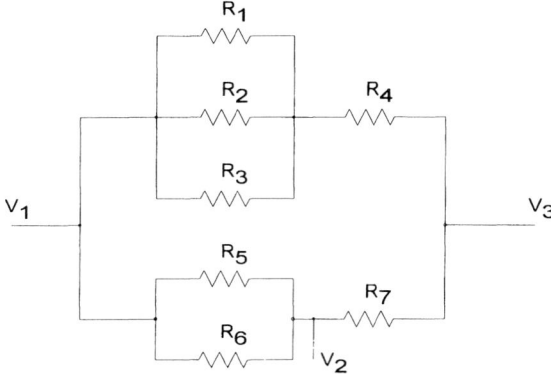

Fig. D-5 A series–parallel circuit.

7. For the wind vane circuit shown in fig. D-3, let the pot resistance be $10\ k\Omega$ and the voltage reference be 1 V. Calculate the voltage output for three different load resistances in the following table.

Direction	V_D		
	$R_L = 10^5\ \Omega$	$R_L = 10^6\ \Omega$	$R_L = \infty\ \Omega$
0°			
60°			
120°			
180°			
240°			
300°			
360°			

What is the maximum error expressed as wind direction due to circuit loading?

8. A #22 wire has a diameter of approximately 0.65 mm. If the wire is copper, what is its resistance per meter? How many meters of wire are required to make one ohm?

9. Contrast the performance of a tungsten filament and a carbon filament in a light bulb. Note that initial power dissipation causes heating of the filament that affects the filament resistance.

Appendix E

Geophysical Coordinate System

Wind direction, as with other geophysical directions, is expressed in the geophysical coordinate system. A problem sometimes arises when computing wind directions on a computer or calculator that incorporates the usual mathematical coordinate system. This appendix shows how to compute wind directions on a computer and resolves the conflict between the two coordinate systems.

E.1 Geophysical versus Mathematical Coordinate System

In the mathematical coordinate system, built into computers and calculators, angles rotate counterclockwise from the positive x-axis whereas in the geophysical coordinate system angles rotate clockwise from the positive y-axis, as shown in fig. E-1.

E.2 Mathematical Coordinates

In the mathematical coordinate system, $x = R\cos\theta$, $y = R\sin\theta$ and $\theta = \tan^{-1} y/x$. Computers and calculators provide an arctangent function, atan ($\tan^{-1}$ on calculators), that returns the angle in quadrants I and IV, $\pm 90°$ (or $\pm\pi/2$ radians). To find the angle over the whole range of $\pm 180°$, one must apply a quadrant correction; see table E-1.

The programming languages C, C++, Java, and FORTRAN, among others, have a function called atan2 that returns the angle over the range $\pm\pi(\pm 180°)$, and no correction is required.

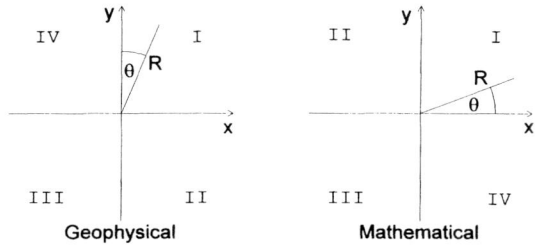

Fig. E-1 Geophysical and mathematical coordinate systems.

E.3 Geophysical Coordinates

In the geophysical coordinate system, $x = R\sin\theta$, $y = R\cos\theta$ and $\theta = \tan^{-1} x/y$. From these expressions and from fig. E-1, one can see that the transformation from one coordinate system to the other is achieved by exchanging the x- and y-axes. When using the atan function, the correction is given in table E-2.

Since the x- and y-axes are interchanged, the atan2 function can be made to work in the geophysical sense by exchanging x and y. Thus, $\theta = \text{atan2}(x, y)$ and no correction is required unless one wants the angle to range from $0°$ to $360°$ instead of $\pm 180°$. Then the correction required is

$$\text{IF } \theta < 0, \ \theta = \theta + 360°$$

Table E-1 Correction to angle returned by the atan function to cover $\pm 180°$.

Quadrant	X-Range	Y-Range	Correction
I	$x \geq 0$	$y \geq 0$	No correction
II	$x < 0$	$y \geq 0$	$\theta = \theta + 180°$
III	$x < 0$	$y < 0$	$\theta = \theta - 180°$
IV	$x \geq 0$	$y < 0$	No correction

Table E-2 Correction to angle returned by the atan function to cover $\pm 180°$.

Quadrant	X-Range	Y-Range	Correction
I	$x \geq 0$	$y \geq 0$	No correction
II	$x \geq 0$	$y < 0$	$\theta = \theta + 180°$
III	$x < 0$	$y < 0$	$\theta = \theta - 180°$
IV	$x < 0$	$y \geq 0$	No correction

Further, to reverse the angle, as in computing the direction from which the wind is blowing,

$$\theta = \theta + 180°$$

$$\text{IF } \theta > 360, \ \theta = \theta - 360$$

A sample calculation is shown in table E-3.

Table E-3 Sample calculation.

X	Y	$\theta \tan^1(x/y)$	Quadrant	θ^1	θ^2	θ^3
1.00	5.67	10	I	10	10	190
5.67	1.00	80	I	80	80	260
5.67	−1.00	−80	II	100	100	280
1.00	−5.67	−10	II	170	170	350
−1.00	−5.67	10	III	−170	190	10
−5.67	−1.00	80	III	−100	260	80
−5.67	1.00	−80	IV	−80	280	100
−1.00	5.67	−10	IV	−10	350	170

[1] Angle after applying the quadrant correction listed in table E-2.
[2] Angle after converting to 0° to 360° range.
[3] Angle after reversing the direction.

Appendix F

Instrumentation Glossary

absolute instrument—An instrument whose calibration can be determined by means of simple physical measurements on the instrument, such as a mercury barometer.
absolute pressure–Pressure relative to a vacuum. Most barometers, including mercury and aneroid barometers, are absolute pressure sensors. These instruments incorporate a reference chamber that contains a high-quality, but not perfect, vacuum.
absorption hygrometer—A humidity measuring instrument utilizing a water-vapor-absorbing chemical.
accuracy—Accuracy of an instrument is the conformity of the measurements made with that instrument to fact, or their lack of error. In common usage, the specified accuracy of an instrument is actually its inaccuracy because a smaller number is associated with less inaccuracy, not less accuracy.
active sensors—An active sensor has an external source of power, usually electrical, which supplies a major part of the output power while the input signal supplies only an insignificant portion. See passive sensor.
aerovane—A wind instrument that measures both wind speed and wind direction. It includes a vane to indicate direction and a propeller to measure wind speed.
Alter shield—A type of rain-gauge shield consisting of freely hanging, spaced slats, arranged circularly around the gauge.
altimeter—An instrument that indicates the altitude of an object above a fixed level. Pressure altimeters use an aneroid barometer with a scale graduated in altitude instead of pressure.
analog signal—(or analogue) A signal (usually electrical), some component (such as amplitude or frequency) of which is continuously proportional to another variable (temperature, wind speed, etc. in meteorology).

analog-to-digital converter—A device which quantizes the information content of an analog signal, that is, transforms it into a finite set of discrete output states, and assigns a digital code to each output state. Essential elements are an analog input signal with a defined input range, a reference quantity, and an output expressed as a finite number of states each representing a discrete input quantum. The most common form converts an analog voltage (information content in the amplitude) to a binary or decimal (binary-coded decimal) number. Conversion is usually performed at discrete time intervals (sampling time) rather than continuously.

anemometer—The general name for instruments designed to measure the speed (or force) of the wind. These instruments may be classified according to the physical principle employed: wind force (cup, propeller, drag-sphere and pitot-static tube anemometers), heat dissipation (hot-film and hot-wire anemometers), or speed of sound (sonic anemometer).

anemometer bivane—An instrument designed to measure the three-dimensional wind vector. It employs a vane that can rotate about both the vertical and horizontal axes providing the wind direction azimuth and elevation. A propeller anemometer is mounted on the front of the vane, in place of a counterweight, to measure the speed.

anemometer, cup—(see cup anemometer).
anemometer, dynamic—(see dynamic anemometer).
anemometer, hot-film—(see hot-film anemometer).
anemometer, hot-wire—(see hot-wire anemometer).
anemometer, propeller—(see propeller anemometer).
anemometer, sonic—(see sonic anemometer).
anemometer, vane—(see vane anemometer).

aneroid barometer—An instrument for measuring atmospheric pressure. (Aneroid means devoid of liquid.) Essential elements are a diaphragm partially enclosing a vacuum chamber, a spring (which is usually part of the diaphragm) to keep the diaphragm from collapsing due to the force of air pressure on it, and a transducer to determine the amount of diaphragm deflection. The transducer may be mechanical, as in a barograph, or electrical, producing a variable resistance, voltage, or frequency.

automatic weather stations—A station designed to monitor meteorological variables, comprising sensors, mounting platform, and a data logger (or data collection platform). It includes a means of local data storage and/or data transmission to another site. Data are sampled at intervals of the sampling period and may be locally processed to produce, for example, averages over the averaging period.

averaging period—Time period over which measurement samples are collected and averaged. See sampling period.

barograph—A recording barometer typically incorporating an aneroid device as the sensing element. A mechanical transducer is used to position an ink pen on a cylindrical chart drum.

barometer—An instrument that measures atmospheric pressure. The two most common barometers are the *aneroid barometer* and the *mercury barometer*. A hypsometer is less commonly used as a barometer.

bias—An instrument error that is uniform over the input range. Bias is normally removed by applying the calibration equation that includes a term to compensate for bias detected at the time of calibration. Instrument bias may change over time due to drift, thus necessitating another calibration.

bimetallic-strip thermometer—A thermometer formed from two strips of metal, having different coefficients of expansion, that are bonded together. As the temperature changes, one strip will expand more than the other, causing the compound strip to bend. Temperature is indicated by the curvature of the strip.

binary quantum—The weight assigned to a change in the least significant bit of a binary output analog-to-digital converter, expressed in terms of the input. For example, the binary quantum of a 10-bit binary analog-to-digital converter with a 5-volt input range would be $5/2^{10} = 4.883$ mV.

bivane— A wind vane designed to rotate about both a vertical and a horizontal axis to measure both the azimuth and elevation angles of the wind vector.

Bourdon tube—A closed curved tube of noncircular cross-section used in some thermometers and barometers. The curve may be C type, spiral, twisted, or helical. The Bourdon-tube thermometer consists of a Bourdon tube completely filled with a liquid. Expansion of the liquid due to temperature change causes an increase in the radius of curvature of the tube. The Bourdon-tube barometer consists of an evacuated Bourdon tube and operates in a similar manner. In both cases the curvature is a measure of the difference between the pressure inside the tube and that outside.

calibration—A process in which the sensor input(s) is held constant (static calibration) or changed in some precisely defined way (dynamic calibration) and the sensor output is measured. The objective is to determine the sensor input–output relation that is called a transfer equation (for the static case) or a transfer function (for the dynamic case) and to determine the coefficients of the calibration equation.

calibration equation—An equation whose independent variable(s) is the sensor raw output(s) and the dependent variable is an estimate of the measurand. For example, temperature can be estimated from the resistance of a platinum resistance thermometer by $T_1 = c_0 + c_1 R_T$, where c_0 and c_1 are calibration constants, R_T is the sensor resistance, and T_1 is the estimate of the measurand T. See transfer equation.

calibration standard—An instrument, used to calibrate other instruments, whose calibration can be traced to national or international standards, and ultimately to primary standards used in the SI system.

carbon hygristor—A suspension of finely divided carbon particles in a hygroscopic film. Increasing ambient relative humidity causes dimensional expansion in the hygroscopic film such that the separation distance of the carbon particles increases; therefore resistance increases with humidity. Corrections must be applied for temperature effects. This sensor is very susceptible to surface contamination which causes calibration drift. It is mainly used in radiosondes where it is kept sealed until flight time.

ceiling balloon—A small balloon used to determine the height of the cloud base. The height can be computed from the ascent velocity of the balloon and the time required for its disappearance into the cloud.

ceilometer—An automatic cloud-height indicator. In the rotating-beam ceilometer, a projector rotates a beam of light from near horizontal to vertical. A photocell, at a known distance (baseline) from the projector, detects light reflected from the cloud base. Knowing the angle of the beam above horizontal, at the moment of detection, and the baseline, the cloud height can be detected. The laser ceilometer projects a pulse of light vertically and measures the time required for the pulse to travel from the ground to the cloud base and back to the instrument.

chilled-mirror hygrometer—See dew-point hygrometer.

complete-immersion thermometer—A complete-immersion thermometer is a liquid-in-glass thermometer designed to be immersed in its entirety, bulb and all of the stem including the portion not filled with the thermometric fluid, in the substance whose temperature is to be measured. A thermometer designed to measure air temperature is completely immersed in the air.

correction—A quantity added to a measured value to compensate for an error, such as the gravity correction used with a mercury barometer.

critical damping—Amount of damping in a second-order system just sufficient to prevent overshoot in response to a step-function input.

cup anemometer—A device, designed to measure the speed of the horizontal component of the wind vector, that rotates about a vertical axis due to the wind force acting on shapes that are concave on one side and convex on the other, that is, cup-like. There are usually three or four hemispherical or conical cups mounted symmetrically about the axis of rotation. When the wind speed exceeds a certain threshold value, the rotation rate of the cup wheel (attached to a vertical shaft) is proportional to the wind speed. Typically, transducers are attached to the shaft to convert rotation rate to electrical voltage or frequency.

damped natural frequency—The frequency at which a second-order system will oscillate in response to a step-function input.

damping ratio—Ratio of damping present in a second-order system to that in a critically damped system. A system with a damping ratio of less than 1 will oscillate in response to a step-function input.

data collection platform—Another term for data logger.

data display—Any form of data presentation to a user. This ranges from simple dial displays to complex images presented on a graphics device (monitor, plotter, etc.).

data logger—A compact device which can be readily connected to a variety of sensors, can sample sensor signals, convert them from analog to digital form, perform some digital signal processing according to user coding, store the data, and, at timed intervals or upon external command, transmit the data to another device.

data storage—A mechanism for holding data until the data are needed. This can include disks, tape, integrated circuit memory, and paper.

data transmission—Any mechanism for transmitting data from one location to another, ranging from hand-carried messages to satellite radio transmission.

dew cell—An instrument for measuring humidity utilizing a solution of lithium chloride in water.

dew gauge—See dew cell.

dew-point hygrometer—(also called a chilled-mirror hygrometer) An instrument which measures the dew-point temperature by cooling or heating a front-surface mirror until condensation just appears on the mirror surface. The observed dew point will differ from the thermodynamic dew point, depending upon the nature of the condensing surface, the presence of condensation nuclei, the accuracy of measurement of the temperature of the front surface of the mirror, and the sensitivity of the condensate detection. A mirror is used because optical techniques are used for condensate detection. At sufficiently cold temperatures, frost will form on the mirror, so the instrument is sometimes called a frost-point hygrometer.

differential pressure—Pressure measured relative to another pressure. A differential pressure sensor is used to read the total pressure relative to the static pressure obtained with a pitot-static tube to obtain the dynamic pressure.

digital signal—A digital signal is one that is represented as a finite number of states. In the binary system, a datum is expressed as a fixed number of binary digits (bits) each of which can be either on or off (1 or 0).

distance constant—The distance constant, $\lambda = \tau V$, where V is the wind speed and τ is the time constant of a sensor which can be modeled with a first-order, ordinary differential equation but where the time constant is observed to be inversely proportional to the wind speed. The distance constant is the appropriate dynamic performance parameter for sensors such as cup and propeller anemometers.

drag coefficient—A dimensionless coefficient whose magnitude is a function of the shape of a body exposed to the wind. It is defined by the equation $F_d = \frac{1}{2} C_d \rho A V^2$ where ρ is air density, A is the cross-sectional area of the body, and V is the wind speed.

drag-cylinder anemometer—A specialized version of the **drag-sphere anemometer** that responds to the two-dimensional wind vector normal to the cylinder axis.

drag-sphere anemometer—An anemometer which measures the three-dimensional wind vector utilizing the drag force exerted by the wind on a sphere. The sphere surface is roughened to ensure turbulence to make the drag coefficient constant over the design range of the sensor. Displacement transducers, such as strain gauges, are used to measure the small deflections caused by the wind force. A drag sphere responds to wind force, not wind velocity, so the transducer outputs are proportional to the wind force components.

drift—A source of error in instruments caused by physical changes in the instrument subsequent to calibration. For example, leakage of air into an aneroid cell would degrade the vacuum and cause drift.

dropsonde—An instrument package similar to a radiosonde but designed to be dropped by parachute from an aircraft for the purpose of obtaining soundings of the atmosphere below.

dry-bulb temperature—Temperature registered by the dry-bulb thermometer of a psychrometer. It is identical to air temperature.

dynamic anemometer—An instrument for measuring wind speed that responds directly to the dynamic force or pressure of the wind. See drag-sphere anemometer or pitot-static tube.

dynamic calibration— A process in which all secondary inputs are held constant while the primary input is forced to change in some known fashion, e.g. a step function, a ramp or a sinusoid. The output is monitored until it settles to a steady-state pattern. The input–output relations, or transfer function, developed in this way comprise a dynamic calibration valid under the stated conditions. Contrast with static calibration.

dynamic error—Dynamic error, due to a changing input, is the sensor output, after static calibration has been applied, minus the input at any given instant. Commonly used with ramp inputs.

dynamic lag—Dynamic lag, due to changing input, is the time delay after the input reaches a certain level until the output, after static calibration has been applied, reaches the same level. Commonly used with ramp inputs.

dynamic pressure—The kinetic energy, $\rho V^2/2$, of a fluid, where ρ is the fluid density and V is the speed.

dynamic response—The response of a sensor that is due to changing input, as opposed to static response observed while the input is held constant.

electrical resistance thermometers—A thermometer comprising of an electrical element whose resistance changes as a function of temperature and the necessary transducers to convert resistance change to some other form, such as voltage change.

energy storage reservoir—Any physical entity which can store energy in any form (electrical, mechanical, or thermal). The number of energy storage reservoirs in a sensor determines the order of the differential equation required to model its dynamic behavior.

error—At any given instant, error is the observed value (sensor or instrument output) minus the reference value (input as measured with a reference quality instrument). Error may be classified as static, dynamic, drift, and exposure.

exposure error—Error due to improper sensor exposure or failure to provide adequate coupling between the sensor and the measurand.

exposure standards—Standard guidelines for the exposure of meteorological instruments. A wind exposure standard may specify distance from the sensor to obstacles, slope of the land, and character of vegetation in the vicinity.

fiducial point—A reference mark used for comparison purposes. The ivory point in a Fortin barometer is a fiducial point; it is used to set the top of the mercury reservoir.

Fortin barometer—A type of cistern barometer. The principal feature of this instrument is that provision is made to increase or decrease the volume of the cistern so that when a pressure change occurs, the level of the cistern can be maintained at the zero of the barometer scale (the ivory point). This is accomplished by rotating of a screw which operates against the leather lower surface of the cistern.

forward scatter sensor—A device which measures the light intensity scattered out of a beam, by particulates and hydrometeors, in a generally forward direction.

forward-scatter visibility meter—A visibility meter which measures the light intensity scattered out of a beam, by particulates and hydrometeors, in a generally forward direction.

frost-point hygrometer—See dew-point hygrometer.

gauge pressure—Pressure measured relative to ambient air pressure, for example, tire pressure.

goldbeater's skin hygrometer—A hygrometer using goldbeater's skin as the sensitive element. Variations of the physical dimensions of the skin caused by its hygroscopic character indicate relative humidity. Goldbeater's skin is the prepared outside membrane of the large intestine of an ox; it is used in gold beating to separate the leaves of metal.

GPS—Acronym for global positioning system. The GPS is a suite of 24 polar-orbiting navigational satellites, each transmitting an identifying code and a time code sequence. A ground receiver that can receive signals from four or more satellites can determine its position on the earth.

ground truth—Reference measurements made by instruments at ground level. All measurements have multiple error sources that can never be completely corrected, therefore, such measurements are always approximations to the measurands or atmospheric variable of interest. It is always incorrect and can be misleading to use the word 'truth' in connection with any measurement.

hair hygrometer—An instrument that measures relative humidity by means of variation in length of a strand of human hair.

hot-film anemometer—An anemometer which uses the same physical principle as the hot-wire anemometer but which uses a film of platinum deposited on a glass base in place of a small wire. It is more rugged and more stable than a hot-wire anemometer, at the expense of reduced frequency response.

hot-wire anemometer—An anemometer which uses the principle that the convection of heat from a body is a function of its ventilation. This principle is employed in two common forms: the constant-current type and the constant-temperature type. In the constant-current type, the wire attains an equilibrium temperature when the heat generated by a fixed current flowing through a fine wire is balanced by the convective heat loss from the wire surface. This equilibrium temperature, measured in terms of the electrical resistance of the wire, is a function of the flow velocity. In the constant-temperature form, current through the wire is adjusted to keep the wire temperature constant. The current required to keep the wire resistance, and thus its temperature, constant is a measure of flow velocity. A hot-wire anemometer is used when very high frequency response, up to 10 000 Hz, is needed.

hygristor—A humidity element used in radiosondes. The resistance of the sensing element is a function of relative humidity. Two types have been used: the carbon hygristor and, in older radiosondes, the lithium chloride element.

hygrometer—Any instrument which measures the water vapor present in the atmosphere.

hygrometer, hair—See hair hygrometer.

hygrometer, infrared—See infrared hygrometer.

hygrometer, Lyman-alpha—See Lyman-alpha hygrometer.

hygroscopic—Readily absorbing moisture, as from the atmosphere.

hypsometer—Literally, an instrument for measuring height; specifically, an instrument for measuring atmospheric pressure by determining the boiling point of a liquid. The relationship between the boiling point of a liquid and atmospheric pressure is given by the Clausius–Clapeyron equation.

hysteresis—A error, caused by sensor friction, that produces errors dependent upon whether the output signal is increasing or decreasing. If the relative humidity were to increase slowly from 50% to 90% and then decrease slowly to 50%, the sensor error observed at, say, 70% would be different for increasing RH than for decreasing RH.

imprecision—An indication of the uncertainty in measurement error. If measurement errors are random with a Gaussian distribution, then the imprecision is twice the standard deviation of the errors. We cannot know the measurand value exactly but usually can assert that it lies within the bounds defined by the measured value plus or minus the imprecision.

inaccuracy—An indication of the total measurement error including the deterministic component and the random component. If the deterministic component is a simple bias, then the inaccuracy is the bias plus or minus the imprecision.

index correction—An instrument correction, usually associated with mercury barometers, obtained by comparison of the test instrument with a reference instrument or transfer standard.

infrared hygrometer—An instrument designed to measure humidity by determining the amount of absorption due to water vapor in certain infrared bands.

input, primary—See primary input.

input, secondary—See secondary input.

instrument—An instrument consists of a sensor, which interacts with the measurand, and may also include signal conditioning elements, an analog-to-digital converter, data processing elements, data transmission elements, data storage elements, and a mechanism for data display.

instrument exposure—The immediate environment of an instrument that affects the instrument output and may cause the output to deviate substantially from a representative measurement.

instrument shelter—A shield designed to protect a sensor from some aspects of the environment while allowing it to interact with the desired elements. A shelter for an air temperature sensor must protect it from rain and direct solar radiation while allowing air to flow freely around the sensor.

irradiance—Total radiant flux received per unit area in $W\,m^{-2}$.

lag error—A dynamic error due to the failure of an instrument to respond instantaneously. Lag is the same as a time constant if the appropriate conditions apply. See time constant.

laser—(acronym for Light Amplification by Stimulated Emission of Radiation) A laser beam is a highly coherent, directional, and monochromatic light beam.

laser ceilometer— An automatic cloud-height indicator that projects a pulse of light vertically and measures the time required for the pulse to travel from the ground to the cloud base and back to the instrument.

laser hygrometer— An instrument designed to measure humidity by determining the amount of absorption due to water vapor in certain wavelength bands, using a laser as the light source.

linearity—In the static sense, a sensor is linear if the transfer plot is a straight line within the specified range of the instrument. In the dynamic sense, linearity holds if the governing differential equation is linear, that is, if the superposition principle holds.

linear thermistor—A circuit of resistors and two or more thermistors whose output voltage is very nearly linear with temperature.

liquid-in-glass thermometer—A thermometer in which the thermally sensitive element is a liquid contained in a graduated glass envelope. Mercury and alcohol are liquids commonly used.

LORAN-C—A type of navigation aid sometimes used for wind finding in radiosondes, designed primarily for marine navigation, particularly in coastal and continental shelf areas. It operates at 200 kHz.

Lyman-alpha hygrometer—A spectroscopic hygrometer which measures absorption due to water vapor in the Lyman-alpha band in the ultraviolet.

manometer—A U-shaped tube containing a fluid such as water or mercury. When used as a differential pressure sensor, the two pressures are connected to the two open ends of the tube. The difference in height of the fluid in the two columns of the U is related to the difference in pressure applied to the two ends. It can also be used as a gauge sensor where one of the tube ends is left open to the atmosphere.

measurand—Generic term for sensor input(s). Temperature is the measurand of a thermometer.

mercury barometer—Instrument for measuring atmospheric pressure which balances the pressure exerted by the atmosphere against the pressure (force per unit area) exerted by the weight of a column of mercury.

meteorological optical range—Length of path in the atmosphere required to reduce the luminous flux in a collimated beam from an incandescent lamp, at a color temperature of 2700 K, to 0.05 of its original value. The luminous flux is evaluated by means of the photopic luminosity function of the International Commission on Illumination (CIE).

microbarograph—A sensitive, recording aneroid barometer.

natural frequency—Frequency of oscillation exhibited by a second-order, linear system after being stimulated by an impulse function or step function. Usually expressed in radians per second.

natural wavelength—When the natural frequency of a second-order linear system is inversely proportional to wind speed, the natural frequency is usually multiplied by the wind speed to yield natural wavelength, in meters, which should be independent of the wind speed.

navigation-aid signals—(GPS, LORAN-C, OMEGA) Used in radiosondes to determine wind velocity from successive position locations found from navigation signals.

net pyranometer— An instrument used to measure net global solar radiation; the downcoming radiation minus upcoming or reflected solar radiation. Could be a matched pair of pyranometers, one pointing up and one pointing down.

Nipher shield—A rain gauge shield used to prevent formation of vertical wind in the vicinity of the mouth of the gauge thereby reducing wind-induced rain and snow catch losses.

noise—(instrumentation definition) Random component of the output signal not present in the measurand, that is, caused by the instrument itself or by sensor–measurand interaction. All fluctuations of the measurand are considered to be input signal.

OMEGA—A network of atomic-clock-controlled transmitters operating in the very low frequency (VLF) band (10.2 kHz to 13.6 kHz), designed to provide global navigation coverage. The signals propagate around the world since the ionosphere and the Earth's surface act as a waveguide at these frequencies.

optical rain gauge—An instrument for measuring precipitation, based on detection of scintillation caused by hydrometeors falling through a beam of light.

outliers—A gross measurement error caused by a mistake, wrong or careless readings, a temporarily defective measuring device, or strong disturbing influences from outside.

parallax error—An error that can occur when reading a mercury barometer or liquid-in-glass thermometer due to not maintaining the line of sight perpendicular to the scale.

partial-immersion thermometer—A liquid-in-glass thermometer designed to be immersed up to the immersion line scribed on the glass.

passive sensors—Passive sensors do not have an external power source; the input signal provides the energy in the output signal. Contrast with an active sensor.

Peltier effect—When current is forced through a thermocouple, one junction will be heated and the other cooled. If current direction is reversed, the junction previously heated will be cooled and the one previously cooled will be heated. This is called the Peltier effect.

performance specification standard—A standard definition defining a test used to measure sensor performance specifications such as time constant and damping ratio.

photodetector—An electronic device that generates a voltage or current; when illuminated by visible or infrared light.

photovoltaic detector— An electronic device that generates a voltage when illuminated by visible or infrared light.

piezoresistive—A resistor whose resistance changes in response to applied mechanical stimuli. Used in some integrated circuit barometers to measure the deflection of the diaphragm.

pilot balloon—A balloon without a payload that is tracked optically, with a theodolite, to determine the wind vector.

pitot-static tube—A sensor for measuring air speed when oriented into the wind. It consists of two concentric tubes. The inner tube is open at the upstream end and closed at the other to measure to total wind pressure, while the outer tube is closed at both ends but is perforated along the side wall to measure the static pressure. Dynamic pressure, the difference between total and static pressure is proportional to the square of the wind speed. Thus $V = \sqrt{2(p_{\text{total}} - p_{\text{static}})/\rho}$ where ρ is the air density.

platinum-resistance thermometer—A platinum film or wire whose resistance is a function of temperature. Resistance at 0°C may be anywhere from 25 to 1000 ohms. Used to measure temperature.

potentiometer—(sometimes called a pot) A variable electric resistor.

precipitation gauge—An instrument used to measure precipitation: rain, snow and other hydrometeors.

precision—A measure of repeatability. Commonly expressed numerically as imprecision so that a smaller numerical value indicates less imprecision.

pressure, absolute—Pressure measured relative to a vacuum. Aneroid and mercury barometers use a near vacuum and so are absolute pressure sensors.

pressure anemometer—An instrument for determining the wind speed from the dynamic wind pressure, $C_d\rho V^2/2$. See pitot-static tube.

pressure, differential—Pressure measured relative to some other pressure. A differential pressure sensor is used to measure the dynamic pressure, which is the difference between the total and the static pressure in a pitot-static tube.

pressure, gauge—Pressure measured relative to ambient atmospheric pressure. Gauge pressure is the pressure in an automobile tire.

pressure-tube anemometer—An anemometer which derives wind speed from measurements of the dynamic wind pressure, $C_d\rho V^2/2$. See pitot-static tube.

primary input—The measurand, the desired input as opposed to secondary inputs which usually are unwelcome.

primary standard—A measurement standard that has the highest accuracy in the country.

procedural standard—A procedural standard defines procedures used in data acquisition such as sampling time and averaging time.

propeller anemometer—A rotation anemometer which has a horizontal or vertical axis upon which helicoidal-shaped vanes are mounted. When the wind speed exceeds the threshold value, the rotation rate of the propeller (and its shaft) is proportional to the wind speed. Typically, transducers are attached to the shaft to convert rotation rate to electrical voltage or frequency.

propeller-vane—A propeller anemometer mounted on a wind vane in place of the usual counterweight. The combination measures speed and direction or the horizontal component of the wind vector.

psychrometer—An instrument to measure humidity utilizing dry-bulb and wet-bulb thermometers.

psychrometer coefficient—The coefficient used in the psychrometer formula: $e = e_s - Ap(T - T_w)$. The psychrometer coefficient is typically $A = 0.00062/°C$.

psychrometer formula—A formula, $e = e_s - Ap(T - T_w)$, used to obtain vapor pressure given the wet-bulb and dry-bulb temperatures and the pressure.

psychrometry—The science of measuring humidity using a psychrometer.

pyranometer—An instrument used to measure global solar radiation, that is both the direct solar beam and the diffuse sky radiation falling on a horizontal flat surface.

pyrgeometer—An instrument used to measure both long-wave earth and atmospheric radiation.

pyrheliometer—An instrument used to measure only the direct solar beam.

pyrradiometer— An instrument used to measure both long-wave earth and short-wave solar radiation.

radar—(from radio detection and ranging) An electronic instrument that transmits microwave energy, usually in pulses, and detects energy scattered back to the receiver from targets (objects, particles, and hydrometeors) in the beam path. It measures the strength of the return, reflectivity, the distance to the target, range, and sometimes the radial velocity of the target.

radio direction-finder—(radio theodolite, radio goniometer) An instrument for determining the direction from which radio waves approach a receiving antenna.

radiometer—An instrument for measuring radiant energy.

radiosonde—An instrument package carried aloft by a balloon, commonly used to measure temperature, humidity, and the wind vector.

radio-theodolite—Same as radio direction-finder.

rain gauge—(sometimes gage) An instrument to measure rainfall. It may be less efficient at measuring snow and other hydrometeors.

rain gauge shield—A wind shield around a rain gauge designed to minimize wind-induced precipitation loss. See Alter shield and Nipher shield.

random error—Random errors are those which cannot be predicted on an individual basis but for which a statistical method can yield information about the mean value of the error and the width of the distribution. The distribution is often assumed to be Gaussian but this is not always the case.

range—Defines the minimum and maximum measurand values a sensor is designed to measure. For example, a temperature sensor may be defined to perform with a given accuracy for temperatures in the range $-30°C$ to $50°C$.

rawinsonde—A balloon-borne radiosonde tracked by radar or by a radio direction-finder, thus measuring wind speed and direction in addition to pressure, temperature, and relative humidity measured by the sonde itself.

reference instrument—A high-quality instrument used to calibrate operational instruments. A reference instrument is stable and its inaccuracy is much smaller than that of the operational instrument.

reference measurement—A measurement made with a high-quality instrument.

repeatability—Error caused by the inability of a sensor to represent the same value under identical conditions.

reproducibility—See repeatability.

residual error—Remaining error after the calibration equation has been applied. Usually due to random error or nonlinearities too small to model.

residual nonlinearity—Non-random residual errors that are too small to model, and thus remove, using the calibration equation.

resistance temperature detectors—(RTD) A temperature sensor in which the resistance of the sensing element, usually platinum, varies with temperature.

resistance thermometer—An instrument for measuring temperature whose sensing element is an electrical resistance that is a function of temperature.

resolution—The smallest increment of the measurand which can be sensed. An analog output from an instrument free of friction has continuous or infinitesimal resolution (sometimes erroneously referred to as "infinite resolution").

root-mean-square error—(RMS) The square root of the average of the residual errors, squared.

$$\text{RMS} = \sqrt{\frac{1}{N}\sum_{n=1}^{N} \varepsilon_n^2}.$$

root-sum-square error—(RSS) A method of combining independent errors to obtain the overall error.

$$RSS = \sqrt{\varepsilon_1^2 + \varepsilon_2^2 + \varepsilon_3^2 + \cdots + \varepsilon_N^2}.$$

rotating-beam ceilometer—An instrument to measure cloud height using a beam of light that is rotated from the horizontal to the vertical. A detector is used to sense when the beam illuminates the cloud deck overhead.
RTD—See resistance temperature detector.
runway visual range—(RVR) Visual range measured along a runway.
sampling period—Time interval between measurements. A measurand may be sampled (measured) at intervals of the sampling period and these data averaged over the averaging period.
secondary inputs—Other sensor inputs in addition to the primary input or measurand. Secondary inputs are usually unwanted. Temperature is a secondary input to an aneroid barometer.
secondary standard—A standard whose calibration is determined by comparison with a primary standard.
Seebeck effect—When two junctions of a thermocouple are maintained at different temperatures, a voltage will be developed between the two junctions. This is called the Seebeck effect.
self-heating error—When current is forced through a resistor, especially a resistance temperature sensor, some of the current will be converted to heat which heats the sensor and causes a temperature error.
sensor—The primary transducer in an instrument. It is the one that interacts with the measurand.
sensor equivalent visibility—Visibility as measured by a sensor. Despite attempts to calibrate visibility sensors to human visibility, some differences remain.
sensor input—The measurand, the variable being measured.
sensor response—The change in sensor output observed when the input is carefully controlled.
signal—A fluctuating quantity, typically mechanical (distance, orientation, rotation rate, etc.) or electrical (voltage, current, frequency, etc.), whose variations convey information about the measurand. The information content may or may not be coded. A voltage proportional to temperature is a signal, as is a sequence of binary bits that represents the digitally coded value of the temperature.
signal conditioning—The process of amplifying, filtering, or otherwise deliberately altering a signal to make it easier to use or to enhance some characteristic of the signal.
significant figures—The digits in a result considered to be meaningful.
siphon rain gauge—A rain gauge that accumulates water in a reservoir and uses the action of a siphon to empty the gauge when the water inside reaches a certain level.
snow gauge—An instrument used to measure snow depth.

snow pillow—An instrument used to measure snow depth from the pressure that accumulated snow exerts on an inflated membrane at ground level.

sonic anemometer—A device used to measure wind velocity by determining the effect of the wind on the transit time of sound pulses.

sorption humidity sensor—A humidity sensor that sorbs (adsorbs and/or absorbs) water vapor to produce some measurable physical change such as length, capacitance, or resistance.

span—The algebraic difference between the upper and lower range limits.

spectral pyranometer— An instrument that measures the spectral distribution of the intensity of global solar radiation.

spectroscopic hygrometer—An instrument that measures humidity by determining the absorption of light caused by the presence of water vapor. Commonly uses water vapor absorption bands in the ultraviolet or infrared portions of the spectrum.

stability—The degree to which sensor characteristics remain constant over time.

static calibration—A process in which all secondary inputs are held constant while the primary input is varied over some range of constant values. That is, the primary input is held constant until the output settles down and the input and output measurements are made. Then the primary input is changed to the next constant value. The input–output relations, or transfer equation, developed in this way comprise a static calibration valid under the stated conditions.

static error—At any given instant, static error is the observed value (sensor or instrument output) minus the reference value (input as measured with a reference quality instrument) while the input is held constant and while the sensor output is steady.

static port—A shield designed to expose a pressure sensor to static pressure while minimizing the effect of wind.

static pressure—Pressure that a sensor would measure if the wind were not blowing.

static sensitivity—Slope of the transfer equation or curve. It is a measure of the sensitivity of a sensor in the static sense. Also the derivative of the raw output with respect to the input.

station pressure—Pressure measured by an instrument after application of the calibration equation (or adding corrections) but before adjusting to a reference elevation such as sea level.

steady-state solution—The solution of a linear differential equation can be expressed as the sum of a transient solution and a steady-state solution. The transient solution decays with time and will eventually go to zero, usually exponentially. What is left is the steady-state solution.

step-function input—An input, mechanical or electrical (or mathematical), to a system (or a differential equation) which rises rapidly (steps) from one constant level to another. Used to test the dynamic performance of a system. In the real world, an input is considered to be a step input if the transition time from one level to the next is small compared to the time constant.

systematic error—Systematic errors are those which can be predicted, from past knowledge of the operation involved, on an individual measurement basis. Calibration removes most, if not all, systematic error. Remaining systematic error is called residual nonlinearity.

temperature scale—A scale for expressing temperature such as the Kelvin scale or the Celsius scale. In these scales, the freezing point of water is 0°C or 273.15K and the boiling point of water is 100°C or 373.15K.

tethered balloon—A balloon, usually flying an instrument package similar to a radiosonde, that is constrained by a line, the tether, to the ground.

theodolite—An optical instrument consisting of a small mounted telescope rotatable in horizontal and vertical planes, used to measure elevation and azimuth angles to a balloon, for the purpose of wind finding.

thermistor—A temperature-sensitive semiconductor. Produces a very large, nonlinear change in resistance for a change in temperature.

thermocouple—A junction, more commonly a pair of junctions, formed by bonding two dissimilar metals to form the junction(s).

thermoelectric—Pertaining to thermocouples. When the thermocouple junctions are held at different temperatures, a thermoelectric voltage is developed at the output.

thermoelectric thermometer—A temperature instrument that uses a thermocouple as the sensor.

thermometer—Any instrument designed to measure temperature.

thermometer screen—Same as an instrument shelter.

threshold—Instruments with internal friction, such as rotational anemometers, require an input to increase from zero to some finite value, called the threshold, before responding.

time constant—Dynamic performance constant of a first-order, linear, ordinary differential equation and, by extension, of physical systems whose dynamic performance can be reasonably modeled by such an equation. It is the time required for such a system to make 63.2% of its final response to a step-function input.

tipping-bucket rain gauge—An instrument designed to measure rainfall rate. Water accumulates in a small bucket which, when full, tips, spilling out the water, rotating another bucket into place to catch water and momentarily closing a relay. Rain rate is proportional to the frequency of tipping or the frequency of contact closure.

total-immersion thermometer— A liquid-in-glass thermometer designed to be immersed such that the bulb and all of the thermometric fluid in the column are immersed in the fluid whose temperature is to be measured.

transducer—A transducer is a device which converts energy from one form to another. All sensors are transducers. In addition to the sensor, transducers are used in an instrument to convert the raw sensor output to a more useful form, often voltage.

transfer equation— An equation whose independent variable(s) is the measurand(s) and the dependent variable is the sensor raw output. For example, the raw output of a platinum resistance thermometer, resistance, is given by $R_T = R_0(1 + aT + bT^2)$ where a, b, and R_0 are constants dependent upon sensor physics and T is the measurand. See calibration equation.

transfer standard—An instrument used to calibrate other instruments. It, in turn, is calibrated by reference to a calibration standard. Necessary qualifications of a transfer standard are high stability and inaccuracy an order of magnitude less than that of the instruments to be calibrated using the transfer standard. A transfer standard is usually kept in a protected laboratory environment where it will not be exposed to shock, temperature extremes, or other forms of abuse typical of a field environment.

transient solution—The solution of a linear differential equation can be expressed as the sum of a transient solution and a steady-state solution. The transient solution decays with time and will eventually go to zero, usually exponentially.

transmissometer—An instrument designed to measure visibility, based on the absorption of light over a path (typically 75 m or 450 m) from a source to a detector.

true value—Theoretical and unobtainable value resulting from a measurement free of all error.

uncertainty—a measure of the residual error, remaining after applying the calibration equation, in a measurement. See also imprecision.

undamped natural frequency—The frequency at which a second-order system would oscillate in response to a step-function input if the system had no damping.

visual range—Visibility as determined by a trained human observer with good or corrected eyesight.

weighing rain gauge—An instrument designed to measure rainfall accumulation by the weight of water accumulated in a container.

wet-bulb temperature—Temperature of a wet-bulb thermometer, cooled by evaporation of pure water into the air stream with adequate ventilation.

wet-bulb thermometer—A thermometer with a cotton "sock" wrapped around the bulb or sensitive portion of the thermometer. Pure water is used to keep the sock wet.

wind run—Average of wind speed irrespective of wind direction, a scalar average.

wind vane—An instrument designed to indicate the direction from which the wind is blowing. The vane rotates about a vertical axis to point into the wind.

wiresonde—A balloon borne instrument package, similar to a radiosonde, that is constrained by a wire to the ground. The wire transmits output data to a ground station.

world radiometric reference—(WRR) Accepted by the World Meteorological Organization (WMO) in 1979 as representing the physical units of total irradiance with an inaccuracy of less than $\pm 0.3\%$.

Index

Absolute humidity, 89, 103
Absolute pyrheliometer, 193
Absorptance, 189
Accuracy, see Inaccuracy
Aerodynamic torque, 136
Aircraft, see Instrument platforms
Albedo, 191–192, 197
Albedometer, 192, 197
Alias, 234–235, see also Frequency folding
Alter rain gauge shield, 179, 181
Amplifier, voltage, 69, 71, 232
Amplitude ratio, 123, 155, 158
Analog signal, see Signal
Analog-to-digital converter, ADC, 231–232, 235–239
Anemometer, 130–143, see also Balloon
   cup, 118, 130–135
   drag cylinder, 137
   drag sphere, 137–138
   hot-film and hot-wire, 140–141
   pitot-static tube, 139–140
   propeller, 130–135
   sonic, 141–143
Aneroid barometer, 29–33, 38, 226, 228, see also Barometer
   Bourdon tube, 33
   characteristics, 38
   diaphragm, 29–33
   error sources, 32–33

   integrated circuit, 31, 34
Assmann psychrometer, see Psychrometer
Atmospheric Radiation Measurement, ARM, 9

Balloon, 215–222
   ascent rate, 215–219
   drag force, 216
   lift force, 216
   pilot, 215, 219–220
   wind finding, 219–222
Barium chloride, $BaCl_2$, 109
Barometer, 22–43, see also Aneroid, Hypsometer, and Mercury barometer
   calibration, 42–43
   comparisons, 37–39
   exposure, 39–40
Beer's law, 101, 202
Bias, 55
Bimetallic strip, 63–64
   static sensitivity of, 64
Binary digit, 236–237
Bridge resistance, 68, 71–72, 83
Buck's equation, 87–89

Calibration, 3, 49, 52, 55–57, 108–109
   laboratory, 16
   static, example of, 56–57
Calibration equation, 53

Calibration equation (*contd*)
 for accumulation gauge, 168
 for aneroid barometer, 29–30
 for drag sphere, 138
 for humidity sensors, 94
 for hypsometer barometer, 36
 for mercury barometer, 25
 for pitot-static tube, 139
 for sonic anemometer, 143
 for transmissometer, 205–206
Capacitance sensor, *see* Hygrometer, sorption
Carbon disulfide, CS_2, 35
Carbon hygristor, 99
Carbon tetrachloride, CCl_4, 35
Ceilometer, 207–210
 laser, 208–210
 rotating beam, RBC, 207–208
 rotating receiver, RRC, 207
Celsius temperature scale, 62–63
Characteristic equation, 152, 154
Chilled mirror dewpoint sensor, 104–106
Clausius–Clapeyron equation, 35, 86
Communication systems, 8–10
Computer, 74, 232–233
Contrast in visibility, 203
Cosine response in anemometers, 132, 134–135
Critical contrast, 203, 205
Cup anemometer, *see* Anemometer

Dalton's law, 87
Damped natural frequency, 160–161
Damped natural period, 161
Damped natural wavelength, 160–161
Damping ratio, 154, 156
Data display, 3, 47
Data logger, 2–4, 231, 233
Data storage, 48
Data transmission, 48
Dewcel, 106–107
Dew-point temperature, 89
Dielectric constant of water, 97, 99
Differential equations, 118–124, 151–154, 156, 160
 first-order, 118–124, 132, 220
 *n*th-order, 151
 second-order, 136, 153–154, 156, 160
Digital signal, *see* Signal
Distance constant, 133–134, 140
Documentation, 18
Doppler frequency shift, 214, 234
Downhill simplex method, 164
Drag coefficient, 130, 133, 137, 216
Drag force, 130, 137

Drift error, 5, 48, 231, 233
 barometer, 33
 drag cylinder or sphere, 137–138
 sorption sensor, 98
Dry-bulb temperature, *see* Temperature
Dynamic error, 5, 122, 158
Dynamic lag, 122, 158
Dynamic performance, 117–127
 cup anemometer, 132–134
 propeller anemometer, 132–134
Dynamic performance parameters, 119, 154
 experimental determination of, 125–126, 162–164

Earth radiation, 190–191
Effective reflectivity factor, 183
Electrical resistance sensor, 70, 76
 conductive sensor, 70–74, *see also* Resistance temperature detector
 semiconductors, 74–76, *see also* Thermistors
Energy storage reservoirs, 152–153, 159, 232
Enhancement effect, 87
Equivocation, 241
Error, definition, 27
Expansion coefficients, 63–64
Exposure, 5–7
 barometric sensor, 39–40
 chilled-mirror, 111
 humidity sensors, 110–111
 precipitation sensor, 178–179
 psychrometer, 93–94, 110
 radiation, 199–200
 radiosonde, 225–228
 sorption, 110
 spectroscopic, 111
 temperature sensor, 77–82, 226–228
 wind sensor, 144

Fahrenheit temperature scale, 63
Field intercomparisons, 17
Figure of merit for temperature sensors, 77
First-order systems, 118, *see also* Linear systems
Fortin barometer, *see* Mercury barometer
Forward scatter meters, 206–207
Freon, 35–36
Frequency folding, 233–234
Frequency response, 123–125, 158–159
Frost-point temperature, 89
Functional model, 1–4, 63, 94, 98, 132, 139, 167–168, 232

Gaussian distribution, 54

Index   287

Geostationary Operational Environmental Satellite, GOES, 10
Glass transmissivity, 195
Global Positioning System, GPS, 214, 224–225
Gold-beater's skin, 100
Ground truth, 182
Gust wavelength, 134
 amplitude, 145
 duration, 145
 frequency, 145
 lull speed, 145
 magnitude, 145
 peak speed, 145

Hair hygrometer, 100, 102
Harmonic mean, 206
Heat transfer, 78–81
 coefficient, 118
 conduction, 78–81
 convection, 78–81
 radiation, 78–81
Helium, He, 215–216
Hot-film and hot-wire anemometers, 140–141
Human aspects, 11–15
Human visual efficacy, 204
Humidity, 86–116, see also Hygrometer
Hydrogen, H, 215–216
Hygrometer, 93–107
 calibration of, 108–109
 capacitive, 98
 chemical reaction, 107
 chilled-mirror, 104–106, 108
 choice of, 110–111
 dew- and frost-point, 104–106
 dewcel, 106–107
 exposure of, 110–111
 infrared, 101–104
 laser, 102
 Lyman-alpha, 103
 mechanical, 100
 psychrometer, 93–96, 107
 resistive, 98–100
 sorption, 96–100, 107–108, 227
 spectroscopic, 101–104, 108
 ultraviolet, 103, 108
Hygroscopicity, 97
Hypsometer, 34–38, see also Barometer
 characteristics, 38–39
Hysteresis errors, 51–52
 barometer, 32

Imprecision, 55, 240
Inaccuracy, 55
Information content or entropy, 239–241
Infrared radiation, 190
Initial conditions, 119–122, 155
Input, primary and secondary, 48
Instrument, 1, 48
Instrument platforms, 8, 213–215
 airplanes, 214–215
 balloons, 215–220
 rockets, 213
 satellites, 214
Instrumentation amplifier, 71
Irradiance, 189
Irrelevance, 241
Irreversible effect, 231

Kelvin temperature scale, 62–63
Kinetic energy, 118, 152
King's law, 141

Least-squares, determining coefficients of, 52–53
Linear systems, 118, 153
Linearity, 50–51, 117
Liquid-in-glass thermometer, 65–67
 complete immersion, 65–66
 maximum thermometer, 65
 minimum thermometer, 65
 partial immersion, 65–66
 total immersion, 65–66
Lithium chloride, LiCl, 107, 109
LORAN, 223, 226
Lyman-alpha, 103–104

Magnesium chloride, $MgCl_2$, 109
Magnesium nitrate, $MgNO_3$, 109
Manometer, 24
Marshall–Palmer drop size distribution, 176
Measurand, 48, 240–241
Mercury barometer, 24–29, 38, 40–42, see also Barometer
 characteristics, 38
 correction for local gravity, 28–29
 Fortin, 24–29, 40–42
 mercury as the barometric fluid, 24–25
Microcomputer, see Computer
Mixing ratio, 90
Multiplexer, MUX, 73, 226, 232

Navigation aids, 223–225
Net pyrradiometer or radiometer, 192, 198
Nipher rain gauge shield, 179, 182
Noise, 51–52, 125–127, 164, 241
Nonlinearity error
 barometer, 32
 bridge, 72–73
 linear thermistor, 75, 78

Nonlinearity error (*contd*)
  thermocouple, 69
Normalized frequency, 158–159
Nyquist frequency, 231, 233–234

Oklahoma Mesonet, 9, 235, 238
OMEGA, 223–224, 226
Optical rain gauge, *see* Precipitation gauge
Overestimate of wind speed, 132, 134–135
Overshoot, 156, 162

Peltier effect, 105
Performance characteristics, 2
  dynamic, 2, 119
  static, 2, 131–132
Phase shift, 123–124, 155, 158
Photosynthetically active radiation, PAR, 190
Pitot-static tube, 139
Planck's law, 190
Platinum, *see* Resistance temperature detectors
Poisson's ratio, 29
Polymer, 98
Potassium acetate, $KC_2H_3O_2$, 109
Potassium carbonate, K_2CO_3, 109
Potassium chloride, KCl, 109
Potassium nitrate, KNO_3, 109
Potassium sulfate, K_2SO_4, 109
Potential energy, 118, 152–153
Power source, 10
Precipitation gauge, 167
  accumulation, 167–176, 179–180
  calibration, 177–179
  error sources, 179–182
  exposure, 178–179
  optical, 174–177
  pressure, 169–181
  radar, 182–185
  siphon, 169–171, 181
  snow, 174
  tipping-bucket, 171–174, 179–180
  weighing, 169, 179–180
Precipitation rate, 166
Pressure, 22–23
  absolute, 23
  differential, 23, 139, 169
  direct measurement of, 23–33
  dynamic, 23
  gauge, 23
  indirect measurement of, 33–37
  static, 22–23
Propeller anemometer, *see* Anemometer
Psychrometer, 93–96, 107
  accuracy of, 95–96

Assmann, 95
  error source, 93–94
  formula, 94
Pyranometer, 79, 192, 194–196
Pyrgeometer, 192, 197
Pyrheliometer, 192–194, 196
Pyrradiometer, 192, 197–198

Quality assurance, 15–19
  data monitoring, 17–18
  documentation, 18
  field intercomparisons, 17
  independent review, 18–19
  laboratory calibrations, 16–17
  publication of data quality assessment, 19
Quantization, 231, 235, 237

Radar, 182–185, 214, 223
Radiance, 189
Radiant flux, 189
Radiant intensity, 189
Radiation sensor errors, 198–199
  absolute calibration, 198
  azimuth, 198
  cosine, 198
  exposure, 199–200
  hysteresis, 198
  linearity, 198
  long-term stability, 198
  response time, 198
  spectral response, 198
  temperature coefficient, 198
  user, 199
  wind speed, 199
Radiation sensor footprint, 199–200
Radiation sensors, 192
  photovoltaic, 192, 196–197
  thermal, 192–195
Radiation shield, 80–82, 228
Radiosondes, 80, 222, 225–228
Rain gauge, *see* Precipitation gauge
Ramp input, 122, 157–158
Random error, 51–52
Range, 50
Range gate, 209–210
Range time, 184, 208
Reference junction, 67, 69
Reflectance, 189
Relative humidity, 90, *see also* Hygrometer
Remote sensors, 182, 214
Removal of water vapor, 93
Residual error, 233
Resistance temperature detector, RTD, 70–74, 105
  power dissipation in, 71–72

sensitivity of, 71
Resistivity of copper, 74
Resolution, 50
Reynolds number, Re, 216
Root-Mean-Square, RMS, 55
Root-Sum-Square, RSS, 57

Sampling error, 179, 231, 233–235
Saturated salt solution, 106–107, 109
Saturation vapor pressure, see Vapor pressure
Second order systems, 154
Seebeck effect, 67
Self-heating, 72–73
Sensor, 1–2, 48
Shadow disk or ring, 196
Signal, 2–3, 48, 232, 237
   analog, 2–3, 47, 237
   averaging, 232, 235
   conditioning, 48, 232, 235
   digital, 4, 48, 237
Significant figures, 58–60
Silicon dome or window, 197
Sinusoidal input, 123–124, 158–159
Sodium chloride, NaCl, 109
Solar radiation, 78–82, 189–192
   diffuse, 190, 196
   direct, 189
   global, 189, 197
Sonic anemometer, 141–143
Sorption sensor, 98–100
   capacitive, 98–99, 228
   resistive, 99–100
Span, 50
Specific humidity, 90
Spectroscopic hygrometer, 100–104
Stability, 51
Standard atmosphere, 216–217
Standard temperature and pressure, STP, 216
Standards, 6–7
   calibration, 6
   exposure, 6–7
   performance, 6
   procedural, 7
Static calibration, see Calibration
Static error, 5, 55
Static pressure port, 39–40
Static sensitivity, 50, 53
   aneroid barometer, 30
   bimetallic strip, 64
   liquid-in-glass thermometer, 66
   resistance temperature sensor, 71
Steady-state, 119, 122–123, 152, 154
Step function, 119–120, 125–126, 154–157
Superposition principle, 117–118, 153

Temperature (dry-bulb), 93–96
   ambient, 90, 118, 126
   virtual, 90
   wet-bulb, 90, 93–96
Temperature errors, barometer, 26–28, 30–32, 38
Temperature scales, 62–63
   Celsius, 62–63
   Fahrenheit, 62–63
   thermodynamic, 62–63
Temperature sensors, 63–76, 161
   bimetallic, 63–64
   comparison of, 76
   differential, 67
   exposure of, 77–80
   liquid-in-glass, 65–67
   resistance, 70–74
   semi-conductor or thermistor, 74–76
   thermoelectric, 67–70
Terminal fall speed, 175, 184
Theodolite, 220–222
Thermal conductivity, 78
Thermal expansion, 63
   coefficients for various materials, 64
Thermistor, 74–76, 105, 226
Thermocouple, 67–70, 194
   copper–constantan, 68–70
Thermometer, 118
Thermopile, 194
Threshold, 51, 131, 135, 140
Time constant, 119–125, 133, 220
Transducer, 2, 48, 131, 136, 138–140, 232
Transfer equation for, 52
   accumulation gauge, 167
   aneriod barometer, 30
   hot-film and hot-wire anemometer, 141
   hypsometer barometer, 36
   infrared hygrometer, 102
   linear thermistor, 75
   mercury barometer, 26
   platinum resistance temperature sensor, 70
   thermistor, 74
   thermocouple, 69
Transfer plot, 50–51
   cup anemometer, 133
   hot-film, 140
   hypsometer, 37
   pitot-static tube, 140
   rotating beam ceilometer, 209
   transmissometer, 206
Transient solution, 119, 122, 152, 154–155
Transmissivity, 205
Transmissometer, 205–206
Transmittance, 189
Trigonometric identities, 123

Turf wall rain gauge site, 178–179

Ultraviolet radiation, 190
Undamped natural frequency, 154, 161
Undamped natural period, 161
Undamped natural wavelength, 160–161
Undetermined coefficients, 119, 152

Vapor pressure, 86–89
  ambient, 89
  saturation, 86–89
Ventilation, 93
Vernier scale, 26
  how to read, 24
Visibility, 202–204
  meteorological optical range, MOR, 204, 206
  meteorological visibility, 202
  runway visual range, RVR, 205
  sensor equivalent visibility, SEV, 205–206
  visual range, 203, 205
Visible radiation, 190

Water, liquid, 35
Water vapor absorption, 102–103
Water vapor pressure, 86–89
Weather radar equation, 182
Wet-bulb temperature, 90
Wexler's empirical equation, 87
Wien's displacement law, 190
Wind force, 130, 137, 219
Wind profiler, 214
Wind run, 140
Wind speed unit conversion, 130
Wind tunnel, 16, 139, 144
Wind vanes, 135–137, 159–161
World Radiation Reference, WRR, 193